Topology and Category Theory in Computer Science

Topology and Category Theory in Computer Science

EDITED BY

G. M. Reed
Oxford University Computing Laboratory

A. W. Roscoe
Oxford University Computing Laboratory

and

R. F. Wachter
U.S. Office of Naval Research, Washington D.C.

CLARENDON PRESS · OXFORD
1991

Oxford University Press, Walton Street, Oxford OX2 6DP
Oxford New York Toronto
Delhi Bombay Calcutta Madras Karachi
Petaling Jaya Singapore Hong Kong Tokyo
Nairobi Dar es Salaam Cape Town
Melbourne Auckland
and associated companies in
Berlin Ibadan

Oxford is a trade mark of Oxford University Press

Published in the United States
by Oxford University Press, New York

© *Oxford University Press, 1991*

All rights reserved. No part of this publication may be reproduced,
stored in a retrieval system, or transmitted, in any form or by any means,
electronic, mechanical, photocopying, recording, or otherwise, without
the prior permission of Oxford University Press

A catalogue record for this book is
available from the British Library

Library of Congress Cataloging in Publication Data
Topology and category theory in computer science / edited by
G. M. Reed, A. W. Roscoe, and R. F. Wachter.
Includes bibliographical references.
1. Computer science—Mathematics. 2. Topology. 3. Categories
(Mathematics) I. Reed, George M. II. Roscoe, A. W. III. Wachter, R. F.
QA76.9.M35T66 1991
004'.01'51—dc20 91-16461
ISBN 0-19-853760-3

Text captured by Alan Jeffrey, Oxford University

Printed by Bookcraft Ltd, Midsomer Norton, Avon

Preface

The Oxford Topology Symposium was held on 27–30 June, 1989. The organizing committee consisted of P. J. Collins and G. M. Reed, both of St Edmund Hall, Oxford and A. J. Ostaszewski of the London School of Economics.

Since techniques from topology and category theory have been increasingly used by theoretical computer scientists in recent years, it was decided to hold a special session at the symposium which would be devoted to the applications of these topics in computer science. By holding this session in the context of the topology symposium the organisers hoped to achieve a cross-fertilization between the communities they brought together — giving one a source of new problems with a more practical flavour, and the other a source of solutions and ideas. The session itself proved successful, attracting a large audience of mathematicians as well as computer scientists. The following gave invited presentations at the special session:

- S. Abramsky, Imperial College
- J. W. de Bakker, Amsterdam
- J. Goguen, Oxford
- C. A. R. Hoare, Oxford
- P. T. Johnstone, Cambridge
- T. Y. Kong, Queen's College, CUNY
- J. Lawson, LSU
- M. Mislove, Tulane
- C. Mulvey, Sussex
- T. C. Przymusinski, Texas at El Paso
- H. Przymusinski, Texas at El Paso
- A. W. Roscoe, Oxford
- A. Rosenfeld, Maryland
- D. S. Scott, Carnegie-Mellon
- M. B. Smyth, Imperial College

and there were 12 contributed presentations.

The organizing committee decided to produce two separate proceedings for the conference: the other is to be a special issue of *Topology and Its Applications* consisting of papers from the 'pure topology' side, with Peter Collins as a co-editor. All those who had presented papers, plus a very few others, were invited to submit papers for these proceedings of the special session.

The papers in this volume have been arranged to maximize continuity. The first paper provides an introductory survey of topology in computer science.

We would like to thank all those who assisted in reviewing papers for this volume. Thanks are also due to Michael Goldsmith, Jeremy Gibbons, and especially Alan Jeffrey, for their assistance in the typesetting of this book. Alan contributed enormously to this volume, by playing the major role in developing its typographical style, by coercing some authors, by doing large amounts of editing on the work of others so that all papers followed this style, and in other ways too numerous to mention.

The symposium would not have been possible without the financial support of the U.K. Science and Engineering Research Council, who funded its general aspects, and the U.S. Office of Naval Research, who funded the special session. We are very grateful to both organizations.

The editors would finally like to thank Peter and Margot Collins, the staff of the Oxford University Computing Laboratory, the Mathematical Institute and various colleges, and everybody else whose efforts made the symposium such a success.

Oxford and Washington D.C. G. M. R.
August 1990 A. W. R. R. F. W.

Contents

1 Topology, computer science, and the mathematics of convergence
A. W. Roscoe

1.1	Introduction	1
1.2	Proving properties of fixed points	5
1.3	Induction over state-spaces	12
1.4	Conclusions	25

2 The soundness and completeness of axioms for CSP processes
Stephen Blamey

2.1	Introduction	29
2.2	The classic failures model	30
2.3	The infinitary model	37

3 Classifying unbounded nondeterminism in CSP
Geoff Barrett and Michael Goldsmith

3.1	Introduction: unboundedly nondeterministic CSP	57
3.2	Predeterministic processes	60
3.3	Restriction and the metric space	63
3.4	Topological properties of process implementations	66
3.5	The branching order of closed processes	69
3.6	Conclusions	74

4 Algebraic posets, algebraic cpo's and models of concurrency
Michael W. Mislove

4.1	Introduction	75
4.2	Algebraic posets and algebraic cpo's	77
4.3	A trace semantics for CSP	88
4.4	A failures model for CSP	96
4.5	Making hiding continuous	101
4.6	Summary	109

5 Concurrency semantics based on metric domain equations
J. W. de Bakker and J. J. M. M. Rutten

5.1	Introduction	113

5.2	Metric spaces and domain equations	116
5.3	Concurrency semantics	123
5.4	Labelled transition systems and bisimulation	146

6 On topological characterization of behavioural properties
Marta Z. Kwiatkowska

6.1	Introduction	153
6.2	Notation and basic notions	155
6.3	The domain of traces	157
6.4	Asynchronous transition systems	159
6.5	Non-interleaving semantics	164
6.6	Properties and systems	165
6.7	The 'good' and 'bad' things	166
6.8	Uniform behavioural properties	167
6.9	Safety properties in Σ	169
6.10	Liveness properties in Σ	171
6.11	Fairness properties	173
6.12	Conclusion	174

7 Order and strongly sober compactifications
J. D. Lawson

7.1	Introduction	179
7.2	Compactly embeddable ordered spaces	180
7.3	Upper and lower semicontinuous functions	183
7.4	Quasi-uniform spaces	189
7.5	Strongly sober compactifications	193
7.6	The Künzi–Brümmer quasi-uniformity	196
7.7	Continuous posets	200
7.8	The Fell compactification	202

8 Totally bounded spaces and compact ordered spaces as domains of computation
Michael B. Smyth

8.1	Introduction	207
8.2	Preliminaries	208
8.3	Totally bounded spaces	211
8.4	Compact ordered spaces	217
8.5	A generalized power domain	223
8.6	Discussion	226

9 A characterization of effective topological spaces II
Dieter Spreen

9.1	Introduction	231

9.2	Strongly effective spaces	234
9.3	Mal'cev topologies	237
9.4	On compatibility	239
9.5	The domain case	245
9.6	The metric case	247
9.7	Effective operators	249

10 The importance of cardinality, separability, and compactness in computer science with an example from numerical signal analysis
Klaus E. Grue

10.1	Introduction	257
10.2	An optimization problem	258
10.3	Further work	263
10.4	Conclusion	268
Appendix	An example from numerical signal analysis	269

11 Digital topology: a comparison of the graph-based and topological approaches
T. Y. Kong and A. Rosenfeld

11.1	Introduction	273
11.2	Graph-based digital pictures; conventional digital pictures	274
11.3	Topological digital pictures	278
11.4	The fundamental group of a digital picture; continuous analogues of graph-based digital pictures	281
11.5	Graph-based equivalents of topological digital pictures	283
11.6	Khalimsky's Jordan curve theorem and a generalization	286
11.7	Concluding remarks	287

12 Tiling the plane with one tile
D. Girault-Beauquier and M. Nivat

12.1	Introduction	291
12.2	Preliminaries	293
12.3	Combinatorics and curves	300
12.4	Tilings	304
12.5	Half-periodicity	309
12.6	Surroundings of exact tiles	330

13 An algebraic axiomatization of linear logic models
Narciso Martí-Oliet and José Meseguer

13.1	Introduction	335
13.2	Dualizing objects and linear categories	337
13.3	Linear logic models	340

13.4 Cancellative linear logic 341
13.5 Girard algebras and quantale models 346
13.6 Categorical combinators 351
13.7 Concluding remarks 351

14 Types as theories
Joseph A. Goguen

14.1 Introduction 357
14.2 Types as theories 361
14.3 Object-oriented concepts 373
14.4 Polymorphism is natural 379
14.5 Programming in the large 382
14.6 Summary 385

Contributors

J. W. DE BAKKER. Centre for Mathematics and Computer Science, P. O. Box 4079, 1009 AB Amsterdam, The Netherlands.

GEOFF BARRETT. Inmos Ltd, 1000 Aztec West, Almondsbury, Bristol BS12 4SQ, U.K.

STEPHEN BLAMEY. Oxford University Computing Laboratory, 11 Keble Road, Oxford OX1 3QD, U.K.

D. GIRAULT-BEAUQUIER. LITP Université Paris VI et VII, France.

JOSEPH A. GOGUEN. Oxford University Computing Laboratory, 11 Keble Road, Oxford OX1 3QD, U.K. and SRI International, Menlo Park CA 94025, U.S.

MICHAEL GOLDSMITH. Oxford University Computing Laboratory, 11 Keble Road, Oxford OX1 3QD, U.K.

KLAUS E. GRUE. DIKU, University of Copenhagen, Universitetsparken 1, DK-2100 Copenhagen, Denmark.

T. Y. KONG. Deparment of Computer Science, Queens College, CUNY, Flushing, NY 11367, U.S.

MARTA KWIATKOWSKA. Department of Computing Studies, University of Leicester, Leicester LE1 7RH, U.K.

J. D. LAWSON. Department of Mathematics, Louisiana State University, Baton Rouge, LA 70803, U.S.

NARCISCO MARTÍ-OLIET. SRI International, Menlo Park, CA 94025, U.S., and Center for the Study of Language and Information, Stanford University, Stanford, CA 94305, U.S.

JOSÉ MESEGUER. SRI International, Menlo Park, CA 94025, U.S., and Center for the Study of Language and Information, Stanford University, Stanford, CA 94305, U.S.

MICHAEL W. MISLOVE. Department of Mathematics, Tulane University, New Orleans, LA 70118, U.S.

M. NIVAT. LITP Université Paris VI et VII, France.

A. W. ROSCOE. Oxford University Computing Laboratory, 11 Keble Road, Oxford OX1 3QD, U.K.

A. ROSENFELD. Center for Automation Research, University of Maryland, College Park, MD 20742, U.S.

J. J. M. M. RUTTEN. Centre for Mathematics and Computer Science, P. O. Box 4079, 1009 AB Amsterdam, The Netherlands.

MICHAEL SMYTH. Imperial College of Science and Technology, Department of Computing, University of London, 180 Queen's Gate, London SW7 2B7, U.K.

DIETER SPREEN. Fachbereich Mathematik, Theoretische Informatik, Universität–GH Siegen, Hölderlinstr. 3, D–5900 Siegen, Germany.

1
Topology, computer science, and the mathematics of convergence

A. W. Roscoe*

Abstract

At first sight, computer science and topology seem unlikely bedfellows. We show that they are linked by the idea of convergence: in computer science we are often interested in processes which converge, often in spaces which are, mathematically speaking, rather unconventional. And point-set topology can be thought of as the study of convergence in general spaces. The main example we use to illustrate this is the way topology can be used to prove properties of the fixed points used in definition of recursive programs. This is developed in the final section for the case of CSP. We show there that it is sometimes possible to use fixed point induction to prove properties of a process which is not defined as a recursion, by synthesizing a recursion of which it is the least fixed point.

1.1 Introduction

That topology and computer science should be closely linked is evidenced by the number, and indeed the variety, of chapters in this volume. But at first sight it is an unlikely combination: computer science is a practical discipline, while topology, in particular point-set topology, has long been regarded as one of the 'purest' parts of pure mathematics. In this chapter we aim to show topologists how their ideas are being applied by computer scientists, and to show theoretical computer scientists how concepts long studied in the abstract by topologists can provide insight into their own problems.

Most of mathematics is rooted in one way or another in the natural, real, or complex numbers. Even those parts which do not study them or structures over them are devoted to the study of objects — for example,

---

*The author gratefully acknowledges that the work reported in this chapter was supported by grant number N00014-87-G-0242 from the Office of Naval Research.

groups, rings, and fields — which generalize them. And, of course, topology is no exception. For point-set topology can be said to be an extension to general spaces of the study of concepts such as convergence, proximity, completeness, continuity, and connectedness from $\mathbb{R}^n$.

Perhaps the most fundamental of these concepts is convergence, for it is in this that the basic idea of an open or closed set arises in $\mathbb{R}^n$: a set X is closed if any convergent sequence of points from X has its limit in X, and open if any sequence of points converging to a point in X must intersect X. Thus the other concepts can all be restated in terms of convergence: for example, $X \subseteq \mathbb{R}^n$ is connected if and only if there are no non-empty Y and Z such that $X = Y \cup Z$ and no sequence of points in either Y or Z converges to a point in the other.

In theoretical computer science one often constructs mathematical models for computer-related objects: usually the semantic values of programs or the data that is stored in a computer. These often look very unlike the real numbers, since, by the nature of computers, they are usually rather discrete. Topology can be used to study how these values relate to each other. Sometimes, on the other hand, as in the study of image processing, signal processing, or numerical analysis, we use topology, or discrete generalizations of it, to help us understand how discrete data stored inside a computer relates to the continuous real world. There is, for example, a large literature on *digital topology* where continuous topological concepts are reinterpreted and exploited in images consisting of a number of grid points in two, three, or more dimensions (see Kong and Rosenfeld's chapter in this volume, for example).

When modelling the behaviour of computer programs we are often interested in limits of one sort or another, usually related to characterizing a program's general behaviour as the limit of approximations — usually finite, or finitely observable, behaviours. This comes most to the fore when deciding on the values of recursively defined programs, which are (either explicitly or implicitly) defined to be the solutions to fixed point equations such as

$$P = F(P)$$

where the function (or program context) F represents the body of the recursion, which makes recursive calls by using its argument. The results used to guarantee the existence of these fixed points in the underlying mathematical spaces all require the topological concepts of completeness and convergence: we set up a sequence of values that ought to converge to the fixed point and use completeness to ensure that it exists. The success of point-set topology is that it allows us to make sense of, and reason about, these concepts in spaces which are very different from those found in traditional mathematics.

The most commonly used results are versions of Tarski's theorem: a monotone function f from a complete partial order X to itself has a least fixed point. Here, a complete partial order (cpo) is one with a least element $\bot$ and where every directed set Δ (Δ is defined to be directed if every finite subset has an upper bound in Δ) has a least upper bound $\bigsqcup \Delta$. The convergent sequence here is found by defining $f^\alpha(\bot)$ for all ordinals α:

$$\begin{aligned} f^0(\bot) &= \bot \\ f^{\alpha+1}(\bot) &= f(f^\alpha(\bot)) \\ f^\lambda(\bot) &= \bigsqcup_{\alpha \in \lambda} f^\alpha(\bot) \end{aligned}$$

This forms an increasing sequence which cannot, by Hartog's theorem, go on increasing strictly for ever. Thus some $f^\alpha(\bot)$ is a fixed point (and it turns out to be the least). If one additionally asserts that f is *continuous* (preserves the limits, that is, least upper bounds, of directed sets), then the least fixed point is always reached by $f^\omega(\bot)$.

Also, an increasing number of computer scientists are making use of the contraction mapping theorem over complete metric spaces to generate their fixed points — see De Bakker and Rutten's chapter in this volume, for example. This theorem, of course, is also proved by setting up a sequence that converges to the (in this case, unique) fixed point. It is interesting to note that, unlike the case of cpos where there was $\bot$, there is no distinguished element in a complete metric space from which to begin converging to the fixed point. Anyone familiar with the proof of the contraction mapping theorem will know that it starts off with an arbitrary point x and establishes that the sequence $\langle x, f(x), f^2(x), \ldots \rangle$ converges to the fixed point. This last observation will be important in the next section.

In both cases the key to setting up the fixed point theorem was to get an appropriate idea of convergence (with related completeness). Of course, this gives us a link with topology. We shall find shortly that we can — in both the partial order and the metric space cases — use topology to analyse the fixed points.

The topology in metric spaces is obvious, but less so in partial orders. Nevertheless there are several different topologies which can be used usefully over cpos. The one which ties best in with the way that the process described above converges on the least fixed point — through the order from below — is known as the Scott topology (for a longer discussion of this topology and its properties, see (Gierz et al. 1980)). It defines a set U to be open if and only if it is upwards closed ($x \in U$ and $x < y$ implies $y \in U$) and such that, if Δ is directed with $\bigsqcup \Delta \in U$, then $U \cap \Delta \neq \emptyset$. A set is Scott-closed if and only if it is downwards-closed and contains the limits of all its directed subsets. It is easy to show that a function between cpos is continuous if and only if it is continuous with respect to the Scott topology. Like some other topologies used in computer science, the

Scott topology is T_0 but not T_1, and indeed the order used to generate the topology can be recovered from it: $x \leqslant y$ if and only if every open set that contains x also contains y. This last property is not one shared by metrics, where many different metrics give rise to the same topology, or by other order topologies that we shall meet later.

The need to understand convergence in mathematically unconventional spaces has been at the root of most of the uses of point-set topology in domain theory and semantics. It has been used not only to study convergence within domains that had been created by other means, but also as a tool for deciding which might be useful spaces to look at and for creating domains themselves. There have been attempts, in particular Smyth's work ((Smyth 1988) and his chapter in this volume), to unify the metric space and order-theoretic approaches by the use of *quasi-metric* spaces (where the symmetry axiom of metric spaces is dropped) and their generalization *quasi-uniform* spaces. A topologically based analogue of completeness that has been much used recently is provided by *sober* spaces — see, for example, (Johnstone 1982) and Lawson's chapter in this volume. The *sobrification* of a T_0 space (see Mislove's chapter in this volume) has a good claim to giving the natural definition of completion over partial-order-based spaces.

Topology has frequently been united with category theory in the search for classes of domains over which to solve *domain equations*. A domain equation is an identity/isomorphism such as

$$D \cong \mathcal{P}(A \times D)$$

(which means D is isomorphic to the powerset of the Cartesian product of A and D.) Domains which, like this one, are defined to be isomorphic to some construction involving themselves, are often required in programming language semantics. Topological considerations apply at several different levels when setting up the theories which allow us to solve them.

- We have to remember that the domains we finally end up with will be used to define programming language semantics. Thus the category of spaces over which one solves equations should have appropriate topological structure to allow us to find fixed points and analyse them. The majority of approaches have used either some class of cpos or a representation of one. However, it has also been shown, for example, (America and Rutten 1988) and see de Bakker and Rutten's chapter in this volume, that it is possible to solve domain equations over classes of complete metric spaces.

- The domain constructors used in the equations, such as $\mathcal{P}$ (powerset or *powerdomain*) and $\rightarrow$ (function space) sometimes appear inconsistent with our desire to solve the equations, because of cardinality considerations (that is, Cantor's theorem). Topological ideas, motivated by the idea of what is computable — (as the limit of a series of finite observations), are used to pare down spaces like $\mathcal{P}(D)$ and $D \to E$ so that the equations do become soluble.

Typically this means restricting ourselves to closed sets and continuous functions (both with respect to appropriate topologies). For example, the function space over cpo models is usually interpreted as the set of Scott-continuous functions, while the powerdomain of a metric domain is usually identified with the set of all closed subsets under the Hausdorff metric.

- Finally, it is necessary to solve the equations themselves. A wide variety of classes of domain have been proposed which allow this, but the method of solution is almost always based, explicitly or implicitly, on inverse limits.

In the next two sections we shall show how the topologies are used to characterize convergence to fixed points in particular domains. In particular we shall examine ways of using these topologies to justify methods for proving properties of the fixed points. In the next section we provide a general survey of these methods, while in the final section we show how one of the techniques can be used in a novel way to prove properties of programs that were not defined explicitly as a fixed point; typically they will be parallel networks of simple CSP processes.

1.2 Proving properties of fixed points

In the introduction we identified convergence within semantic models as the driving force behind topology in theoretical computer science. To illustrate this we will show in this section how topology can be exploited to prove properties of fixed points. In this section we have deliberately kept details of semantic domains and the languages they model to a minimum, in an attempt to keep it accessible to the topological audience.

Our discussions will be based around the following proof rule, whose validity or otherwise depends on the choice of healthiness conditions applied to the body $F(\cdot)$ of the recursion and to the predicate $R(\cdot)$ we are trying to establish.

Rule 1. (Fixed point induction) *Suppose that the property $R(\cdot)$ is satisfiable (that is, not uniformly false) and satisfies healthiness condition H_R, and that $F(\cdot)$ satisfies healthiness condition H_F. Then, if*

$$\forall P\,.\,R(P) \Rightarrow R(F(P))$$

(that is, we can prove R holds of the body of the recursion on the assumption that it holds of all recursive calls), then infer $R(\mu P.F(P))$ (the value of the recursion — namely the appropriate fixed point of the function F).

(This principle is also known as *recursion induction*.)

Here, the assumption that the predicate is satisfiable corresponds to the usual inductive base case: without it, we could prove *false* true of anything. Intuitively, the condition H_F should be strong enough (under the assumption of satisfiability) to prove that there is some sequence of points, all satisfying R, converging to $\mu P.F(P)$ in some sense; and H_R should ensure that this is enough to establish $R(\mu P.F(P))$. H_R will thus usually state that $\{P \mid R(P)\}$ is closed in the topology determined by the mode of convergence used by H_F. By studying different types of space and different modes of convergence we shall be able to establish a variety of pairs (H_F, H_R) that suffice.

The best-known form of this rule is based on the Scott topology which we met in the Introduction (and is known as Scott induction), and naturally applies to functions F from a cpo X to itself. All we have to assume of the function F is that it is monotone (that is, just what is required to ensure a fixed point). A strong condition is required of R though: $R(\bot)$ must hold (reinforcing the satisfiability constraint) and, if Δ is a directed set all of whose members satisfy R, then $R(\bigsqcup \Delta)$ holds[1]. The assumption $R(P) \Rightarrow R(F(P))$ in the rule, plus the assumptions about R, combine to prove inductively that all of the $F^\alpha(\bot)$ that converge to, and include, the fixed point satisfy R.

The unfortunate feature of this form of induction is that the property must hold at $\bot$ — frequently the worst-behaved element of the models computer scientists build. One way of getting around this is to use not one predicate, but a series of strengthening ones. For a continuous F one would try to construct inclusive predicates R_n such that $R_0(\bot)$ holds, and $R_{n+1}(P) \Rightarrow R_n(P)$ and $R_n(P) \Rightarrow R_n(F(P))$ both hold for all n and P. It is then straightforward to prove that $R_n(F^m(\bot))$ holds whenever $m \geqslant n$, so that $\forall n . R_n(\mu P . F(P))$ holds.

Scott induction is based very clearly on the mode of convergence used in the proof of Tarski's theorem. There is another version which is based in the same way on the contraction mapping theorem, and was introduced by the author in (Roscoe 1982, Brookes *et al.* 1984).

If we can view the space X as a complete metric space, the body F of our recursion as contraction mapping, and $\{P \mid R(P)\}$ as a closed subset, then the rule is valid, as is very simple to demonstrate. We know,

---

[1] Any R such that $\{P \mid R(P)\}$ is non-empty and Scott-closed satisfies both these constraints. A predicate R satisfying this directed limit condition is called called *inclusive* or sometimes *continuous*.

by the contraction mapping theorem, that F has a unique fixed point in the space of all processes; but, as F preserves R, it is also a contraction mapping on the complete metric space $\{P \mid R(P)\}$ (a closed non-empty subset of a complete metric space is itself a complete metric space). As F has a fixed point in the subspace, the unique fixed point in the whole space must lie in the subspace.

An attractive feature of this induction is that the base case is totally arbitrary: all we need know is that there is some object satisfying R. (This is related to the observation we made earlier about convergence in a complete metric space being able to start from anywhere.) Even in models which are usually viewed as cpos, such as the untimed models of CSP (Brookes and Roscoe 1985, Brookes et al. 1984), it is frequently useful to introduce a metric structure specifically to apply this version of recursion induction. In the case of the CSP models this metric is usually based on the number of communications which it takes to tell processes apart:

$$d(P,Q) = \inf\{2^{-n} \mid P{\downarrow}n = Q{\downarrow}n\}$$

where $P{\downarrow}n$ captures P's behaviour for its first n steps. The condition that a recursion represents a contraction map then corresponds to its being constructive: the $n+1$-step behaviour of $F(P)$ must depend only on the n-step behaviour of P. Almost all reasonable well-formed recursions which CSP 'programmers' devise turn out to be constructive. Chiefly because of the hiding operator $P\backslash X$, which makes all occurrences of events in X internal and so shortens traces, there are, however, many recursions one can write down which are not constructive. Thus the metric fixed-point theory is not, in this case, as general as the cpo one[2].

Notice that if we can prove a given partial-order-based recursion contractive with respect to a metric, then we have proved that the least fixed point is actually the only (unique) fixed point. This can often be a very useful principle in itself. If P is any fixed point of the function F, then the property $R(Q) \equiv (Q = P)$ clearly inducts through F (that is, $\forall Q \,.\, R(Q) \Rightarrow R(F(Q))$). Since only one of these predicates is actually true of the least fixed point, the inductive principle cannot be valid for proving equality with a single point unless F has a unique fixed point[3]. Thus the property of having a unique fixed point is actually closely tied in with the validity of our induction.

These metrics based on restriction operators (usually indexed by the

---

[2] Notice that, by looking at the same space both as a cpo and as a complete metric space, it has proved possible to deduce things from each view that were not obvious from the other.

[3] Notice also that if the condition H_R on predicates is the same as $\{P \mid R(P)\}$ being closed in some T_1 topology, then equality with a single point must be admissible.

natural numbers but sometimes on larger subsets of the real numbers[4], where restriction operators with higher index are strictly more discriminating and all distinct objects are distinguished by one of them) are very common in computer science. They have a number of special properties, the most notable of which is that they are *ultra-metrics* (sometimes called *non-Archimedian* metrics) which satisfy the following stronger version of the triangle inequality:

$$d(x,z) \leqslant \max(d(x,y), d(y,z))$$

Among the interesting properties of ultra-metrics is that all open balls $B(x,r) = \{y \mid d(x,y) < r\}$ are closed, and no two of these balls (with the same or different radii) intersect without one being contained in the other. We shall return briefly to the theory of these metrics later. It turns out that the closed sets (that is, the ones admissible for recursion induction) are, in the case of a metric based on restriction operators, just the ones whose failure (for a given object) is detectable by one of the restrictions, or formally

$$\neg R(x) \;\Rightarrow\; \exists n \,.\, \forall y \,.\, (x{\downarrow}n = y{\downarrow}n \Rightarrow \neg R(y))$$

Fortunately it usually turns out that the predicates thus admissible include most of those one seeks to prove of processes in practice.

Two further examples where metrics based on restrictions have been used with this type of recursion induction are timed CSP (Reed 1988, Reed and Roscoe 1986, 1988), where there is a restriction operator for every non-negative real, and finitely branching synchronization trees (Brookes et al. 199?), frequently used to model the operational semantics of languages like CSP and CCS. These two examples turn out to be well-suited for similar reasons. In real-time CSP it turns out that every operator is non-expanding and every recursion is a contraction because no operator can see into the future, which would be necessary for it to push a pair of processes further apart.

Where, as in timed CSP, *every* recursion is a contraction map, the only conditions for a recursion induction to be valid are that the predicate represents a non-empty and closed set in the metric topology. In real-time CSP a predicate represents a closed set if and only if its failure (on any given process) is detectable by some finite time (which can vary from process to process). In other words, there must be some time t such that no process with the same t-restriction as the given one can satisfy the predicate. Thus any predicate which states that all of a process's finitely observable behaviours must satisfy some specification represents a closed set, and so is

---

[4] Occasionally restriction operators are used indexed by more exotic orders such as the countable ordinals, but these do not give rise naturally, if at all, to metrics.

admissible for induction if it is satisfiable. Schneider (1990) has shown how in models like those of timed CSP where processes are identified with sets of behaviours one can often, when using the metric-space version of recursion induction, eliminate the need for proving satisfiability for these *behavioural predicates* among the space of processes by expanding the models to contain non-processes, including one with no behaviours, thereby making satisfiability trivial. (The price one pays for this generalization is that one cannot assume that P is a process when proving $\forall P . R(P) \Rightarrow R(F(P))$, but this rarely proves much of a handicap in practice.)

Over synchronization trees, natural operators usually have the similar property that they never forget actions (though they may make them internal); the n-step behaviour of the input determines the n-step behaviour of the output, so to be constructive it is enough that some stage of the definition of a function adds a step of behaviour. Interestingly, the major use of the induction rule in (Brookes et al. 199?) was one over spaces of operators (functions from trees or tuples of trees to trees), where the operators were defined by a form of recursion, to show that the operators being defined were themselves non-expanding or contractions. For the properties of being a constructive or non-destructive operator are admissible since any failure would show up in a finite number of steps. (There would be two inputs with identical n-step behaviours, the results of which, when one applied the offending operator, would have different $(n+1)$- or n-step restrictions as appropriate.)

Before leaving the subject of the version of the rule based on ultrametric spaces, it is worth pointing out another topological idea which can be of use in its application. This is that of *zero dimensionality*. A T_1 space is said to have weak inductive dimension zero if, whenever a point x belongs to open U, we can find a clopen (closed and open) C such that $x \in C \subseteq U$. Every ultra-metric space trivially has weak inductive[5] dimension zero because, as we have already remarked, all open balls $B(x,r) = \{y \mid d(x,y) < r\}$ are also closed. A space is said to have *strong* inductive dimension zero if it satisfies the stronger property that any pair E, F of disjoint closed sets can be separated by a clopen set: that is, there is a clopen set C such that $E \subseteq C$ and $F \cap C = \emptyset$. Notice how these definitions mimic the definitions of regular and normal spaces. Even though every metric space is normal, there is a famous example due to Roy (1969) which shows that, in general, a metric space can have weak induction zero without having strong inductive dimension zero. However, it is easy to show that every ultra-metric space does have the stronger property. Given E and F

---

[5] We do not need to consider spaces with dimension greater than zero, and so do not define them. However, as is suggested by the names, the definitions we gave here of weak and strong inductive dimension zero are just the base cases of inductions for defining what it means for a space to have weak or strong inductive dimension n.

as above, we can set $C = \bigcup\{B(x,r) \mid x \in E \land B(x,r) \cap F = \emptyset\}$. This set (which is obviously open) can be shown to be closed using the fact that if $y \in B(x,r)$ then $B(x,r) = B(y,r)$.

Why is this interesting when it comes to applying our rule? It is because of the condition which says that any allowable predicate must be satisfiable, which means that the set of all things satisfying it is non-empty. For we sometimes have to deal with predicates of the form $R_1 \land R_2$, where both R_1 and R_2 are satisfiable and represent closed sets, but where it is not clear that there is any object satisfying them both. What the above result says is that if there is none then there is a predicate S such that both S and $\neg S$ are admissible and where $R_1 \Rightarrow S$ and $R_2 \Rightarrow \neg S$, which can be useful[6].

We have already discovered that, on the assumption that equality with a given process is to be an acceptable predicate R, the 'well-definedness' condition H_F on recursions must at least imply uniqueness of fixed points. Interestingly, over complete lattices, it turns out to be possible to establish a version of fixed point induction that uses no stronger assumption. Over any partial order we can define the *interval topology* to be the smallest topology in which all the 'closed intervals' $[x,y] = \{z \mid x \leqslant z \leqslant y\}$ are closed sets. Now the interval topology over a complete lattice is compact (see (Gierz et al. 1980), for example), so that the following theorem can always be applied. (In its original reference cited below, this theorem was proved for a specific complete lattice, namely the traces model for CSP.)

Theorem 2. (Roscoe 1982) *Suppose $\langle X, \leqslant \rangle$ is a complete lattice and that $\mathcal{T}$ is a compact topology on X which extends the interval topology. Suppose also that the monotonic function $f : X \to X$ has a unique fixed point x. Then if $Y \subseteq X$ is non-empty, closed in $\mathcal{T}$, and such that $f(Y) \subseteq Y$, we can deduce $x \in Y$.*

Proof. We shall prove by transfinite induction that the sets $[f^\alpha(\bot), f^\alpha(\top)] \cap Y$ (which are closed) are all non-empty, where $\bot$ and $\top$ are respectively the least and greatest elements of X. ($f^\alpha(\bot)$ is defined as earlier, and $f^\alpha(\top)$ is defined by a symmetric iterative process starting from $\top$.)

- When $\alpha = 0$, the set equals Y, which is non-empty by assumption.

- If $x \in [f^\alpha(\bot), f^\alpha(\top)] \cap Y$ then

$$f^{\alpha+1}(\bot) \leqslant f(x) \leqslant f^{\alpha+1}(\top)$$

---

[6]Of course, if both R_1 and R_2 individually induct through the body of the recursion, that is, $\forall P . R_i(P) \Rightarrow R_i(F(P))$ for $i \in \{1,2\}$, then it is not necessary to go through this analysis, since we then know that $R_1(\mu p . F(p)) \land R_2(\mu p . F(p))$ holds by two separate applications of the rule. Also, we have already discussed Schneider's technique for removing the satisfiability constraint from *behavioural* predicates.

since $x \in [f^\alpha(\bot), f^\alpha(\top)]$ and by monotonicity of f. And $f(x) \in Y$ since $f(Y) \subseteq Y$. Hence

$$[f^{\alpha+1}(\bot), f^{\alpha+1}(\top)] \cap Y$$

is non-empty, for it contains $f(x)$.

- Finally, if each $[f^\alpha(\bot), f^\alpha(\top)] \cap Y$ is non-empty for $\alpha < \lambda$ (a limit ordinal) then, since these sets form a descending chain of closed sets in a compact space, their intersection is non-empty. But clearly

$$\bigcap_{\alpha < \lambda} ([f^\alpha(\bot), f^\alpha(\top)] \cap Y) = (\bigcap_{\alpha < \lambda} [f^\alpha(\bot), f^\alpha(\top)]) \cap Y$$

and it is not hard to see that

$$\bigcap_{\alpha < \lambda} [f^\alpha(\bot), f^\alpha(\top)] = [\bigsqcup_{\alpha < \lambda} f^\alpha(\bot), \bigsqcap_{\alpha < \lambda} f^\alpha(\top)] = [f^\lambda(\bot), f^\lambda(\top)]$$

Hence $[f^\lambda(\bot), f^\lambda(\top)] \cap Y$ is non-empty, completing the induction.

Now from the proof of Tarski's theorem in the Introduction we know that there is some α such that $f^\alpha(\bot)$ is the least fixed point. And by symmetry there is a β such that $f^\beta(\top)$ is the greatest fixed point. But, since we have assumed that f has a *unique* fixed point, we know $f^\alpha(\bot) = f^\beta(\top)$. If γ is the greater of α and β we know that $[f^\gamma(\bot), f^\gamma(\top)]$ contains only the fixed point. Since its intersection with Y is non-empty, the fixed point is contained in Y. □

This proof depends crucially on the existence of a top element. Indeed, the theorem is not true of more general partial orders. One need only consider the flat truth value domain $\{true, false, \bot\}$ and the continuous function $\neg$. This has unique fixed point $\bot$ and maps the closed set $\{true, false\}$ into itself.

This is unfortunate, since a great many of the partial orders one meets in computer science fail to have top elements. For a large class of these orders, moving up through the order corresponds to becoming more and more defined in some way, either losing nondeterministic choices or places where a program might loop. They usually contain many maximal elements, which are the values of fully defined programs. For example, the standard failures–divergences model of CSP can (Roscoe 1988a) be ordered in two ways, one of which places all divergence-free processes (ones that can never loop) on top[7], and the other in which only the fully deterministic processes are maximal. In orders like these one can often deduce by

---

[7] A similar 'alternative order', for Dijkstra's language of guarded commands, has been developed independently in (Nelson 1987). It seems quite possible that strong orders of this type can be developed for other application as well.

some means that the least fixed point of a function is maximal (for example, because it is divergence free). Notice that if the least fixed point of f is maximal, then it is necessarily the only fixed point of f. There is a variant of Theorem 2 which allows us to prove properties of this type of fixed point.

The *lower* topology on any cpo X is the smallest one in which all sets of the form $x\!\uparrow = \{y \mid y \geqslant x\}$ are closed. (Notice that, on an order with a $\top$, the interval topology refines this one.) It is easy to show that it is compact for every consistently complete X[8].

Theorem 3. (Roscoe 1988a) *Let X be a complete partial order and let $\mathcal{T}$ be a compact extension of its lower topology. Then, if $f : X \to X$ is a monotonic function whose least fixed point x is maximal (and hence unique) and Y, closed in $\mathcal{T}$, is such that $f(Y) \subseteq Y$, we have $x \in Y$.*

Proof. This is very similar to that of Theorem 2. Transfinite induction establishes that the (closed) sets $(f^\alpha(\bot))\!\uparrow \cap Y$, where $\bot$ is the least element of X, are all non-empty. But the fact that the least fixed point of f is maximal implies that $(f^\alpha(\bot))\!\uparrow$ eventually (that is, for large enough α) contains only the fixed point. The result follows immediately. □

It is mathematically pleasing to have characterizations of when the recursion induction rule is valid that are independent of a specific metric. Even though the great majority of recursions included in CSP programs are constructive, it proves useful to have these abstract forms of the rule when the function is derived from the process and not the other way round: this is discussed in the next section.

1.3 Induction over state-spaces

In this section we concentrate on CSP, a notation which originally motivated much of the work in the last section. The fact that the semantics of CSP has traditionally been given in abstract mathematical models with well-defined notions of convergence has frequently led to its being studied via topology. A number of references which relate directly to the subject of this chapter have already been given. Two rather different applications of topology to CSP may be found in the chapters by Barrett and Goldsmith and by Mislove in this volume.

The application of the rules in the last section to processes defined explicitly by recursion has been well documented in the literature. What we seek to do here is to show how they can sometimes be used to prove properties of processes which have not been defined by recursion, typically

---

[8] A cpo is consistently complete if every set with an upper bound has a least upper bound.

networks of processes communicating with their internal communications hidden.

This section assumes familiarity with the notation of CSP and its semantics in the failures–divergences semantics. A knowledge of operational semantics is also a prerequisite.

A simple example which we shall use to illustrate and develop our techniques is provided by the following. If $B_1 = in?x \to out!x \to B_1$, consider the process

$$B_M = B_1 \gg B_1 \gg \cdots \gg B_1$$

where there are M copies of B_1. We shall assume that $M > 1$. (The piping, or chaining, operator $P \gg Q$ operates on two processes with alphabet $in.T \cup out.T$, and has the effect of putting P and Q in parallel with the out communications of P synchronized with the in communications of Q, and hidden. The in communications of P and the out communications of Q become the visible communications of the combination. The operator is associative.)

It should be obvious that B_1 is a one-place buffer and that B_M is an M-place buffer, in the sense that a k-place buffer never refuses an input if it contains less than k items, and never refuses to output (the oldest input it has not yet output) when it contains any items. For an input to the leftmost B_1 will filter across to the rightmost via hidden communications irrespective of what else the system does or does not do. There are two types of result one might want to prove about the network B_M. Firstly, one might wish to prove it met some specification defined abstractly over a semantic model — we shall be using the failures–divergences model for this example — and secondly one might want to prove that it either equalled or refined some other process.

A typical specification one might want to prove would be that it was a buffer, where a buffer (of type T) is defined to be a process whose failures and divergences $\langle F, D \rangle$ satisfy (i) $D = \emptyset$ (that is, the process is divergence free) and (ii) for all $(s, X) \in F$

$$s{\downarrow}out \leqslant s{\downarrow}in \qquad (1)$$
$$s{\downarrow}out = s{\downarrow}in \Rightarrow X \cap in.T = \emptyset \qquad (2)$$
$$(s{\downarrow}out)\langle a \rangle \leqslant s{\downarrow}in \Rightarrow out.a \notin X \qquad (3)$$

Here, $s{\downarrow}in$ and $s{\downarrow}out$ are respectively the sequences of values in T communicated over the channels in and out within s. Condition 1 says that the outputs are always an initial subsequence of the inputs: in other words no data is lost and order is preserved. Condition 2 says that when a buffer is empty it must accept any input, and condition 3 says that a non-empty buffer cannot refuse to output the next item which is due. This specification has been well studied before, for example in (Roscoe 1982) and (Brookes et al. 1984), and indeed the law established in (Roscoe 1982) that, if P and Q are buffers then so is $P \gg Q$, proves B_M satisfies it.

(Roscoe 1982) also proved that B_M equalled $B^M_{\langle\rangle}$, where the processes B^M_s for $s \in T^*$ with $|s| \leqslant M$ are defined as follows:

$$B^M_{\langle\rangle} = in?x \to B^M_{\langle x \rangle}$$
$$B^M_{s\langle y\rangle} = (in?x \to B^M_{\langle x\rangle s\langle y\rangle}) \,\square\, (out!y \to B^M_s) \quad \text{if } |s| < M - 1$$
$$B^M_{s\langle y\rangle} = out!y \to B^M_s \quad \text{if } |s| = M - 1$$

We are going to use these well-tried examples, namely proving that B_M is a buffer and that it equals $B^M_{\langle\rangle}$, to illustrate the techniques we develop.

In order to be able to apply our results about fixed point induction we must find a recursive definition that is provably equivalent to the network or other CSP term we are considering. One convenient way of doing this, and the method we shall concentrate on, is based on operational semantics.

The operational semantics of CSP (Brookes et al. 199?) describes how a CSP term is modified by rewrite rules as it performs various actions: essentially it provides an abstract interpreter for the notation. We can think of the finite or infinite set of terms which is reachable from a given term P as being the state-space of P. It will be denoted $S(P)$.

The set of all CSP terms with a given alphabet Σ forms a labelled transition system under the operational semantics — transitions being drawn from $\Sigma \cup \{\tau\}$, where τ represents internal, invisible actions. Clearly $S(P)$ is a subtransition system for any term P, which cannot be distinguished from the whole set if we are only allowed to start our transitions from P.

Given any labelled transition system it is easy to define functions which construct the sets of finite or infinite traces (sequences of visible actions) an element can perform, as well as the set of divergences (traces on which the process can perform an infinite sequence of internal actions) and failures — pairs (s, X) such that the process can perform the finite trace s and then come into a stable state (one where no internal action is possible) where no element of X is possible. Provided we close up as appropriate under divergence in defining these functions, we can form abstraction functions which map the transition system into any of the CSP models based on these types of behaviour. As a consequence of the congruence theorems of (Brookes et al. 199?) and (Roscoe 1988b) we know that the result of doing this to an element of the transition system of CSP terms formed by the operational semantics is exactly the same value as predicted in the appropriate denotational semantics in the case (i) where we are using finitely nondeterministic CSP and the failures–divergences model and (ii) where we are allowed to use unboundedly nondeterministic constructs but use the failures–divergences–infinite traces model. Of course, exactly the same statement holds good of a CSP process P when we think of P as an element of the transition system $S(P)$ rather than the transition system of the whole language.

We shall concentrate on the failures–divergences model and only allow finitely nondeterministic CSP. In particular this means that we must restrict ourselves to hiding only finite sets of events. On the positive side it means we can use the standard version $\mathcal{N}$ of the failures–divergences model with the compactness axiom

$$(\forall Y \in \wp(X).(s,Y) \in F) \Rightarrow (s,X) \in F$$

which improves the topological properties of the model (see (Roscoe 1988a) for details of what happens without this axiom). ($\wp(X)$ is the set of all finite subsets of X.) Under the definedness order $\leqslant$ (introduced in (Roscoe 1988a)), this model has as its maximal elements precisely the divergence free processes.

Suppose $\Phi(\cdot)$ is the abstraction map that maps an arbitrary finitely nondeterministic transition system into $\mathcal{N}$. It is easy to see that, for any $P \in T$, we have

$$\Phi(P) = x : P^0 \to \sqcap \{\Phi(P') \mid P \xrightarrow{x} P'\}$$

if P is stable in the sense discussed above. If P is not stable, we get instead

$$\Phi(P) = (x : P^0 \to \sqcap \{\Phi(P') \mid P \xrightarrow{x} P'\}) \triangleright \sqcap \{\Phi(P') \mid P \xrightarrow{\tau} P'\}$$

where $P \triangleright Q$ is a shorthand for $(P \square Q) \sqcap Q$, and P^0 is the set of visible actions P can perform on its first step. (The uses of the $\sqcap$ operation are well defined because the sets over which they have been applied are always non-empty and finite.)

If we replace all occurrences of $\Phi(Q)$s in the above equations by V_Q, for a process variable V_Q (Q varies over T), we get a mutual recursion defined over vectors of processes indexed by the transition system T. We know from the above that the vector of $\Phi(P)$ as P varies is one solution to this recursion, which we shall henceforth refer to as the *state recursion*. In fact, it is the least solution, which is a corollary to part of the proof of the congruence theorem in (Roscoe 1988b)[9]. The assumptions we have made about finite nondeterminism ensure that the function from $\mathcal{N}^{S(P)}$ to itself associated with the state recursion is continuous (like all functions definable in this dialect of CSP).

---

[9] That reference constructs a sequence of approximate abstraction functions Φ_α which are just the iterations of the above recursion, and shows that this sequence eventually reaches the usual abstraction function Φ.

16   *The mathematics of convergence*

In the case where the transition system T is the one produced by the operational semantics (on the whole set of CSP terms or on an $S(P)$), we know by the congruence theorem that the vector of $\Phi(Q)$s is identical to the vector of denotational semantic values. Thus, given an arbitrary CSP term P, we have constructed an indexing set $S(P)$ and a mutual recursion over vectors indexed by it whose least solution has a component equal to P. We have thus achieved one part of what we set out to do in this section: to synthesize a recursion from a process term. We can now use this recursion and our fixed point induction rules to prove properties of $S(P)$. It is worth making two observations about the state recursion.

- It will almost always be non-constructive, because any internal action causes an unguarded recursive call. It will, however, always be non-destructive since it is formed purely from non-destructive operations (the two forms of choice and prefixing).

- The particularly simple form of the recursion is likely to make the checking of individual components of $R(\underline{V}) \Rightarrow R(F(\underline{V}))$ easy. However, there may be a large finite number, or even an infinite number, of such components.

Because of the first of these observations and the fact that the failures/divergence model $\mathcal{N}$ has no top element, only two of the versions of recursion induction discussed in the last section are potentially applicable. Scott induction can be used to prove properties such as $R(\underline{V}) \equiv \forall Q.V_Q \sqsubseteq C_Q$ for any vector $\underline{C}$ of processes. (Here, $\sqsubseteq$ is the standard nondeterminism order on $\mathcal{N}$.) This is not the sort of property which is very useful to have if we are seeking to use (a component of) the vector $\underline{V}$, but it may well be useful if we are seeking to use $\underline{C}$. In other words, we can prove that one vector refines another (which can be thought of as the specification) by doing induction on the specification.

The other form of induction which is applicable, and the one we shall concentrate on, is that established by Theorem 3. Of course we shall use the stronger definedness order on CSP along with this, since it gives us far more maximal elements. However, we shall have to assume that our process P has been proved divergence free before we start this part of our analysis. It should be easy to see that this implies that all elements of $S(P)$ are also divergence free and hence that the vector consisting of the semantic values of all of these processes is maximal, under $\leqslant$, in the product space $\mathcal{N}^{S(P)}$.

It is certainly worth noting that, under this divergence freedom assumption, the vector of semantic values is not only the least, but actually the only, fixed point of the derived recursion. Thus if one can find any vector of values which is a fixed point, the two must be equal. This gives us a unique fixed point principle which can be used in place of the commonly used metric one that is applicable to constructive state-based mutual recursions. We shall shortly apply a version of it to our buffer example.

The major obstacle to the application of the constructions given above is the observation made earlier about the number of cases that need to be checked. Realistically we can only hope to employ them (at least by hand) if we can find some reasonably concise classification of the states — and perhaps a useful notation for them — which will allow us to combine all of the cases into a few pieces of analysis. From now on we shall restrict ourselves to a particular class of CSP processes, one which is interesting in itself and which will allow us to introduce some specific ideas to help codify and reduce this complexity.

This class will be the parallel compositions of a number of deterministic processes with a single hiding operation at the outside. Usually we can expect this hiding to conceal the internal communications of the system, leaving visible just those which are only in the alphabet of a single process. As has been observed elsewhere (for example, (Roscoe and Dathi 1987)) this syntax can, using the various laws of CSP and suitable renaming of hidden communications, cover any composition of (in this case deterministic) processes using interleavings of the parallel and hiding operators. For any hiding that takes place at inner levels can be moved to the outside by means of the law

$$(P\backslash X \parallel Q) = (P \parallel Q)\backslash X \quad \text{if } X \cap \alpha Q = \emptyset$$

after renaming of the hidden communications which appear in αQ. This type of example shows up our difficulty with number of states very well, since even if all of the processes we are combining in parallel are finite state (say with the same number of states) then the potential number of states of the combination grows exponentially with the number of processes.

The advantage of this type of process from our point of view is that its structure is static under the operational semantics. In other words if the initial state is $(P_1 \parallel \cdots \parallel P_n)\backslash X$, then every element of $S(P)$ has the form $(P_1' \parallel \cdots \parallel P_n')\backslash X$ where each P_i is deterministic and the alphabet of P_i' is the same as that of P_i. We shall write Σ_i for the alphabet of the ith component of the parallel, and $\Sigma = \bigcup \{\Sigma_i \mid 1 \leqslant i \leqslant n\}$ for the alphabet of the system before hiding.

Even though the processes P_i we are composing together are assumed to be deterministic, this does not mean that they are free of internal (τ) actions. Indeed, because of the operational semantics of recursion it is impossible to build a non-terminating CSP term whose standard operational semantics is free of τs. These internal τs will almost always be an inconvenience, since they add to the size and complexity of the state-space without, in this instance, adding anything to the semantics. The analysis of our restricted class of processes would be easier if we could think of the P_i as generating only visible actions, so we could restrict our attention to how the parallel and hiding operators combine these and turn some of them into τs.

It is easy, for each alphabet Σ, to construct a transition system with one element for each deterministic process with alphabet Σ and where there are no τ actions. If we think of the deterministic processes as prefix-closed non-empty sets of traces, we shall have $P \xrightarrow{a} Q$ if and only if $\langle a \rangle \in P$ and $Q = \{s \mid \langle a \rangle s \in P\} = P/a$, say. The abstraction map $\Phi(\cdot)$ applied to this system will produce exactly the values we would hope. And we note that, because all states in this transition system are now stable and there is never more than one arc with a given label from any node, the states all satisfy the equation

$$\Phi(P) = x : P^0 \to \Phi(P/a)$$

In determining the state-space of our system, and thus in subsequent computations, we can substitute objects with this nicer transition behaviour for the standard operational semantics of the deterministic processes without destroying any of the analysis carried out earlier. In deriving the transitions of the network $(P_1 \| \cdots \| P_n) \backslash X$ from the transitions of the P_i, we shall assume that the P_i are drawn from the canonical systems of deterministic processes (with appropriate alphabets). This substitution is easily justified in terms of the work in (Brookes et al. 199?) and (Roscoe 1988b) where *operational environments* are used. The congruence theorems proved in those chapters extend to the cases where the CSP terms we consider have free variables, and the values of those free variables are determined by environments which assign elements of transition systems to the variables.

With this simplification the state-space $S((P_1 \| \cdots \| P_n) \backslash X)$ of the system is now a set of objects of the form $N \backslash X = (P'_1 \| \cdots \| P'_n) \backslash X$ where the P'_i are drawn from the canonical deterministic transition systems. The transitions of these states, and hence the mutual recursion derived from them, are easy to deduce from those of the network N before hiding. For a given element $N \backslash X = (Q_1 \| \cdots \| Q_n) \backslash X$ of the state-space there is exactly one transition for any $a \in \Sigma$ such that $\forall i.a \in \Sigma_i \Rightarrow a \in Q_i^0$, and then

$$N \backslash X \xrightarrow{a} N' \backslash X \quad \text{if } a \notin X$$
$$N \backslash X \xrightarrow{\tau} N' \backslash X \quad \text{if } a \in X$$

where $N' = (Q'_1 \| \cdots \| Q'_n)$, for $Q'_i = Q_i/a$ if $a \in \Sigma_i$ and $Q'_i = Q_i$ otherwise. In either case we have $N \xrightarrow{a} N'$. Clearly the state is stable if and only if there is no $a \in X$ such that $a \in N^0$.

The clauses of the recursion as the state varies can now be derived from our earlier discussion. We index the recursion by the N as above:

$$V_N = a : N^0 \to V_{N/a} \qquad \text{if } N^0 \cap X = \emptyset$$
$$= a : (N^0 - X) \to V_{N/a}$$
$$\triangleright \sqcap \{V_{N/a} \mid a \in N^0 \cap X\} \qquad \text{otherwise}$$

Now, on the assumption that our original network $(P_1 \parallel \cdots \parallel P_n) \backslash X$ was divergence free, we know that the vector $\underline{V}$ of semantic values of these states is the unique (and maximal in the product space) fixed point of this recursion. This result can be used directly, since by demonstrating that some other vector is a fixed point we can prove it equals $\underline{V}$. It can also be used, via Theorem 3, to prove results about $\underline{V}$ by fixed point induction.

There is another way one can derive the state recursion of one of our parallel–hiding combinations without thinking about operational semantics at all. We have already observed that any deterministic process can be written uniquely in the form

$$a : A \to P(a)$$

where each $P(a)$ is deterministic. If we write out all of the P'_i in a typical state $(P'_1 \parallel \cdots \parallel P'_n) \backslash X$ in this way, then we can apply (i) the *expansion* laws of $\parallel$ to bring one step of communications out of the parallel operator and then (ii) one of the laws

$$(a : A \to P(a)) \backslash X = a : A \to (P(a) \backslash X) \qquad \text{if } X \cap A = \emptyset$$
$$(a : A \to P(a)) \backslash X = a : (A - X) \to (P(a) \backslash X)$$
$$\triangleright \sqcap \{P(a) \backslash X \mid a \in A \cap X\} \qquad \text{otherwise}$$

This will derive exactly the same equation among the semantic values of our state-space as was produced by the constructions above.

We shall now return to our example of M one-place buffers piped together. This example is not actually in exactly the same form as the one we have been considering, since instead of using parallel and hiding operators separately we have used the hybrid operator $\gg$. The main consequence of this is that the hiding gets done at intermediate stages as we put the network together rather than all at once at the end. However, with some renaming of internal communications and applications of the law mentioned above, this system becomes equal to

$$(B_c^{in} \parallel B_d^c \parallel \cdots \parallel B_{out}^h) \backslash X$$

where $c, d, \ldots, h$ are a sequence of distinct channel names we have picked different from in and out, $X = c.T \cup \cdots \cup h.T$, and

$$B_b^a = a?x : T \to b!x \to B_b^a$$

In fact, not only the denotational semantics but also the transitions of the operational semantics of this system are identical to those of the original one

$$B_1 \gg B_1 \gg \cdots \gg B_1$$

so that in fact by exploring the state-space of one system we are exploring that of the other. The state-spaces of these systems are particularly easy to describe: each component process either holds a value $x \in T$ or is empty, and furthermore any combination of these is possible. A typical state can thus be denoted by an M-tuple of elements of $T \cup \{\epsilon\}$, where ϵ or $x \in T$ in the ith place denotes that the ith process is empty or contains x. The transitions of the state $\sigma \in (T \cup \{\epsilon\})^M$ (a set which we shall abbreviate as S) are precisely those described below.

- If $\sigma = \langle \epsilon \rangle \sigma'$, then there is an action labelled $in.x$ for every $x \in T$ to the state $\langle x \rangle \sigma'$.

- If $\sigma = \sigma' \langle x \rangle$ for $x \in T$, then there is an action labelled $out.x$ to the state $\sigma' \langle \epsilon \rangle$.

- If $\sigma = \sigma_1 \langle x, \epsilon \rangle \sigma_2$, then there is a τ action to the state $\sigma_1 \langle \epsilon, x \rangle \sigma_2$ corresponding to the communication from one element of the chain to the next.

Thus a state is stable precisely when no ϵ follows an element of T in the M-tuple.

We can tabulate the various cases of the state recursion, defining a process V_σ for every state of the type discussed above. We give the stable cases first. If $\sigma = \langle \epsilon, \epsilon, \ldots, \epsilon \rangle = \epsilon^M$, we get

$$V_\sigma = in?x : T \to V_{\langle x \rangle \epsilon^{M-1}}$$

If $\sigma = \epsilon^k \sigma' \langle y \rangle$, where $k > 0$ and σ' contains no ϵ, then

$$V_\sigma = (in?x : T \to V_{\langle x \rangle \epsilon^{k-1} \sigma' \langle y \rangle}) \,\square\, (out!y \to V_{\epsilon^k \sigma' \langle \epsilon \rangle})$$

If $\sigma = \sigma' \langle y \rangle$ contains no ϵ, then

$$V_\sigma = out!y \to V_{\sigma' \langle \epsilon \rangle}$$

Secondly, we give the unstable cases. An unstable state of the form $\langle x \rangle \sigma' \langle \epsilon \rangle$ has no external actions, so we get

$$V_\sigma = \sqcap \{V_{\sigma^*} \mid \sigma \xrightarrow{\tau} \sigma^*\}$$

One of the form $\langle \epsilon \rangle \sigma' \langle \epsilon \rangle$ can input, so we get

$$V_\sigma = (in?x : T \to V_{\langle x \rangle \sigma' \langle \epsilon \rangle}) \,\triangleright\, (\sqcap \{V_{\sigma^*} \mid \sigma \xrightarrow{\tau} \sigma^*\})$$

Induction over state-spaces 21

One of the form $\langle z\rangle\sigma'\langle y\rangle$ can perform an output, so we get

$$V_\sigma = (out!y \to V_{\langle z\rangle\sigma'\langle\epsilon\rangle}) \triangleright (\sqcap\{V_{\sigma^*} \mid \sigma \xrightarrow{\tau} \sigma^*\})$$

And finally, one of the form $\langle\epsilon\rangle\sigma'\langle y\rangle$ can do either:

$$V_\sigma = (in?x : T \to V_{\langle x\rangle\sigma'\langle y\rangle}) \square (out!y \to V_{\langle\epsilon\rangle\sigma'\langle\epsilon\rangle})$$
$$\triangleright (\sqcap\{V_{\sigma^*} \mid \sigma \xrightarrow{\tau} \sigma^*\})$$

A prerequisite of using our methods on this system is proving it free of divergence. This is an easy consequence of the following rule, which applies to any system of the form $(P_1 \parallel P_2 \parallel \cdots \parallel P_n)\backslash X$, where X consists of all communications that are in the alphabets of at least two of the P_i. Versions of this rule have previously appeared in several references (for example (Roscoe 1982, Dathi 1989)).

Lemma 4. *Suppose we can place a partial order between the nodes P_i such that (i) each process is comparable in the order with all its neighbours (that is, P_j such that $\alpha P_i \cap \alpha P_j \neq \emptyset$) and (ii) no process communicates infinitely with neighbours less than itself without either communicating externally or with a neighbour greater than itself. Then the system is divergence free.*

The partial order we use on the current system is the obvious linear order where the leftmost process is the least.

We can thus deduce that the state recursion has a unique and maximal fixed point. This fact can be used to prove that, for all states σ, we have

$$V_\sigma = B^M_{|\sigma|}$$

where B^M_s is as described earlier and $|\sigma|$ is the σ with all ϵs removed (that is, the actual contents of the multi-stage buffer). For consider the vector $\underline{W}$, indexed by S, where $W_\sigma = B^M_{|\sigma|}$. It is easy to show that this is a fixed point of the state recursion. For example, no internal action changes $|\sigma|$, and so all of the unstable clauses of the application of the function F associated with the recursion have the form

$$F(\underline{W})_\sigma = (x : A \to B^M_{f(\sigma,x)}) \triangleright B^M_{|\sigma|}$$

where in every case $A \subseteq (B^M_{|\sigma|})^0$ and, for all $a \in A$, $B^M_{f(\sigma,a)} = B^M_{|\sigma|}/a$. The fact that these clauses all work thus follows from the fact that, if $A \subseteq B$ and $\forall a \in A . P(a) \sqsupseteq Q(a)$, then

$$(a : A \to P(a)) \triangleright (b : B \to Q(b)) = (b : B \to Q(b))$$

After a little more analysis of this type we would conclude that the two vectors $\underline{V}$ and $\underline{W}$ are equal, so that in particular the network B_M equals the process $B^M_{\langle\rangle}$ as claimed earlier. We shall show shortly how this type of argument can be neatly summarized in an easily stated principle.

We may well want to prove more abstract properties of our system than equality with some other process, and besides it may not always be as easy as it was here to find an alternative description of the processes represented by the state-space. As an example to show how to prove such properties, we shall show how the recursion derived from the network B_M can be used to prove directly that it is a buffer in the sense described earlier.

We need a predicate which applies to the whole product space $\mathcal{N}^S$, which is closed in a compact topology, which is satisfiable, which inducts through the state-recursion, and which implies that the ϵ^M-state is a buffer. The property we shall use is

$$R(\underline{W}) \equiv \forall \sigma . (W_\sigma \text{ is a } |\sigma|\text{-buffer})$$

where, for $w \in T^*$, a w-buffer is a divergence free process $\langle F, \emptyset \rangle$ such that, for all $(s, X) \in F$,

$$s{\downarrow}out \leqslant w(s{\downarrow}in) \qquad (4)$$
$$s{\downarrow}out = w(s{\downarrow}in) \Rightarrow X \cap in.T = \emptyset \qquad (5)$$
$$(s{\downarrow}out)\langle a\rangle \leqslant w(s{\downarrow}in) \Rightarrow out.a \notin X \qquad (6)$$

A w-buffer is simply a buffer that initially contains the sequence w. Obviously, a $\langle\rangle$-buffer is precisely a buffer in the sense that we specified earlier.

One can put topologies on $\mathcal{N}$ and its product spaces in several different ways, though when one is considering (as we are, for reasons discussed earlier) the case of finite alphabets, several of these actually coincide. The author introduced a sequence of topologies termed (α), (β), (γ) and (δ) in (Roscoe 1988a), which coincide in the case of finite alphabets and finite product spaces. The α-topology is just the topology defined, as described in the last section, with respect to the standard length-of-trace metric. The others were defined as follows, over the product space $\mathcal{N}^\Lambda$. (Below, the failures or divergences of the λ-component of the vector $\underline{P}$ are denoted D^P_λ or F^P_λ respectively.)

(β) A set C is closed iff whenever $\underline{P} \notin C$ then there are finite sets T of traces and Ψ of indices such that, for any $\underline{Q}$, if

$$s \in D^P_\lambda \Leftrightarrow s \in D^Q_\lambda \text{ and } \{X \mid (s, X) \in F^P_\lambda\} = \{X \mid (s, X) \in F^Q_\lambda\}$$

for all $s \in T$ and $\lambda \in \Psi$, then $\underline{Q} \notin C$.

(γ) C is closed iff whenever $\underline{P} \notin C$ then there are finite sets Ψ of indices, T of traces, and F of failures such that, for any $\underline{Q}$, if whenever $\lambda \in \Psi$

$$s \in D_\lambda^P \Leftrightarrow s \in D_\lambda^Q$$

for all $s \in T$ and

$$(s, X) \in F_\lambda^P \Leftrightarrow (s, X) \in F_\lambda^Q$$

for all $(s, X) \in F$, then $\underline{Q} \notin C$.

(δ) C is closed iff whenever $\underline{P} \notin C$ then there are finite sets Ψ of indices, T of traces, and F of finite failures (failures (s, X) with X finite) such that, for any $\underline{Q}$, if whenever $\lambda \in \Psi$

$$s \in D_\lambda^P \Leftrightarrow s \in D_\lambda^Q$$

for all $s \in T$ and

$$(s, X) \in F_\lambda^P \Leftrightarrow (s, X) \in F_\lambda^Q$$

for all $(s, X) \in F$, then $\underline{Q} \notin C$.

These topologies are successively weaker. The following result from (Roscoe 1988a) lists some of their basic properties.

Lemma 5.

1. The conditions β, γ, and δ are equivalent if the underlying alphabet Σ is finite. If it is infinite, δ is strictly stronger than γ and γ is strictly stronger than β. α is equivalent to β if the alphabet is finite and the index-set Λ is finite. Otherwise β is strictly stronger than α.

2. For all $\xi \in \{\beta, \gamma, \delta\}$ the ξ-topology on $\mathcal{N}^\Lambda$ defined above is well-defined. The topology generated by ξ on $\mathcal{N}^\Lambda$ is the same as the product topology where each copy of $\mathcal{N}$ is given the topology generated by ξ.

3. The δ-topology on $\mathcal{N}^\Lambda$ is compact, Hausdorff, and zero-dimensional.

4. If $\underline{P} \in \mathcal{N}^\Lambda$, each of the sets $\{\underline{Q} \mid \underline{Q} \geqslant \underline{P}\}$, $\{\underline{Q} \mid \underline{Q} \leqslant \underline{P}\}$, $\{\underline{Q} \mid \underline{Q} \sqsupseteq \underline{P}\}$, and $\{\underline{Q} \mid \underline{Q} \sqsubseteq \underline{P}\}$ are closed in the δ-topology (and hence in the others as well).

Since in our case the set T of values passed over channels is finite, we know the state-space S is as well and hence all of these topologies coincide. But in fact the property of being a w-buffer is (δ)-closed even in the general case, since any failure shows up in a process $\langle F, D \rangle$ in one of

1. a failure $(s, \emptyset)$ where s violates (a), or

2. a failure $(s, \{in.x\})$ where $s{\downarrow}out = w\,(s{\downarrow}in)$, or

3. a failure $(s, \{out.a\})$ where $(s{\downarrow}out)\langle a\rangle \leqslant w\,(s{\downarrow}in)$

even though, as stated, the predicate is phrased in terms of infinite as well as finite failures. Since the set of processes satisfying R is just the Cartesian product of (δ)-closed sets, it follows by Lemma 5(2) that the set of vectors satisfying R is closed in the product topology.

R is satisfiable, since for any sequence $w = \langle a_1, \ldots, a_r\rangle$ the process

$$out.a_1 \to \cdots \to out.a_r \to B_1$$

is easily shown to be a w-buffer.

All we now have to do is prove that $\forall \underline{W}.R(\underline{W}) \Rightarrow R(F(\underline{W}))$, where F is the function of our state recursion. Much as in the case of the equality proof above, this simply requires some easy case checking, which is assisted by the following observations.

1. Any behavioural specification like w-buffer is distributive, in the sense that if U is a finite set of w-buffers, then $\sqcap U$ is one too.

2. If P and Q are divergence free then so is $P \triangleright Q$, and its failures are $F_Q \cup \{(s, X) \in F_P \mid s \neq \langle\rangle\}$. In particular this implies that, if $a : A \to P(a)$ and Q are both w-buffers, and $B \subseteq A$, then $(a : B \to P(a)) \triangleright Q$ is a w-buffer.

The details of the proof are left to the reader.

We took care above to show that our predicate R was δ-closed, so that Theorem 3 would be applicable even if the alphabet were infinite. In fact, because of the particular structure of the CSP model $\mathcal{N}$, it turns out that, for continuous functions like our state recursion (and all functions definable using this CSP dialect), closure in the β-topology is enough. This is proved in (Roscoe 1988a), via a more detailed analysis of the way the iteration from bottom converges to the least fixed point. One practical benefit of this is that behavioural predicates on $\mathcal{N}$ are always γ-closed, and hence β-closed, but may not be δ-closed if the alphabet is infinite.

The δ-topology actually coincides with another well-known order topology, the *Lawson* topology (or Λ-topology). It can be defined to be the coarsest common refinement of the two T_0 topologies we have already met, namely the Scott topology and the lower topology. For detailed discussion of this topology, see (Gierz et al. 1980). For a consistently complete algebraic[10] cpo it is always compact Hausdorff. If the alphabet is finite then

---

[10] An algebraic cpo is one where the set of *finite* elements less than or equal to x is directed with limit x. A finite element is one e such that, whenever Δ is directed with $\bigsqcup \Delta \geqslant e$, then $\exists y \in \Delta . y \geqslant e$. The failures–divergences model $\mathcal{N}$ is algebraic and consistently complete under both $\leqslant$ and $\sqsubseteq$.

the δ-topology (and hence the others as well, as described in Lemma 5) is equivalent to the Lawson topology with respect to both orders. If the alphabet is infinite then it is the same as the $\sqsubseteq$-Lawson topology. The $\leqslant$-Lawson topology is then different from all the others we have met. The relationship of these two compact Hausdorff topologies will be an interesting topic for further work: it may be possible to take advantage of them in combination just as we did with the pair of orders.

1.4 Conclusions

In this chapter we have seen how convergence links topology and computer science, and have in particular studied conditions under which fixed point induction is valid.

In Section 1.3 we showed how to take advantage of the topological structure of the CSP model $\mathcal{N}$ and construct a sequence which converged on (the state-vector of) an arbitrary process, and then use fixed point induction to establish properties of the process itself. While we showed how the state recursion could be derived for an arbitrary CSP term, we concentrated on parallel–hiding combinations of deterministic processes. The method we evolved for proving properties of the process using this recursion can be summarized in the following general principle.

Rule 6. (State induction) *Suppose we have a divergence free CSP process with state-space $S(P)$, and to each $Q \in S(P)$ we associate a satisfiable and β-closed predicate R_Q. Then if we can prove each $R_Q(Q)$ holds simply on the assumption that $R_{Q'}(Q')$ holds of all states Q' reachable in one step from Q using the Q-clause of the state recursion, we can infer $\forall Q \in S(P) . R_Q(Q)$.*

One particular example to which this was applied was where each R_Q specified equality with a particular process W_Q. Equally valid would be the case where each R_Q specified that the Q-component refined (in either order) W_Q. Thus, for example, if P is divergence free and we can prove that $F(\underline{W}) \sqsupseteq \underline{W}$, then the state-vector of P also refines $\underline{W}$. (Since we have used the definedness order $\leqslant$ crucially in proving this, it provides an interesting case of the two orders complementing each other.)

We have seen how the state recursion technique applies to one fairly simple and regular example. The combinatorial explosion which arises from parallel composition means that we are only likely to be able to apply the techniques by hand to examples where there is enough regularity to allow us to describe the state-space reasonably compactly, and to find a classification of states which allows us to deal with only a few essentially different cases. The work we have described here is clearly closely related to the very active field of exploring state-spaces mechanically, known as model checking. Some references to this are (Browne et al. 1986, Burch et al. 1990, Kurshan and McMillan 1989). (Indeed, the work we have carried out here could be viewed as providing the mathematical basis for using model-checking techniques for proving properties of CSP programs specified over $\mathcal{N}$.

References

America, P., and Rutten, J. (1988). *Solving reflexive domain equations in a category of complete metric spaces.* Proceedings of MFPLS87, Springer LNCS 298 pp 254-288.

Brookes, S. D. (1983). *A model for communicating sequential processes.* Oxford University D.Phil. thesis.

Brookes, S. D., Hoare, C. A. R., and Roscoe, A. W. (1984). *A theory of communicating sequential processes.* Journal of the Association for Computing Machinery, **31**, 3, pp 560-599.

Brookes, S. D., and Roscoe A. W. (1985). *An improved failures model for communicating processes.* Springer LNCS 197, pp 281-305.

Brookes, S. D., Roscoe A. W., and Walker, D. J. (199?). *An operational semantics for CSP.* Submitted for publication.

Browne, M. C., Clarke, E. M., and Grumberg, O. (1986). *Reasoning about networks with many identical finite state processes.* Proc. 5th ACM Symposium on Principles of Distributed Computing.

Burch, J. R., Clarke, E. M., Dill, D. L., and Hwang, L. J. (1990) *Symbolic model checking: 10^{20} states and beyond.* Proc. 5th IEEE Annual Symposium on Logic in Computer Science, IEEE Press.

Dathi, N. (1989). *Deadlock and deadlock-freedom.* Oxford University D.Phil. thesis.

Gierz, G., Hofmann, K. H., Keimel, K., Lawson, J. D., Mislove, M., and Scott, D. S. (1980). *A compendium of continuous lattices*, Springer-Verlag.

Hoare, C. A. R. (1985) *Communicating sequential processes.* Prentice-Hall International.

Hoare, C. A. R., Brookes, S. D., and Roscoe, A. W. (1981). *A theory of communicating sequential processes*, Oxford University Computing Laboratory, Programming Research Group, Technical Report PRG-16.

Johnstone, P.T. (1982) *Stone spaces.* Cambridge University Press.

Kurshan, R. P., and McMillan, K. (1989). *A structural induction theorem for processes.* Proc. 8th ACM Symposium on Principles of Distributed Computing, Edmonton.

Nelson, G. (1987). *A generalisation of Dijkstra's calculus.* Research report 16, Digital Systems Research Center.

References

Reed, G. M. (1988). *A uniform mathematical theory for real-time distributed computing.* Oxford University D.Phil. thesis.

Reed, G. M., and Roscoe, A. W. (1986). *A timed model for communicating sequential processes.* Proceedings of ICALP'86, Springer LNCS 226, pp 314-323.

Reed, G. M., and Roscoe, A. W. (1988). *Metric spaces as models for real-time concurrency.* Proceedings of MFPLS87, Springer LNCS 298 pp 331-343.

Roscoe, A. W. (1982). *A mathematical theory of communicating processes.* Oxford University D.Phil. thesis.

Roscoe, A. W. (1988a). *An alternative order for the failures model.* in 'Two Papers on CSP', Technical Monograph PRG-67, Oxford University Computing Laboratory.

Roscoe, A. W. (1988b). *Unbounded nondeterminism in CSP.* in 'Two Papers on CSP', Technical Monograph PRG-67, Oxford University Computing Laboratory.

Roscoe, A. W. and Dathi, N. (1987). *The Pursuit of deadlock freedom.* Information and Computation **75**, 3.

Roy, P. (1969). *Failure of equivalence of dimension concepts for metric spaces.* Bull. Amer. Math. Soc. **68** pp609-613.

Schneider, S. (1990) *Correctness and communication in real-time systems* Oxford University D. Phil Thesis.

Smyth, M. B. (1988). *Quasi-uniformities: reconciling domains with metric spaces.* Proceedings of MFPLS87, Springer LNCS 298 pp236-253.

2
The soundness and completeness of axioms for CSP processes

STEPHEN BLAMEY*

Abstract

A model for CSP—for example the classic failures model (Brookes et al. 1984)—is standardly presented as the totality of mathematical structures, of a specific type, that satisfy certain 'axioms'. These structures are taken to represent processes and are used to interpret terms of CSP. Here, the value of looking for a more basic 'implementational' definition of a given model is stressed. Any explicit set of axioms will then be required to be sound and complete with respect to this definition.

To illustrate the idea, the classic failures model will be given an implementational definition, and the usual axioms will be proved sound and complete. But the bulk of this chapter is about a model for 'unbounded non-determinism', which can handle the infinite non-deterministic composition of processes. The model is infinitary in a way that would make justifying an appropriate set of axioms a very murky business, were there not an implementational definition to hand—to keep track of what is going on and to make comparisons with the original finitary model. The axioms actually proposed include an infinite formula: it is intuitively natural, but it is of a logical complexity that calls for a game-theoretic interpretation. This is spelt out in some detail, and so is a proof that the axioms are indeed sound and complete.

2.1 Introduction

This chapterpursues, a little way, the thought that the failures model for CSP, and any modification or extension of that model, should be provided with a precise 'implementational' definition delineating which structures of the relevant logical type count as representations of processes—a definition

---

*The author gratefully acknowledges that the work reported in this chapter was supported by grant number N00014-87-G-0242 from the ONR.

which directly reflects an intuitive understanding of the modelling, and against which a set of explicit axioms should be required to be sound and complete.

The standard axioms for the classic failures model (Brookes et al. 1984, Brookes and Roscoe 1985) will be proved sound and complete for a suitable implementational definition (Section 2.2). But the guts of this chapter (Section 2.3) concern structures modelling 'unbounded nondeterminism' in CSP. The classic model is finitary in that it gives the identity of a process in terms merely of finite items of behaviour, but, as we shall see, this makes it unable to handle the infinite nondeterministic composition of processes in a fully adequate way; we need to model processes by structures which make non-trivial reference to 'infinite items of behaviour'. Roscoe's (1988) pioneering essays provide an axiomatic characterization of a model for unbounded nondeterminism, but one of the axioms—the most interesting and important—can in fact be seen to import the idea of an implementational definition. This is something of a mixing of levels (and it creates some redundancy). But there is an explicitly implementational definition that turns out to capture what is, extensionally speaking, precisely the same model as Roscoe's. And it is also possible to provide a natural set of axioms—which are quite independent of the implementational definition, but which can be shown to be sound and complete with respect to it.

In a longer version of this chapter, one could take full account of a 'divergence' component in the modelling of processes. But here we shall ignore divergence—or, better, we shall adopt the assumption that all the processes we are concerned with in fact exhibit no divergent behaviour. Some of the manouevring in Section 2.2 will feel familiar to anyone acquainted with work like that in (Brookes 1983, Chapter 3), which pursues the idea that a process is the union—the 'nondeterministic composition'—of the deterministic processes which it contains—which are 'above' it in the nondeterminism ordering. Indeed, one way roughly to describe our theme might be to say that it converts this idea from a provable result about an axiomatically defined domain of processes into an independent definition, which can then be used to justify the axioms. We should, however, record a caveat to glossing the union of a (possibly infinite) family of deterministic processes, as its 'nondeterministic composition'. So long as we stay with the finitary models, this gloss is not an entirely happy one: in fact, we have an instance of the problem which motivates moving to infinitary models.

2.2 The classic failures model

The classic failures model for CSP is described by Brookes et al. (1984) and Brookes and Roscoe (1985). Let us think in terms of the set up in this second paper, where there is a fixed non-empty background set Σ of communications. Processes are then represented by particular sets of pairs (s, X), where s is an element of Σ^* (the set of finite sequences of elements of Σ) and X is a finite subset of Σ. And which sets represent processes is stipulated by a set of axioms.

An intuitive understanding of this model is that a set F of pairs gives the identity of the process it represents by containing all and only those pairs modelling 'observable' items of behaviour that the process 'might' exhibit. The items of behaviour are called *failures*, since, as an element of a process-representation F, a pair (s, X) is to be understood to model the communication of the sequence s followed by the refusal to communicate any element of X. And the axioms that F must satisfy are supposed to pick out exactly those sets of pairs which can be viewed, in this way, as representations of processes. But how might the informal account of the model be spelt out in a precise enough way for us to be able to verify that the axioms do indeed pick out the right sets?

So long as we restrict attention to the classic failures model, we can shelve any questions about the notion of observability in play when we talk about 'observable' items of behaviour. What crucially requires explication is the modality involved in saying that a process *might* exhibit a given failure. There are, of course, two strands that must first be distinguished: on the one hand, the behaviour of a process is conditional on what communications are presented to it to share with its environment; on the other hand, its behaviour may be nondeterministic. Let us consider nondeterminism. This is frequently glossed in mentalistic vocabulary: it is said, for example, that a process can '*decide*' to communicate one thing rather than another from a range of '*choices*'. But this is no help, if we are aiming for conceptual clarification. It would seem better to direct attention firmly towards nondeterminism as a way of being non-committal about (deterministic) implementation; then we can consider spelling out the 'might' of nondeterminism simply in terms of quantification over possible implementations. Hence, taking account also of the conditional strand in the naive modality, I suggest that 'process P might exhibit failure (s, X)' be spelt out as follows:

> There is an implementation of P such that there is a presentation of communications (which could perhaps be thought of as offered by an experimenter trying to elicit 'observations') such that the sequence s is communicated and then all of X refused.

Observe that if we adopt this explication, then the 'might' of nondeterminism has wider scope than the 'might' of conditional dependence on the presentation of communications, so that the basic notion of communication is taken to be communication with a process *under a given implementation*. Maybe this is to ride rough-shod over some theoretical subtleties—in particular, because it obscures the idea that nondeterministic possibilities might be seen as arising stage by stage as a process progresses—but at least it is readily intelligible, and it will certainly provide a formally adequate foundation for the representation of a CSP process P as the set F of all and only those failure-pairs that P 'might' exhibit.

And so, if we are to make this explication mathematically formalizable, we have incurred the obligation first to provide a suitable (abstract) representation of implementations for CSP processes, secondly to provide a definition matching our account of an implementation's (conditionally) yielding a failure—thinking of it as the implementation of some process or other—and thirdly to spell out what it is for an implementation to be the implementation of a particular process (as represented by a particular set of failure-representing pairs). This third obligation arises because the range of the existential quantifier in terms of which we want to spell out the 'might' of nondeterminism is the domain of precisely the implementations of the process in question. But how are we going to avoid circularity? All this is supposed to be leading up to a definition of what sets of failure-representing pairs represent processes, yet we just seem to have presupposed that we already know. In fact there is no circularity. The point is that we are assuming that the identity of a process is to be given by a set of failures, and we can say what it is to be an implementation of a process in a way which appeals to this fact and no more; we shall then be in a position to characterize process representations as all and only those sets of failures which can be seen to give the identity of a process in the intended way—that is to say, which contain all and only those failures which they *might* exhibit, when 'might' is spelt out in terms of the definitions we have framed. Or almost. For, if we want to fit in with the usual definitions, then we should rather say that this characterization provides a fundamental necessary condition for when a set of failures is a process representation; it can be taken to leave room for other special constraints. And there will be one further condition we need to impose in order to capture the standard CSP model.

The implementational definition

The obvious way to represent implementations is to use trees—which we can take to be subsets T of Σ^* satisfying the conditions

$$\langle\,\rangle \in T$$
$$s\langle a\rangle \in T \;\Rightarrow\; s \in T$$

An implementation tree T is to be thought of as containing the range of possible traces that it would impose on a process as its implementation—possible, that is, only in the sense that they are conditional on communications offered. There is no nondeterminism here.

The classic failures model

Next we have to define what it is for (s, X) to be a failure of T—to model the idea that (s, X) is a possible failure of a process under the implementation T. This must be when T can yield s (if presented with suitable communications) and then refuse to yield c for any $c \in X$; in other words, when $s \in T$ but $s\langle c\rangle \notin T$ for any $c \in X$. Hence

$$f(T) = \{(s, X) \in \Sigma^* \times \mathsf{F}(\Sigma) \mid s \in T \wedge \forall c \in X . s\langle c\rangle \notin T\}$$

gives the set $f(T)$ of all possible failures of T—where $\mathsf{F}(\Sigma)$ is the set of finite subsets of Σ.

Recall that our strategy is to define what it is to be an implementation of a process in a way which presupposes only that processes can be represented by some set or other of failures. If, then, we had a general definition of what it is for T to implement an arbitrary set of failures, we would be guaranteed what we want. And such a general definition is, in any case, not an unreasonable thing to provide. For specifications which are formulated as 'predicates' of (observable) behaviour, after the approach of (Hoare 1985), for example, will have sets of failures as their extension—sets which are not themselves expected to be process representations—and it then makes sense to raise the direct question of how such specifications might be (deterministically) implemented. In fact, the question of what sets of failures are to count as processes in our modelling could be seen as precisely the question of which predicate-specifications already have a (possibly nondeterministic) process as their extension.

Following this idea, the obvious definition of what it is for T to be an implementation of an (arbitrary) set F of failures is that T yields no failure that is not in F; if F is thought of as the extension of a specification, this just means that T does not go beyond the constraints that the specification imposes. Hence we can define

$$\mathcal{I}(F) = \{T \mid f(T) \subseteq F\}$$

to give the set $\mathcal{I}(F)$ of all possible implementations of F.

And now we are in a position to mathematicize the point that nondeterministic 'choice' is to be spelt out in terms of existential quantification over implementations—and thereby to pick out which sets of failures represent processes. Trivially, if (s, X) is a failure of some implementation of F, then $(s, X) \in F$. But, to represent a process, F must satisfy the converse also: (s, X) must be contained in F (if and) *only if* (s, X) is a failure of some implementation. This means that F must be determined by its implementations in the sense that

$$F = \bigcup \{f(T) \mid T \in \mathcal{I}(F)\}$$

Here, then, is a necessary condition of F's representing a process. But is it sufficient? Observe that this condition allows F to be empty—and it entails that F is empty if and only if $\mathcal{I}(F)$ is empty. Do we want $\emptyset$ to represent a process—'the impossible process', as it were, which has no implementations? If only because the definition of CSP operations would be messy if we allowed it, $\emptyset$ is standardly ruled out. And so I propose that an implementational definition should pick out as process-representations all and only those sets F of failures such that $\mathcal{I}(F) \neq \emptyset$ and such that the above identity obtains.

Observe that the way F is determined by $\mathcal{I}(F)$ in this formulation can be generalized to define a set of failures $\mathcal{P}(M)$ from an arbitrary set M of implementation trees:

$$\mathcal{P}(M) = \bigcup \{f(T) \mid T \in M\}$$

We can then use $\mathcal{P}$ to provide the following formal definition for when, according to our modelling, a set F of failures counts as a process: .

Definition 1. F is a process iff $\mathcal{I}(F) \neq \emptyset$ and $F = \mathcal{P}(\mathcal{I}(F))$.

And the operator $\mathcal{P}$ can also be used to provide an alternative characterization of process-representations, which does not involve $\mathcal{I}$. For it turns out that any non-empty totality of implementations will determine a process.

Lemma 2. F is a process iff $F = \mathcal{P}(M)$, for some $M \neq \emptyset$.

Proof. The implication from left to right is immediate. And the converse obtains because, for any M, $M \subseteq \mathcal{I}(\mathcal{P}(M))$ and $\mathcal{P}(M) = \mathcal{P}(\mathcal{I}(\mathcal{P}(M)))$. □

The proof of this lemma reveals a little bit about how $\mathcal{I}$ and $\mathcal{P}$ fit together. But we might also record that both these operators are monotonic with respect to set-inclusion; matching what we had above, $\mathcal{P}(\mathcal{I}(F)) \subseteq F$ and $\mathcal{I}(F) = \mathcal{I}(\mathcal{P}(\mathcal{I}(F)))$. And the compound operators $M \mapsto \mathcal{I}(\mathcal{P}(M))$ and $F \mapsto \mathcal{P}(\mathcal{I}(F))$ are of some interest. It is easy to check that the first one is a closure operator: that is to say, (i) it is monotonic with respect to set-inclusion, (ii) it maps to supersets, and (iii) it is the identity function on its own range—in other words, $\mathcal{I}(\mathcal{P}(\mathcal{I}(\mathcal{P}(M)))) = \mathcal{I}(\mathcal{P}(M))$, for any M. The closed sets, viz. those M such that $M = \mathcal{I}(\mathcal{P}(M))$, are then precisely the sets $\mathcal{I}(F)$, for some F, and non-empty closed sets are in a natural one–one correspondence with processes—the correspondence given by $M \mapsto \mathcal{P}(M)$, with $F \mapsto \mathcal{I}(F)$ as its inverse. We could, then, think of non-empty closed sets of implementation trees as themselves representations of processes: the (non-variable-binding) CSP operators could be directly interpreted by operations over these sets, and the nondeterminism ordering $\sqsubseteq$—which could be invoked in the usual way to guarantee an interpretation for the fixed point operator—is just the converse of set inclusion. We would then have a sort of powerdomain semantics for CSP: the identity of a process would be given directly in terms of the range of its implementations. This, we should observe, is precisely the idea arrived at by a different route, and worked out in detail, in (Brookes 1983, Chapter 3).

The operator $F \mapsto \mathcal{P}(\mathcal{I}(F))$, on the other hand, could be thought of as a 'co-closure' operator (or an 'interior' operator, though *not* in the topological sense). This is because (i) it is monotonic with respect to set inclusion, (ii) it maps to *sub*sets, and (iii) it is the identity function on its own range. And the class of non-empty 'co-closed' sets of failures then turns out to be precisely the domain of processes. This domain is easily seen to be closed under arbitrary unions, and, provided that $\mathcal{I}(F)$ is non-empty, $\mathcal{P}(\mathcal{I}(F))$ will then be the 'largest' process contained in F; not only is it contained in F itself, but it is also the union of all the processes contained in F. This is noteworthy, because it means that if F is the extension of a specification, then either $\mathcal{P}(\mathcal{I}(F)) = \emptyset$, in which case the specification cannot be satisfied, or else $\mathcal{P}(\mathcal{I}(F))$ is the most nondeterministic process satisfying the specification.

However this may be, we now have a definition against which to assess any collection of axioms proposed for the failures model. At the outset we introduced notions of 'soundness' and 'completeness', which are defined in the following way: a collection of axioms will be sound just in the case when any set of failures which is a process (as defined above) satisfies all of them; conversely, it will be complete just in the case when any set of failures which satisfies all of the axioms is a process. And what we want is a collection of axioms which is both sound and complete.

The soundness and completeness of the standard axioms

The following axioms are those used in (Brookes and Roscoe 1985) to stipulate which sets F of failure-pairs are to be taken to represent processes:

(A0) $(\langle\,\rangle, \emptyset) \in F$

(A1) $(t, \emptyset) \in F \wedge s \leqslant t \Rightarrow (s, \emptyset) \in F$

(A2) $(s, X) \in F \wedge Y \subseteq X \Rightarrow (s, Y) \in F$

(A3) $(s, X) \in F \wedge \forall c \in Y . (s\langle c\rangle, \emptyset) \notin F \Rightarrow (s, X \cup Y) \in F$

In fact logically simpler formulations of (A1), (A2), and (A3) are available, but they would not generalize so straightforwardly to Section 2.3, and so we shall stick to these. But observe that, in the presence of (A1) and (A2), (A0) is equivalent to the condition that $F \neq \emptyset$.

And now we can turn directly to the main result of this section.

Theorem 3. *Axioms (A0) to (A3) are sound and complete.*

Proof. Soundness—that every process F satisfies these axioms—is routine. Conversely, to show completeness, say F satisfies (A0)–(A3); then we have to show (i) that $\mathcal{I}(F) \neq \emptyset$, and (ii) that $F = \bigcup \{f(T) \mid T \in \mathcal{I}(F)\}$. But in (ii), inclusion from right to left is trivial; what remains to be shown is that if $(s, X) \in F$, then there exists a $T \in \mathcal{I}(F)$ such that $(s, X) \in f(T)$, and this will also be sufficient for demonstrating (i), since, by (A0), $(\langle\,\rangle, \emptyset) \in F$, and so the existence of something in $\mathcal{I}(F)$ will be guaranteed.

So, take $(s, X) \in F$. Then we can define T as follows:

$$T = \{t \in \Sigma^* \mid (t, \emptyset) \in F \land \forall c \in X \,.\, s\langle c\rangle \not\leqslant t\}$$

By (A0) and (A1), T will clearly be an implementation tree, and, by (A2), $(s, X) \in f(T)$. Now we just have to check that $T \in \mathcal{I}(F)$—in other words, that $f(T) \subseteq F$. Assume, then, that $(t, Y) \in f(T)$; we have to show that $(t, Y) \in F$. But, by definition, our assumption means that the following conditions obtain:

$$(t, \emptyset) \in F$$
$$\forall c \in X \,.\, s\langle c\rangle \not\leqslant t$$
$$\forall d \in Y \,.\, [(t\langle d\rangle, \emptyset) \notin F \lor \exists c \in X \,.\, s\langle c\rangle \leqslant t\langle d\rangle]$$

Hence Y is the union of two sets Y_0 and Y_1 given by:

$$Y_0 = \{d \in Y \mid \exists c \in X \,.\, s\langle c\rangle \leqslant t\langle d\rangle\}$$
$$Y_1 = \{d \subset Y \mid (t\langle d\rangle, \emptyset) \notin F\}$$

And then, if $Y_0 = \emptyset$, $(t, Y) = (t, Y_1) \in F$ by (A3). On the other hand, if $Y_0 \neq \emptyset$, we can deduce that $s = t$ and $Y_0 \subseteq X$, and so, since $(s, X) \in F$, $(t, Y_0) = (s, Y_0) \in F$ by (A2); hence, by (A3), $(t, Y) = (t, Y_0 \cup Y_1) \in F$. □

To round off our discussion of the classic model, we should say something about deterministic processes. The familiar definition of when a process F is deterministic is as follows:

$$(s, X) \in F \Rightarrow \forall c \in X \,.\, (s\langle c\rangle, \emptyset) \notin F$$

But we might also record the following simpler formulation, which (granted that F is indeed a process) is equivalent:

$$(s, \{c\}) \in F \Rightarrow (s\langle c\rangle, \emptyset) \notin F$$

On the other hand, a natural definition would be to say that a process F is deterministic if and only if it has exactly one implementation, and clearly this is if and only if $F = \mathcal{P}(\{T\})$—in other words, $F = f(T)$—for some particular implementation tree T. But happily it is easy to check that the definitions are equivalent. Observe, moreover, that the function $T \mapsto f(T)$ provides a natural one–one correspondence between implementation trees and deterministic processes, whose inverse is given by $F \mapsto \{s \mid (s, \emptyset) \in F\}$. This correspondence can be proved by working entirely with the axiomatic characterization of processes and using the explicit definition of 'deterministic'. Therefore, if we had started out with such a characterization and then proved that any process (thus defined) contains some deterministic process, and moreover that it is the union of all deterministic processes which it contains, then we would have covered essentially the same ground as the proof of completeness. The correspondence between deterministic processes and implementation trees reveals that we would have proved that any process has at least one implementation and that it is determined by the range of all the implementations it has. But we have made this condition the *definition* of a process, and we have proved that the axiomatic characterization is then sound as well as complete.

The function $F \mapsto \{s \mid (s, \emptyset) \in F\}$ is of course well defined even if F is not deterministic: it takes a process to the implementation T specified in the completeness proof for the case when $(s, X) = (\langle\,\rangle, \emptyset)$. This is the 'most generous' implementation of F, and so, if we go back to a process with the function $T \mapsto f(T)$, then we have defined a retraction on the space of processes which takes F to the deterministic process that definitely *can* communicate anything that F might. Observe that, under the representation of processes as non-empty closed sets M of implementations, this retraction is characterizable as $M \mapsto \{\bigcup M\}$; any such M will be closed under arbitrary unions of constituent trees, and we have just picked out the biggest.

2.3 The infinitary model

In (Roscoe 1988) the *desideratum* of being able to handle the infinite nondeterministic composition of processes—that is, their composition with an infinitary version of ⊓—motivates two modifications to the model we have been discussing. First, failures are to be taken to be possibly infinite objects: they come from $\Sigma^* \times \mathbb{P}(\Sigma)$, where $\mathbb{P}(\Sigma)$ is the unrestricted power set of Σ. Secondly, the representation of processes is to be enriched with a new component: this is a subset I of Σ^ω, which is to be thought of as containing the 'infinite traces' that the process might exhibit. Hence, since we are going to ignore divergence, the idea will simply be to represent processes by pairs $\langle F, I\rangle$, where $F \subseteq \Sigma^* \times \mathbb{P}(\Sigma)$ and $I \subseteq \Sigma^\omega$, and to represent the nondeterministic composition $\bigsqcap_{\alpha \in A}\langle F_\alpha, I_\alpha\rangle$ of any A-indexed family of processes as their co-ordinate-wise union $\langle\bigcup_{\alpha \in A} F_\alpha, \bigcup_{\alpha \in A} I_\alpha\rangle$.

It is important, then, that the domain of processes turns out to be closed under such 'unions'. Of course, the classic failures model was already closed under arbitrary unions; however, some examples we shall consider shortly show that they are not guaranteed to provide an adequate interpretation for infinitary $\sqcap$. In fact, the examples will establish something stronger—that there is no way at all to interpret $\sqcap$ satisfactorily in this finitary model. To handle $\sqcap$ we have to give the identity of processes in a way which makes non-trivial reference to infinite items of behaviour.

But first we must address the very idea of 'infinite behaviour'. The infiniteness of the refusals component of a failure is of a very different kind from the infiniteness of an infinite trace, yet the same general point emerges concerning both kinds. We can put it like this. So far as we are concerned with a process under a given implementation, then there is no more to the idea that a process can exhibit an infinite item of behaviour than that it can exhibit every finite sub-behaviour, but the idea that a process non-deterministically 'might' exhibit an infinite item of behaviour is something stronger than that it might exhibit each finite sub-behaviour: it is the idea that there is some one implementation under which every finite sub-behaviour can be exhibited, not just that every finite sub-behaviour can be exhibited by some implementation or other.

We can now show that the model of Section 2.2 collapses the contrast we have just explained because of strong closure conditions that its finitary character imposes on the range of possible implementations of a process. The following lemmas make this precise. First, say $\sigma \in \Sigma^\omega$; then we have the following lemma.

Lemma 4. $\forall s < \sigma \,.\, \exists T \in \mathcal{I}(F) \,.\, s \in T \;\Rightarrow\; \exists T \in \mathcal{I}(F) \,.\, \forall s < \sigma \,.\, s \in T$

Proof. Assume the antecedent and then check that $T \in \mathcal{I}(F)$, where $T = \{t \in \Sigma^* \mid (t, \emptyset) \in F\}$. □

And, using '$Y \subseteq_f X$' to mean that Y is a finite subset of X, we have another lemma.

Lemma 5.

$\forall Y \subseteq_f X \,.\, \exists T \in \mathcal{I}(F) \,.\, (s, Y) \inf(T) \;\Rightarrow\; \exists T \in \mathcal{I}(F) \,.\, \forall Y \subseteq_f X \,.\, (s, Y) \in f(T)$

Proof. Assume the antecedent and then show that $T \in \mathcal{I}(F)$, where $T = \{t \in \Sigma^* \mid (t, \emptyset) \in F \land \forall c \in X \,.\, s\langle c\rangle \nleq t\}$. The details are similar to those of the completeness proof in Section 2.2. (X's being infinite does not spoil the overall strategy.) □

Lemma 4 means that if every initial segment s of an infinite sequence σ of communications is a possible trace of a process, then there is a single implementation which can yield every such s, so that σ is a possible 'infinite trace'. Lemma 5 means that if for every finite subset Y of X it is possible that, after communicating s, Y is refused, then there is a single implementation under which every such subset is refused (after s); but, under a given implementation, a process's refusing a set is the same as its refusing each communication in the set individually, and so this condition can be seen to be equivalent to saying that every element of X is refused under some single implementation—or, in other words, that the whole of X is refused. If X is infinite, then this would be an 'infinite refusal' of the process.

And now, to see what this has to do with the infinite nondeterministic composition of processes, consider two examples. First, let $\Sigma = \{a\}$ and put

$$P_0 = STOP$$
$$P_{n+1} = a \to P_n$$

Alternatively, if T_n is the implementation tree $\{a^0, \ldots, a^n\}$, where $a^0 = \langle\,\rangle$ and $a^{n+1} = a^n \langle a \rangle$, then we could characterize these processes directly by saying that $P_n = \mathcal{P}(\{T_n\})$. But now what should the implementations of the infinite nondeterministic composition $\bigcap_{n<\omega} P_n$ be? Obviously all the T_n are included. But according to the first lemma, if we are working in the finitary model, then there must be a single implementation containing all finite sequences of a's; in other words, $T_\omega = \{a^0, a^1, \ldots, a^n, \ldots\}$ also has to be an implementation. Yet this is surely more than we bargained for; in writing down $\bigcap_{n<\omega} P_n$ we might naturally have wanted to mean a process whose *only* implementations were the T_n.

For the second example, let Σ be a denumerable set $\{a_0, a_1, \ldots\}$ and then put

$$Q_n = a_n \to STOP$$

In other words, $Q_n = \mathcal{P}(\{T_n\})$, where $T_n = \{\langle\,\rangle, \langle a_n \rangle\}$. And we can now ask the same question as before: what should the implementations of $\bigcap_{n<\omega} Q_n$ be? Clearly, given any finite subset X of Σ there must be some implementation T such that $(\langle\,\rangle, X) \in f(T)$, since $(\langle\,\rangle, X)$ is a failure of all those (infinitely many) Q_n such that $a_n \notin X$. But by the second lemma, the finitary model then commits us to a single implementation T for $\bigcap_{n<\omega} Q_n$ such that $(\langle\,\rangle, X) \in f(T)$ for every finite subset X of Σ, which can only be the implementation $\{\langle\,\rangle\}$. And this is certainly more than we should bargain for in writing down $\bigcap_{n<\omega} Q_n$; in allowing for an infinite range of initial possibilities, we do not want to be forced thereby into allowing also for the possibility of immediate deadlock.

To overcome the problems that these two examples reveal, what we now have to do is to set things up so that the possibility of infinite behaviour does not automatically follow, in the way the above lemmas describe, from the possibility under some implementation or other of every finite sub-behaviour. And this is remarkably easy if we take pairs $\langle F, I \rangle$ and think of providing an implementational definition of which ones count as representing processes. There is no new notion of implementation to bother about, and definitions can be framed in the obvious naive way to match the pattern of development in Section 2.2. Observe, moreover, that although F and I will themselves contain items of infinite behaviour, we can still ground our account in terms of *finite*—and hence 'observable'—behaviour. This is because, if we follow the same strategy as before, then when we talk about what behaviour a process can (conditionally on a presentation of communications) exhibit, this falls within the scope of the 'might' of nondeterminism, and so we are talking about the behaviour of a process *under a given implementation*; but, under a given implementation, a process can exhibit infinite behaviour if and only if it can exhibit every finite item of sub-behaviour. We shall leave the reader to formulate an explicit account of nondeterministic 'might', if he wants to be rigorous along the same lines as before.

The implementational definition

To represent implementations we just take the trees T defined in Section 2.2. But we have to modify the definition of $f(T)$ to include infinite failures:

$$f(T) = \{(s, X) \in \Sigma^* \times \mathbb{P}(\Sigma) \mid s \in T \wedge \forall c \in X . s\langle c \rangle \notin T\}$$

We also have to define the set $i(T)$ of infinite traces that (a process under) the implementation T can yield:

$$i(T) = \{\sigma \in \Sigma^\omega \mid \forall s < \sigma . s \in T\}$$

Observe that $i(T)$ is just the set of infinite paths through the tree T. Then, for an arbitrary pair $\langle F, I \rangle$ (of the right logical type) the set $\mathcal{I}(F, I)$ of all the implementations of $\langle F, I \rangle$ is given by

$$\mathcal{I}(F, I) = \{T \mid f(T) \subseteq F \wedge i(T) \subseteq I\}$$

And, for an arbitrary set M of implementation trees, we define the pair $\mathcal{P}(M)$ by stipulating that

$$\mathcal{P}(M) = \langle \bigcup \{f(T) \mid T \in M\}, \bigcup \{i(T) \mid T \in M\} \rangle$$

The operators $\mathcal{I}$ and $\mathcal{P}$ then fit together in the same way as before, so that we can give the implementational definition of a process as follows.

The infinitary model

Definition 6. $\langle F, I \rangle$ is a process iff $\mathcal{I}(F, I) \neq \emptyset$ and $\langle F, I \rangle = \mathcal{P}(\mathcal{I}(F, I))$.

And there is an alternative characterization of processes, given as follows.

Lemma 7. $\langle F, I \rangle$ is a process iff $\langle F, I \rangle = \mathcal{P}(M)$, for some $M \neq \emptyset$.

This is easy to check along the same lines as before. It is, of course, trivial that if T is an implementation of $\langle F, I \rangle$ and $(s, X) \in f(T)$, then $(s, X) \in F$, and also that if T is an implementation and $\sigma \in i(T)$, then $\sigma \in I$; hence in the equation $\langle F, I \rangle = \mathcal{P}(\mathcal{I}(F, I))$ (co-ordinate-wise) inclusion from right to left makes no contribution to the definition of a process. The guts of the definition is the other way round—in other words, the following two conditions:

$$(s, X) \in F \Rightarrow \exists T \in \mathcal{I}(F, I) . (s, X) \in f(T)$$
$$\sigma \in I \Rightarrow \exists T \in \mathcal{I}(F, I) . \sigma \in i(T)$$

We now invite the reader to check that, as we require, the co-ordinate-wise union $\langle \bigcup_{\alpha \in A} F_\alpha, \bigcup_{\alpha \in A} I_\alpha \rangle$ of any non-empty family $\{\langle F_\alpha, I_\alpha \rangle\}_{\alpha \in A}$ of processes is itself a process.

Such unions, we claimed, would provide an interpretation for the (possibly infinite) nondeterministic composition of a family of processes. Let us illustrate this with the examples $\{P_n\}_{n<\omega}$ and $\{Q_n\}_{n<\omega}$ considered above. First, P_n was given as $\mathcal{P}(\{a^k \mid k \leqslant n\})$ in the background alphabet $\Sigma = \{a\}$, and so, if we understand this in our present model, $P_n = \langle F_n, I_n \rangle$ where

$$F_n = \{(a^k, \emptyset) \mid k \leqslant n\} \cup \{(a^n, \{a\})\}$$
$$I_n = \emptyset$$

Hence, the co-ordinate-wise union can be described as follows:

$$\langle \bigcup_{n<\omega} F_n, \bigcup_{n<\omega} I_n \rangle = \langle \Sigma^* \times \mathbb{P}(\Sigma), \emptyset \rangle$$

And this works perfectly as an interpretation for $\bigsqcap_{n<\omega} P_n$, since then

$$\mathcal{I}(\bigcup_{n<\omega} F_n, \bigcup_{n<\omega} I_n) = \{\{a^k \mid k \leqslant n\} \mid n < \omega\}$$

Observe that the tree $\{a^k \mid k < \omega\}$ is *not* an implementation, because the emptiness of the infinite traces component rules it out. The trouble with the classic failures model was that this implementation could not be ruled out.

On the other hand, consider $\{Q_n\}_{n<\omega}$. The background alphabet is $\Sigma = \{a_0, \ldots, a_n, \ldots\}$, and $Q_n = \mathcal{P}(\{\langle\,\rangle, \langle a_n\rangle\})$; understanding this in the new model, $Q_n = \langle F_n, I_n \rangle$, where

$$F_n = \{(\langle\,\rangle, X) \mid a_n \notin X \subseteq \Sigma\} \cup \{(\langle a_n\rangle, X) \mid X \subseteq \Sigma\}$$
$$I_n = \emptyset$$

Hence

$$\langle \bigcup_{n<\omega} F_n, \bigcup_{n<\omega} I_n \rangle = \langle \{(\langle\,\rangle, X) \mid X \subset \Sigma\} \cup \{(\langle b\rangle, X) \mid b \in \Sigma \wedge X \subseteq \Sigma\}, \emptyset \rangle$$

And this provides a good interpretation for $\sqcap_{n<\omega} Q_n$, since then

$$\mathcal{I}(\bigcup_{n<\omega} F_n, \bigcup_{n<\omega} I_n) = \{\{\langle\,\rangle\} \cup \{\langle b\rangle \mid b \in X\} \mid \emptyset \neq X \subseteq \Sigma\}$$

This set contains an explosion of implementations, but $\{\langle\,\rangle\}$ is not among them, because $(\langle\,\rangle, \Sigma) \in f(\{\langle\,\rangle\})$, though $(\langle\,\rangle, \Sigma) \notin \bigcup_{n<\omega} F_n$. The tree $\{\langle\,\rangle\}$ was not ruled out in the classic model, because, when failure-pairs are restricted to those with a finite refusals component, then $f(\{\langle\,\rangle\})$ can contain no pair to stop its being an implementation.

These two examples can each be converted into an interesting demonstration that, according to our new definitions, the closure operator $\mathcal{I}(\mathcal{P}(\cdot))$—and it is easy to check that it *is* a closure operator—does not close up as much as the corresponding operator in Section 2.2. Every set of implementation trees which is closed in the old sense is also closed in the new sense, but the converse does not hold: there are now *more* closed sets of implementations. To see this we could consider the set $\{\{a^k \mid k \leq n\} \mid n < \omega\}$ of trees in the alphabet $\{a\}$; to obtain a closed set in the old sense we should have to throw in the tree $\{a^k \mid k < \omega\}$, but in the new sense it is closed already. On the other hand, whenever Σ is infinite, then $\{\langle\,\rangle\}$ must be in the old-sense closure of $\{\{\langle\,\rangle, \langle b\rangle\} \mid b \in \Sigma\}$, but it cannot be in the new-sense closure.

As we would expect, the function $\langle F, I \rangle \mapsto \mathcal{I}(F, I)$, with inverse $M \mapsto \mathcal{P}(M)$, sets up a one–one correspondence between processes and non-empty closed sets of implementation trees, and we can therefore think of the identity of a process as given directly by the totality of its implementations. This is the natural way to think about the identity of *deterministic processes*—which, as before, can be taken to be processes that have exactly one implementation. Obvious examples are the processes P_n and Q_n above. And, since all singleton sets of implementation trees are obviously closed in both the old and the new senses, the class of deterministic processes over a given alphabet Σ will be precisely 'the same' now as it was in Section 2.2.

The infinitary model

But there are some important differences from the set-up in Section 2.2. We have just seen, for example, that the set of implementations of an unboundedly non-deterministic process is not necessarily closed under unions of constituent trees (though it will be closed under *finite* unions). Again, chains of implementation sets—ordered by $\supseteq$—do not behave as nicely as before. An intersection of closed sets will certainly be closed, but the intersection of a chain of non-empty closed sets is not guaranteed to be non-empty. This points up the fact that we cannot carry over to the infinitary model any straightforward fixed point theory based on the non-determinism ordering (and a lot of (Roscoe 1988) is directed towards getting round the problem).

We can, of course, pick out the finitary model as a retract of the infinitary one: the old closure operation applied to sets of implementation trees which are closed in the new sense is the required retraction. This throws in all those additional implementations that cannot be ruled out on the basis of finite items of behaviour. And, if we go back to the representation of a process as a pair $\langle F, I \rangle$, then the retraction can be characterized as the composition of the following projection and injection mappings from the infinitary model to the finitary one and back again:

$$\langle F, I \rangle \mapsto \{(s, X) \in F \mid X \in \mathsf{F}(\Sigma)\}$$
$$F \mapsto \langle \{(s, X) \mid \forall Y \subseteq_f X \, . \, (s, Y) \in F\}, \{\sigma \mid \forall s < \sigma \, . \, \langle s, \emptyset \rangle \in F\} \rangle$$

It is straightforward to check that this definition corresponds to the first one.

To provide another way to get a feel for the difference between the finitary and the infinitary model, we might briefly consider the the standard ultrametric space of (implementation) trees over a given alphabet Σ. In this space the notion of convergence can be characterized directly by saying that a sequence $\langle T_i : i < \omega \rangle$ converges to T if and only if, for any N (however large), there exists an n such that, whenever $n \leqslant m$, then T and T_m contain precisely the same elements from Σ^N. And it is then easy to check that, in the finitary model, the limit of a convergent sequence of implementations of a process is again an implementation of that process so that the set of all implementations of a process will be closed in the metric space sense (though the converse does not hold: after all, simple examples show that the system of implementation sets is not closed under finite unions, and so it is not the system of closed sets of any topology). But things are different in the infinitary model. We can appeal again to the first example above: $\{\{a^k \mid k \leqslant n\} \mid n < \omega\}$ does not contain the limit tree $\{a^k \mid k < \omega\}$. This point in fact has to do with only one of the two kinds of unbounded non-determinism, and we can draw a contrast between them. For, if we liberalized the classic model just by allowing infinite refusals, then the set of implementations of a process would still be closed in the metric space sense.

A set of axioms

We now present a set of axioms which, as we shall demonstrate, is sound and complete for the implementational definition of a process. We shall label them (I0), (I1), (I2), (I3), (I6), and (I8) in order to leave room for additional axioms, which would govern a divergence component if we included it. The labelling will then mesh with that of (Roscoe 1988)—except for (I8), the axioms are actually the same, or only trivially different. The question of what the eighth axiom should be has been one of the most interesting and difficult problems in addressing the model for infinite traces, and, as a halfway stage in the proof of soundness and completeness, we shall in fact offer an alternative to (I8)—again distinct from Roscoe's axiom—which is less appealing but sometimes easier to work with.

Axioms (I0)–(I3) look the same as (A0)–(A3) of Section 2.2, though X and Y are no longer restricted to finite sets:

(I0) $(\langle\,\rangle, \emptyset) \in F$

(I1) $(s, X) \in F \wedge Y \subseteq X \Rightarrow (s, Y) \in F$

(I2) $(s, X) \in F \wedge \forall c \in Y . (s\langle c\rangle, \emptyset) \notin F \Rightarrow (s, X \cup Y) \in F$

Axiom (I6) states an obvious necessary condition of a process's containing an infinite trace—that every initial segment is a (finite) trace. Hence it can be seen as an extension of (I1) to cover the infinite case:

(I6) $\sigma \in I \wedge s < \sigma \Rightarrow (s, \emptyset) \in F$

Axiom (I8), on the other hand, could be seen as an infinite extension of (I3). But it will take a little time to explain how this goes.

Observe first that, if we take $X = \emptyset$, then we can manipulate (I3) into the logically equivalent form

($I_0 3$) $(s, \emptyset) \in F \Rightarrow \forall Y_{(s,Y) \notin F} . \exists a \in Y . (s\langle a\rangle, \emptyset) \in F$

where the notational scheme $\forall X_{\Phi(X)} . \Psi(X)$ means 'for all X such that $\Phi(X)$, $\Psi(X)$', which we could have written $\forall X \in \{X \mid \Phi(X)\} . \Psi(X)$. Axiom ($I_0 3$) itself just states that if s is a trace, then, for any Y, (s, Y)'s not being a failure implies that the process might be prepared to extend s with some communication or other in Y. But we can iterate the principle to any finite stage n in the following way (where we adopt the convention that $\langle a_0, \ldots, a_{n-1}\rangle = \langle\,\rangle$ if $n = 0$, so that ($I_0 3$) is itself an instance of the scheme):

The infinitary model

$$(I_n 3) \quad (s, \emptyset) \in F \Rightarrow \begin{bmatrix} \forall Y_{0\,(s,Y_0) \notin F} \cdot \exists a_0 \in Y_0. \\ \forall Y_{1\,(s\langle a_0 \rangle, Y_1) \notin F} \cdot \exists a_1 \in Y_1. \\ \ddots \\ \forall Y_{n\,(s\langle a_0, \ldots, a_{n-1} \rangle, Y_n) \notin F} \cdot \exists a_n \in Y_n. \\ (s\langle a_0, \ldots, a_n \rangle, \emptyset) \in F \end{bmatrix}$$

And, for all n, this iteration follows from $(I_0 3)$. To see this, assume that $(s, \emptyset) \notin F$; then if you can find a Y_0 such that $(s, Y_0) \notin F$, I can pick an $a_0 \in X_0$ such that $(s\langle a_0 \rangle, \emptyset) \in F$; and so *then*, if you find a Y_1 such that $(s\langle a_0 \rangle, Y_1)$, I can pick an $a_1 \in Y_1$ such that $(s\langle a_0, a_1 \rangle) \in F$; and so..., if you can find a Y_n such that $(s\langle a_0, \ldots, a_{n-1} \rangle, Y_n) \notin F$, I can pick an a_n such that $(s\langle a_0, \ldots, a_n \rangle, \emptyset) \in F$—in other words, $s\langle a_0, \ldots, a_n \rangle$ is a (finite) trace. And what (I8) will do is extend this iteration into the infinite: as it were, (I8) will be $(I_\omega 3)$.

A more cumbersome, though perhaps more revealing, iteration of $(I_0 3)$ would be obtained if we reflected the informal 'you find, I pick' proof of $(I_n 3)$ by restricting the range of $\exists a_k$ not merely to Y_k but to elements of Y_k such that $(s\langle a_0, \ldots, a_k \rangle, \emptyset) \in F$. Or another modification would be to restrict $\forall Y_k$ to non-empty sets; clearly this would be no restriction in the logical force of the formula, since if $(s, \emptyset) \in F$, then you cannot in any case pick $Y_0 = \emptyset$ such that $(s, Y_0) \notin F$, and a similar argument applies at each successive stage. And we could think of extending either of these modifications into the infinite; as we shall show later, the result would be equivalent to the extension of $(I_n 3)$ as it stands.

Modified or not, we should stress that $(I_n 3)$ does not actually guarantee the existence of a full-length trace $s\langle a_0, \ldots, a_n \rangle$ extending s. Consider our proof: if you can't find any Y_k such that $(s\langle a_0, \ldots a_{k-1} \rangle, Y_k) \notin F$, then my last choice of a_{k-1} establishes the iteration by exhibiting the possibility of deadlock after $s\langle a_0, \ldots, a_{k-1} \rangle$. And it might be true only because I have a way of picking elements which must, however you choose, produce the possibility of deadlock at some stage. So then, we might think of $(I_n 3)$ as saying that if I don't have any such way of picking a_k, then I do have a way of picking a_k that may—if your choices let it—extend s to a trace $s\langle a_0, \ldots, a_n \rangle$. But we can switch modalities here and alternatively think of $(I_n 3)$ as saying that if I don't have any way of picking a_k that may—if you let it—produce deadlock, then I do have a way of picking a_k that must, however you choose, extend s to a trace $s\langle a_0, \ldots, a_n \rangle$.

And these ways of thinking can be extended to the infinite axiom too—but concerning an *infinite* extension of s. For (I8) is formulated using an infinite string of alternating universal and existential quantifiers, which can be understood in terms of the choosing and picking story:

46   Axioms for CSP processes

$$(\text{I8}) \quad (s, \emptyset) \in F \ \Rightarrow \ \begin{bmatrix} \forall X_{0\,(s, X_0) \notin F} \cdot \exists a_0 \in X_0. \\ \forall X_{1\,(s\langle a_0 \rangle, X_1) \notin F} \cdot \exists a_1 \in X_1. \\ \forall X_{2\,(s\langle a_0, a_1 \rangle, X_2) \notin F} \cdot \exists a_2 \in X_2. \\ \ddots \\ s\langle a_0, a_1, \ldots \rangle \in I \end{bmatrix}$$

We ought, however, to provide a precise way of modelling the interpretation of the consequent.

This will be a game-theoretic definition that picks up our informal remarks. The general idea of using games to interpret infinite strings of quantifiers is not unfamiliar—see (Kolaitis 1985), for example—but we shall set things up from scratch in a style tailored to our particular example. Truth conditions for the consequent of (I8) will, then, be given in terms of an infinite game $G(F, I, s)$ between two players, ENV and PRO say (which we could think of as the environment and the process respectively). ENV picks sets and PRO picks elements, and the truth of the consequent will be defined to consist in the existence of a winning strategy for PRO. And so we can then take (I8) to mean that

$$(s, \emptyset) \in F \ \Rightarrow \ \text{PRO has a winning strategy in } G(F, I, s)$$

The game $G(F, I, s)$ is defined as follows. At stage n ENV is to move first and then PRO is to move. The moves must obey the following constraints:

ENV's moves	PRO's moves
pick X_0 such that $(s, X_0) \notin F$	pick $a_0 \in X_0$
pick X_1 such that $(s\langle a_0 \rangle, X_1) \notin F$	pick $a_1 \in X_1$
pick X_2 such that $(s\langle a_0, a_1 \rangle, X_2) \notin F$	pick $a_2 \in X_2$
$\vdots$	$\vdots$

The game ends if either player is unable to move at some finite stage; otherwise there is an infinite sequence of moves. We also stipulate that

ENV wins if *either*: PRO cannot go at some finite stage
          or: after infinite play, $s\langle a_0, a_1, \ldots \rangle \notin I$

PRO wins if *either*: ENV cannot go at some finite stage
          or: after infinite play, $s\langle a_0, a_1, \ldots \rangle \in I$

Hence it is obvious that either ENV wins or PRO wins. But axiom (I8) is interpreted to mean that PRO has a *winning strategy* when s is a trace, and so we had better provide a way of modelling such strategies.

The infinitary model

A strategy for PRO has to dictate PRO's moves at each stage as a function of the course of play up to that stage—and a winning strategy has to do this in a way that guarantees a win. To capture this idea, we can take a winning strategy for PRO in $G(F, I, s)$ to consist in a family $\{\phi_i\}_{i<\omega}$ of functions into Σ, where the domain of ϕ_0 contains precisely those subsets X_0 of Σ such that $(s, X_0) \notin F$, where the domain of ϕ_{n+1} contains precisely those $(2n+1)$-tuples $\langle X_0, a_0, \ldots, X_n, a_n, X_{n+1}\rangle$ such that

$$\langle\langle X_0, a_0, \ldots, X_n\rangle, a_n\rangle \in \phi_n$$
$$(s\langle a_0, \ldots, a_n\rangle, X_{n+1}) \notin F$$

and the following conditions are satisfied:

$$\langle\langle X_0, a_0, \ldots, X_n\rangle, a_n\rangle \in \phi_n \Rightarrow a_n \in X_n \quad \text{(a)}$$
$$\forall n . \exists X_0 \ldots \exists X_n . \langle\langle X_0, a_0, \ldots, X_n\rangle, a_n\rangle \in \phi_n \Rightarrow s\langle a_0, a_1, \ldots\rangle \in I \quad \text{(b)}$$

But it is often convenient to identify a winning strategy ϕ with the union $\bigcup_{i<\omega} \phi_i$ of the constituent functions. Since the domains of the ϕ_i are mutually disjoint, ϕ will then itself be a function—and one which it is easy to decompose back into the ϕ_i by restricting its domain to tuples of the right type.

Various comments are now prompted concerning this definition. First, notice that ϕ_n might be empty (and hence also ϕ_m for all $m > n$); if n is the smallest i such that $\phi_i = \emptyset$, then ϕ guarantees a win for PRO at the nth stage, whatever moves ENV makes.

Secondly, we should look back to the discussion of $(I_n 3)$ to see what more a winning strategy for PRO in $G(F, I, s)$ provides than the totality of all these finite iterations of $(I_0 3)$. Clearly, when s is a trace, then for any particular n, $(I_n 3)$ guarantees PRO a strategy for not losing at or before stage n. And (assuming the axiom of choice) an obvious argument strengthens this fact, by infinite picking, to show that PRO has a single strategy for not loosing at any finite stage—in other words, we have a family of functions $\{\phi_i\}_{i<\omega}$ satisfying all the conditions for a winning strategy except possibly (b). But a very simple example shows that (I8) says something stronger. For let $\Sigma = \{a\}$ and let $\langle F, I\rangle$ be defined by

$$F = \{(a^n, \emptyset) \mid n < \omega\}$$
$$I = \emptyset$$

Clearly this pair then satisfies all our axioms aside from (I8). But it does not satisfy (I8). For consider the game $G(F, I, \langle\ \rangle)$, in which ENV has no option but to pick $\{a\}$ all the time and PRO has no option but to pick a, and these pickings are admissible. This means that if there were a winning strategy for PRO, then this inevitable course of play would have to be in accordance with it. However, by the definition of a winning strategy, I would then have to contain the infinite sequence consisting of a at every coordinate, which contradicts I's emptiness, and so there can be no winning strategy for PRO. Hence (I8) is falsified, since $(\langle\,\rangle, \emptyset)$ is certainly contained in F.

Thirdly, we might ask whether it would make any difference if we worked with games $G'(F, I, s)$ which were exactly like $G(F, I, s)$ except that they required PRO to pick a_n not just in X_n but also such that $(s\langle a_0, \ldots, a_n\rangle, \emptyset) \in F$. For the condition that whenever $(s, \emptyset) \in F$, then PRO has a winning strategy in $G'(F, I, s)$, would constitute an interpretation for the modification of axiom (I8) obtained by restricting the range of $\exists a_k$ to those elements of X_k such that $(s\langle a_0, \ldots, a_k\rangle, \emptyset) \in F$—and we claimed earlier that this modification would yield an equivalent axiom. Well, happily, a winning strategy for PRO in $G'(F, I, s)$ is precisely the same thing as a winning strategy in $G(F, I, s)$. We shall leave the reader to check this out, and also to check that it would not make any difference either, if we stipulated that ENV pick non-empty sets.

Fourthly, is the game $G(F, I, s)$ determined? Is it such that either one or other of the players must have a winning strategy? Obviously, if $(s, \emptyset) \notin F$, then ENV has a winning strategy: finish the game at one stroke by picking $\emptyset$ first go. But this partial answer is of little interest. We want to know whether $G(F, I, s)$ is determined when $(s, \emptyset) \in F$. This question naturally arises if we think of contraposing axiom (I8). For then the temptation is to drive negation through the infinite string of quantifiers to produce the formula

(I?8) $\left[\begin{array}{l} \exists X_{0(s, X_0)\notin F} \cdot \forall a_0 \in X_0. \\ \quad \exists X_{1(s\langle a_0\rangle, X_1)\notin F} \cdot \forall a_1 \in X_1. \\ \quad\quad \exists X_{2(s\langle a_0, a_1\rangle, X_1)\notin F} \cdot \forall a_2 \in X_2. \\ \quad\quad\quad \ddots \\ \quad\quad\quad\quad s\langle a_0, a_1, \ldots\rangle \notin I \end{array}\right] \Rightarrow (s, \emptyset) \notin F$

in which it is natural to take the truth of the antecedent to consist in the existence of a winning strategy for ENV in $G(F, I, s)$. But is this equivalent to axiom (I8) as we have interpreted it? It is entailed by (I8), since ENV and PRO cannot both have winning strategies in $G(F, I, s)$; but, the other way round, it would entail (I8) just in case we were guaranteed that one or other of the players must have a winning strategy—assuming that $(s, \emptyset) \in F$.

A winning strategy for ENV is neater to model than a winning strategy for PRO: essentially it is just a tree—which we can represent in the same way as the implementation trees—satisfying the conditions

$$t \in T \Rightarrow (st, \{a \mid t\langle a\rangle \in T\}) \notin F$$

The infinitary model 49

$$\tau \in i(T) \Rightarrow s\tau \notin I$$

where $i(T)$ (as before) is the set of infinite paths through T. A tree T works as a strategy for ENV in the following way: ENV first chooses the set $\{a \mid \langle a \rangle \in T\}$, in which case PRO must pick a particular a_0 such that $\langle a_0 \rangle \in T$, and then ENV chooses $\{a \mid \langle a_0, a \rangle \in T\}$, and so on. Hence, the interpretation we are entertaining for the new infinite formula (I?8) makes it equivalent to the following condition, which quantifies over trees:

$$\exists T . [\forall t \in T . (st, \{a \mid t\langle a \rangle \in T\}) \notin F \wedge \forall \tau \in i(T) . s\tau \notin I] \Rightarrow (s, \emptyset) \notin F$$

Or, contraposing *this* formula (quite unproblematically),

$$(s, \emptyset) \in F \Rightarrow \forall T . [\forall t \in T . (st, \{a \mid t\langle a \rangle \in T\}) \notin F \Rightarrow \exists \tau \in i(T) . s\tau \in I]$$

And it is interesting to observe that here we have precisely a condition reported in (Roscoe 1988) as an early attempt at formulating an eighth axiom.

But the condition is too weak, and Roscoe provides an example to show this. The reader is referred to (Roscoe 1988): it is easy to convert Roscoe's argument into a direct demonstration that, although the example satisfies our reading of (I?8), it is not a process according to the implementational definition, and we can then wait for the completeness theorem to deduce that (I?8) is strictly weaker than (I8). The argument, we should observe, depends on the assumption that $2^{\aleph_0}$ can be well ordered. But, granted this, we now have an answer to the question we started with: $G(F, I, s)$ is not in general determined.

Fifthly, and finally, we should observe that though our modelling of winning strategies for PRO in $G(F, I, s)$ appropriately reflects the overall logical form of the consequent of (I8), in fact a strategy for PRO need only depend, at each stage, on ENV's immediately previous move, together with the sequence of PRO's own moves up to the stage in question, and not on the entire history of ENV's moves at earlier stages. To see this, we can define a notion of *economical winning strategy* for PRO in $G(F, I, s)$ and show that an economical winning strategy exists if and only if a winning strategy exists in the sense we have so far been working with. Let us, then, model an economical winning strategy ϕ as (the union of) a family $\{\phi_i\}_{i<\omega}$ of functions into Σ such that the domain of each ϕ_n is in fact a set of failure-pairs—a subset of $\Sigma^n \times \mathbb{P}(\Sigma)$:

$$\phi_0 : \{(\langle \rangle, X) \mid (s, X) \notin F\} \to \Sigma$$
$$\phi_{n+1} : \{(s\langle a \rangle, X) \mid \exists Y . \langle (t, Y), a \rangle \in \phi_n \wedge (st\langle a \rangle, X) \notin F\} \to \Sigma$$

The following two conditions must be satisfied:

$$\langle (t, X), a \rangle \in \phi \Rightarrow a \in X \quad \text{(a)}$$
$$\forall t . \forall a . [t\langle a \rangle < \sigma \Rightarrow \exists X . \langle (t, X), a \rangle \in \phi] \Rightarrow s\sigma \in I \quad \text{(b)}$$

As before, it does not matter whether we think of ϕ as a family of functions or as one big function: these conditions have been given according to the second way of thinking, but they can easily be modified to suit the other.

Intuitively an economical winning strategy just *is* a winning strategy, but we have to show that any ϕ, as defined above, can be canonically transformed into a corresponding object $\hat{\phi}$ in the original modelling of winning strategies. So then, given ϕ, let us define $\hat{\phi}$ by induction. First, $\langle X, a \rangle \in \hat{\phi}_0$ if and only if $\langle (\langle\ \rangle, X), a \rangle \in \phi_0$. Then, assuming we have defined $\hat{\phi}_n$, $\langle \langle X_0, a_0, \ldots, X_n, a_n, X_{n+1}\rangle, a_{n+1}\rangle \in \hat{\phi}_{n+1}$ if and only if

$$\langle \langle X_0, a_0, \ldots, X_n \rangle, a_n \rangle \in \hat{\phi}_n$$
$$\langle (\langle a_0, \ldots, a_n \rangle, X_{n+1}), a_{n+1}\rangle \in \phi_{n+1}$$

And it is now a routine matter to check that $\hat{\phi}$ meets the required conditions. Hence, if PRO has an economical winning strategy (e.w.s.), then PRO has a winning strategy (w.s.). And, once we have proved the converse, then we shall have established that our canonical reading of axiom (I8), which can itself be labelled (I8), is equivalent to the following slicker variant (I′8):

(I8) $(s, \emptyset) \in F \Rightarrow$ PRO has a w.s. in $G(F, I, s)$

(I′8) $(s, \emptyset) \in F \Rightarrow$ PRO has an e.w.s. in $G(F, I, s)$

But we are about to formulate a completely different alternative to axiom (I8). This too will take the form 'if s is a trace, then $\Phi(F, I, s)$', and we shall be able to demonstrate its equivalence to both (I8) and (I′8) by completing a circle of implications—by showing that if there is a w.s. for PRO in $G(F, I, s)$, then $\Phi(F, I, s)$, and that if $\Phi(F, I, s)$, then there is an e.w.s.

An Alternative Axiom

One way to think of our alternative eighth axiom would be as a statement about implementations. It will, of course, stand on its own as an axiom, independently of the implementational definition of a process, but it talks about trees—subsets of Σ^* containing $\langle\ \rangle$ and closed under shorter sequences—and it has a natural understanding if we think of these as implementation trees. To state the axiom neatly, we first need some notation: if G is any subset of $\Sigma^* \times \mathbb{P}(\Sigma)$ and J is any subset of Σ^ω, let $s {\frown} G$ be the set $\{(st, X) \mid (t, X) \in G\}$ and let $s {\frown} J$ be the set $\{s\sigma \mid \sigma \in J\}$. Then, taking T to range over trees, the alternative axiom is

(I*8) $(s, \emptyset) \in F \Rightarrow \exists T . s{\frown}f(T) \subseteq F \land s{\frown}i(T) \subseteq I$

The infinitary model

And this could be thought of as saying 'whenever s is a trace of $\langle F, I \rangle$, then there is an implementation of $\langle F, I \rangle$-after-s', where $\langle F, I \rangle$-after-s is taken to be the pair $\langle \{(t, X) \mid (st, X) \in F\}, \{\sigma \mid s\sigma \in I\} \rangle$.

However this may be, it is interesting to observe that there is a slight variation of (I*8) whose logical complexity exactly matches a formula we considered in the last section—the formula which quantifies over trees to capture the condition that if s is a trace of $\langle F, I \rangle$, then ENV does not have a wining strategy in $G(F, I, s)$. This, we noted, was strictly weaker than (I8), but we need only switch a couple of quantifiers and shift a negation sign to obtain something strong enough to adopt as an eighth axiom:

(I*'8) $(s, \emptyset) \in F \Rightarrow \exists T.[\forall t \in T.(st, \{a \mid t\langle a \rangle \notin T\}) \in F \wedge \forall \tau \in i(T).s\tau \in I]$

Here we have what is really just a rewriting of (I*8), except that to show the equivalence of $s \wedge f(T) \subseteq F$ and $\forall t \in T.(st, \{a \mid t\langle a \rangle \notin T\}) \in F$ we need to appeal to axiom (I2). We shall leave the reader to check this equivalence.

But let us work with (I*8) in this section, because then the following theorem can be established for arbitrary pairs $\langle F, I \rangle$, independently of any of the other axioms for processes.

Theorem 8. *Axioms (I8), (I'8) and (I*8) are equivalent to each other.*

Proof. As we indicated at the end of the last section, we can ignore the common antecedent $(s, \emptyset) \in F$, and simply prove the following equivalent:

1. there is a w.s. for PRO in $G(F, I, s)$;

2. there is an e.w.s. for PRO in $G(F, I, s)$;

3. there is a tree T such that $s \wedge f(T) \subseteq F$ and $s \wedge i(T) \subseteq I$.

We have already established that (2) implies (1); now we shall show that (1) implies (3) and that (3) implies (2).

To prove that (1) implies (3), assume (1) and take a w.s. ϕ for PRO in $G(F, I, s)$. Then define T_ϕ to be the union $\bigcup_{n < \omega} T_n$, where

$$T_0 = \{\langle \rangle\}$$
$$T_{n+1} = \{\langle a_0, \ldots, a_n \rangle \mid \exists X_0 \ldots \exists X_n . \langle \langle X_0, a_0, \ldots, X_n \rangle, a_n \rangle \in \phi_n\}$$

This is the set of all possible sequences of moves on PRO's part which may arise when PRO plays in accordance with ϕ. And it is easy to check that T_ϕ is then a tree. We claim, moreover, that it satisfies the conditions required in (3). First, to show that $s \wedge f(T_\phi) \subseteq F$, assume that $(t, X) \in f(T_\phi)$; it will then be sufficient to show that the additional assumption that $(st, X) \notin F$ leads to a contradiction. Let us consider two cases separately:

$t = \langle\,\rangle$: Aiming for a contradiction, say $(st, X) = (s, X) \notin F$; then X would be in the domain of ϕ_0, and so there would be an $a \in X$ such that $\langle X, a\rangle \in \phi_0$, but then $\langle a\rangle \in T_1 \subseteq T_\phi$. And, since $t = \langle\,\rangle \in T$, this means that $(t, X) \notin f(T_\phi)$—which is a contradiction.

$t = \langle a_0, \ldots, a_n\rangle$: In this case, since $(t, X) \in f(T_\phi)$, we know that $t \in T_{n+1}$ and $T_{n+1} \subseteq T_\phi$; hence we also know that there exist $X_0, \ldots, X_n$ such that $\langle\langle X_0, a_0, \ldots, X_n\rangle, a_n\rangle \in \phi_n$. Aiming, as before, for a contradiction, say now that $(st, X) \notin F$; then the tuple $\langle X_0, a_0, \ldots, X_n, a_n, X\rangle$ would be in the domain of ϕ_{n+1}, and so there would be an $a \in X$ such that $\langle\langle X_0, a_0, \ldots, X_n, a_n, X\rangle, a\rangle \in \phi_{n+1}$, but then $t\langle a\rangle \in T_{n+2} \subseteq T_\phi$. And, since $t \in T_\phi$, this means that $(t, X) \notin f(T_\phi)$—which is a contradiction.

This completes our proof that $s{\scriptstyle\wedge}f(T_\phi) \subseteq F$. And clearly $s{\scriptstyle\wedge}i(T_\phi) \subseteq I$—because if $\langle a_0, a_1, \ldots\rangle$ is an infinite path through T_ϕ, then ϕs satisfying condition (b) in the definition of a w.s. implies that $s\langle a_0, a_1, \ldots\rangle \in I$.

To prove that (3) implies (2), let us invoke the axiom of choice and assume a function χ which takes every non-empty subset X of Σ to an element of X. And, for convenience, let us take χ to be defined on $\emptyset$, so that $\chi(\emptyset)$ is some element or other of Σ (we do not care what). Then, if we assume (3), we are guaranteed a tree T such that $s{\scriptstyle\wedge}f(T) \subseteq F$ and $s{\scriptstyle\wedge}i(T) \subseteq I$, and we can define a family $\{\phi_i\}_{i<\omega}$ of functions in terms of T and χ to provide an e.w.s. for PRO in $G(F, I, s)$. The definition is by induction, and so first we specify ϕ_0: the domain of ϕ_0 is what the definition of an e.w.s. requires, viz. the set of those failure-pairs $(\langle\,\rangle, X)$ such that $(s, X) \notin F$, on this domain

$$\phi_0 : (\langle\,\rangle, X) \longmapsto \chi(X \cap \{b \mid \langle b\rangle \in T\})$$

Then, assuming that we have defined ϕ_n, we stipulate the domain of ϕ_{n+1} to be the set of those failure-pairs $(t\langle a\rangle, X)$ such that $\langle(t, Y), a\rangle \in \phi_n$, for some Y, and such that $(st\langle a\rangle, X) \notin F$; and on this domain

$$\phi_{n+1} : (t\langle a\rangle, X) \longmapsto \chi(X \cap \{b \mid t\langle a, b\rangle \in T\})$$

Because χ takes a value on $\emptyset$ this is guaranteed to be a well-defined family of functions of the right type to be an e.w.s., but what we now have to check—to establish conditions (a) and (b) in the definition of an e.w.s.—is essentially just that χ does indeed have non-empty sets to select from at each stage. This will be an inductive proof. The actual condition $\Phi(n)$, to be proved for all n, is most conveniently formulated as follows:

$$\forall c\,.\,\langle(t, X), c\rangle \in \phi_n \Rightarrow c \in X \wedge t\langle c\rangle \in T$$

Then condition (a) in the definition of an e.w.s. will follow immediately, and condition (b) can easily be deduced, given that $s{\scriptstyle\wedge} i(T) \subseteq I$.

To begin the induction, say $\langle(\langle\,\rangle, X), c\rangle \in \phi_0$. Then, by definition of the domain of ϕ_0, $(s, X) \notin F$, and so, from the fact that $s{\scriptstyle\wedge} f(T) \subseteq F$, we can conclude that $(\langle\,\rangle, X) \notin f(T)$; since of course $\langle\,\rangle \in T$, this means that $X \cap \{b \mid \langle b \rangle \in T\}$ is non-empty, so that χ must have selected c appropriately—$c \in X$ and $\langle c \rangle \in T$. Hence we have proved $\Phi(0)$.

Next, assuming $\Phi(n)$, say $\langle(t\langle a\rangle, X), c\rangle \in \phi_{n+1}$. The definition of the domain of ϕ_{n+1} this time tells us two things. First, we know that $\langle(t, Y), a\rangle \in \phi_n$ for some Y, so that, by the assumption that $\Phi(n)$ holds, $t\langle a \rangle \in T$. Secondly, $(st\langle a\rangle, X) \notin F$, so that, as before, since $s{\scriptstyle\wedge} f(T) \subseteq F$, we can conclude that $(t\langle a \rangle, X) \notin f(T)$. But, together with the fact that $t\langle a \rangle \in T$, this means that the set $X \cap \{b \mid t\langle a, b\rangle \in T\}$ is non-empty, and so χ must again have selected c appropriately—$c \in X$ and $t\langle a, c\rangle \in T$. Hence, assuming $\Phi(n)$, we have proved $\Phi(n+1)$, and this completes the proof that (3) implies (1). $\square$

And this proof suggests another way of looking at axiom (I*8), which points up its connection with the original (I8). For a tree satisfying condition (3) can be seen to provide what is sometimes called a 'quasi-strategy' (for PRO in $G(F, I, s)$)—a strategy which does not necessarily determine moves uniquely, but which specifies a range of options at each stage.

The soundness and completeness of the axioms

Finally, we shall prove that the set of axioms we started out with is sound and complete with respect to the implementational definition of a process.

Theorem 9. *Axioms (I0), (I1), (I2), (I3), (I6), and (I8) are sound and complete.*

Since (I8) is equivalent to (I8*), we can deduce this as a corollary of the following theorem.

Theorem 10. *Axioms (I0), (I1), (I2), (I3), (I6), and (I8*) are sound and complete.*

Proof. The proof of soundness is routine: it is easy to check that each axiom is satisfied by any pair $\langle F, I \rangle$ which is a process according to the implementational definition. But we might run through the case of axiom (I*8): Say that $\langle F, I \rangle$ is a process: then, if $(s, \emptyset) \in F$, we know that there exists a $T \in \mathcal{I}(F, I)$ such that $(s, \emptyset) \in f(T)$—in other words, $s \in T$—and it is easy to verify that $\{t \mid st \in T\}$ will then be a tree satisfying the conditions required by the axiom.

To establish completeness, let us assume that $\langle F, I \rangle$ satisfies the axioms. Then we have to show (i) that $\mathcal{I}(F, I) \neq \emptyset$ and (ii) that $\langle F, I \rangle = \mathcal{P}(\mathcal{I}(F, I))$. But, as we pointed out earlier, (ii) boils down to the following two conditions:

$$(s, X) \in F \Rightarrow \exists T . (s, X) \in f(T) \land f(T) \subseteq F \land i(T) \subseteq I \qquad \text{(iia)}$$
$$\sigma \in I \Rightarrow \exists T . \sigma \in i(T) \land f(T) \subseteq F \land i(T) \subseteq I \qquad \text{(iib)}$$

And, once we have established (1), then (i) will follow immediately, by (I0). So it is sufficient to prove (1) and (1).

To do this we shall invoke the axiom of choice again. If $(s, \emptyset) \in F$, then (I*8) guarantees the existence of a tree T such that $s \wedge f(T) \subseteq F$ and $s \wedge i(T) \subseteq I$, and in the course of the proof we shall, more than once, want to select such a tree, for each trace in what is, or might be, an infinite range of traces. We shall, then, assume a choice function θ which takes a trace s to a particular tree $\theta(s)$ such that $s \wedge f(\theta(s)) \subseteq F$ and $s \wedge i(\theta(s)) \subseteq I$. And let us introduce a third use for the '$s \wedge$' notation: if S is any subset of Σ^*, then $s \wedge S$ will be the set $\{st \mid t \in S\}$—in particular, $s \wedge \theta(s)$ will be the pseudo-tree obtained by attaching $\theta(s)$ to the stalk s in place of the natural root $\langle \rangle$.

We are now in a position to tackle the proof of (1). Let us assume that $(s, X) \in F$. Then we want to define a tree T meeting the conditions required in (1). This tree is going to be a grand union consisting essentially of $t \wedge \theta(t)$, for various traces t of $\langle F, I \rangle$, and it will be easiest to understand if we provide some preliminary definitions specifying disjoint chunks of the tree. First, then, for each $t < s$, take

$$\Theta_s(t) = \bigcup \{t\langle a\rangle \wedge \theta(t\langle a\rangle) \mid (t\langle a\rangle, \emptyset) \in F \land t\langle a\rangle \not\leqslant s\}$$

Next take

$$\Theta_{(s,X)} = \bigcup \{s\langle a\rangle \wedge \theta(s\langle a\rangle) \mid (s\langle a\rangle, \emptyset) \in F \land a \notin X\}$$

And now glue these sets together by adding in s and its initial segments:

$$T = \{t \mid t \leqslant s\} \cup \bigcup_{t<s} \Theta_s(t) \cup \Theta_{(s,X)}$$

It is then straightforward to verify that T is indeed an implementation tree. And, by construction, it is easy to see that $(s, X) \in f(T)$. But we still have to establish that $f(T) \subseteq F$ and that $i(T) \subseteq I$.

First, to prove that $f(T) \subseteq F$, assume that $(t, Y) \in f(T)$; we have to show that $(t, Y) \in F$. The problem falls naturally into three cases.

$t < s$: Since $(s, X) \in F$, (I1) and (I2) imply that $(t, \emptyset) \in F$, and, by construction of T, $(t\langle c\rangle, \emptyset) \notin F$ for any $c \in Y$; hence, by (I3), $(t, Y) \in F$.

$t = s$: Since $(s, X) \in F$, (I2) implies that $(s, X \cap Y) \in F$, and, by construction of T, $(s\langle c\rangle, \emptyset) \notin F$ for any $c \in Y \setminus X$; but $Y = (X \cap Y) \cup (Y \setminus X)$, hence, by (I3), $(t, Y) \in F$.

$t \not\leqslant s$: By construction of T, there exist $u \leqslant s$, such that for some v and a, $t = u\langle a\rangle v \in u\langle a\rangle {\scriptstyle\wedge} \theta(u\langle a\rangle)$, and, also by construction of T, $(v, Y) \in f(\theta(u\langle a\rangle))$; hence, $(t, Y) \in u\langle a\rangle {\scriptstyle\wedge} f(\theta(u\langle a\rangle)) \subseteq F$.

Now we have to check that $i(T) \subseteq I$. Assume, then, that $\sigma \in i(T)$; we have to show that $\sigma \in I$. But let t be the longest common initial segment of s and σ. (There must always be one, if only $\langle\rangle$ and at the other extreme it might be s itself.) Then $\sigma = t\langle a\rangle \sigma'$, for some (unique) a and σ', and, by construction of T, $t\langle a\rangle u \in t\langle a\rangle {\scriptstyle\wedge} \theta(t\langle a\rangle)$ whenever $t\langle a\rangle u < \sigma$—in other words, $u \in \theta(t\langle a\rangle)$ for any $u < \sigma'$, which means that $\sigma' \in i(\theta(t\langle a\rangle))$. But then $\sigma \in t\langle a\rangle {\scriptstyle\wedge} i(\theta(t\langle a\rangle)) \subseteq I$. And we have finished checking condition (1).

A much simpler construction can be used to establish (1). Given $\sigma \in I$, we can define

$$T = \bigcup_{s < \sigma} s {\scriptstyle\wedge} \theta(s)$$

This is well defined because, by (I6), $(s, \emptyset) \in F$ whenever $s < \sigma$. And T is obviously a tree such that $\sigma \in i(T)$. It remains only to prove that $f(T) \subseteq F$ and that $i(T) \subseteq I$.

To show that $f(T) \subseteq F$, we can argue in a way similar to the case $t \not\leqslant s$ above. For if $(t, Y) \in f(T)$, then, by construction of T, there exists $s < \sigma$ such that, for some u, $t = su \in s {\scriptstyle\wedge} \theta(s)$ and, also by construction of T, $(u, Y) \in f(\theta(s))$; hence $(t, Y) \in s {\scriptstyle\wedge} f(\theta(s)) \subseteq F$.

On the other hand, to show that $i(T) \subseteq I$, assume that $\tau \in i(T)$; we have to check that $\tau \in I$. But, if $\tau = \sigma$, this is immediate. And, if $\tau \neq \sigma$, then it will be sufficient to take the longest (finite) common initial segment t of τ and σ, and show that there is an $s \leqslant t$ such that, for any u, if $su < \tau$, then $su \in s {\scriptstyle\wedge} \theta(s)$. For in that case $\tau = s\tau'$ where $\tau' \in i(\theta(s))$, so that $\tau \in s {\scriptstyle\wedge} i(\theta(s)) \subseteq I$.

Aiming, then, for a contradiction, suppose that for every $s \leqslant t$ there is a u such that $su < \tau$ but $su \notin s {\scriptstyle\wedge} \theta(s)$, and, for each such s, let μ_s be the shortest such u. Now consider the set $\{s\mu_s \mid s \leqslant t\}$; this set is finite, and so we can select its longest element—v say. And we claim that $v \notin T$, which, since v is an initial segment of τ, contradicts our starting assumption that $\tau \in i(T)$.

To establish that $v \notin T$, we have to show that $v \notin s {\scriptstyle\wedge} \theta(s)$ for any $s < \sigma$; we can divide the problem into two cases. First, if $s \leqslant t$, then we know that $s\mu_s \not\in s{\scriptstyle\wedge}\theta(s)$; but $s\mu_s \leqslant v$, and so, since $\theta(s)$ is a tree, $v \not\in s{\scriptstyle\wedge}\theta(s)$. On the other hand, if $t < s$, we can infer that $v \not\in s{\scriptstyle\wedge}\theta(s)$, because otherwise $s \leqslant v$, in which case s would be an initial segment of τ. However, this is impossible, since t is the longest initial segment common to both σ and τ. This concludes our proof of (1). □

Acknowledgements

It will be obvious that the material in Section 2.3 would not have been what it is without close contact with Bill Roscoe's developing work on unbounded nondeterminism. And comments from many other people have been helpful throughout—especially from Alan Jeffrey, David Murphy, and Mike Reed—though no one has yet been able to make me understand 'runtime nondeterminism'.

References

Brookes, S. D., 1983. *A Model for Communicating Sequential Processes*, D.Phil thesis, Oxford University.

Brookes, S. D., Hoare, C. A. R., and Roscoe, A. W., 1984. A theory of communicating sequential processes, *JACM*, 31.

Brookes, S. D., and Roscoe, A. W., 1985. An improved failures model for communicating sequential processes, in *Seminar on Concurrency*, LNCS 197, Springer.

Hoare, C. A. R., 1985. Programs are predicates, in C. A. R. Hoare and J. C. Shepherdson (Eds.) *Mathematical Logic and Programming Languages*, Prentice Hall.

Kolaitis, Ph. G., 1985. Game quantification, in J. Barwise and S. Feferman (Eds.) *Model-Theoretic Logics*, Perspectives in Mathematical Logic, Springer.

Roscoe, A. W., 1988. *Two Papers on CSP*, Technical Monograph PRG-67, OUCL

3
Classifying unbounded nondeterminism in CSP

GEOFF BARRETT AND MICHAEL GOLDSMITH

Abstract

Recent research has provided models for Communicating Sequential Processes (CSP) which allow the compactness axiom implying finite nondeterminism to be relaxed. In particular, arbitrary processes can be identified with the set of *predeterministic* processes which implement them. This chapter considers the topological properties of such sets considered as subsets of the complete ultra-metric space of predeterministic processes under the metric

$$d(Q, R) = \inf \left\{ 2^{-n} \mid Q \downarrow n = R \downarrow n \right\}$$

where $Q \downarrow n$ is the n-step approximation to Q.

The principal result is a bound on the trans-finite index of non-determinacy of a process in terms of the Lindelöf number of some set of predeterministic processes which exhibit all the required behaviours.

3.1 Introduction: unboundedly nondeterministic CSP

Communicating Sequential Processes (Hoare 1985) is a notation for expressing distributed systems which co-operate by synchronizing engagement in events from some alphabet Σ. Traditionally, its full semantics have been given denotationally with process values in the domain of pairs consisting of a set of failures ($F \subseteq \Sigma^* \times \mathbb{P}(\Sigma)$) and a set of divergences ($D \subseteq \Sigma^*$) which satisfy certain axioms. $(s, X) \in F$ means that the process, having successfully engaged in the sequence of events s, may persistently refuse to engage in any of the events in X; thus $(s, \Sigma) \in F$ shows a potential for deadlock, and $(s, \{a\}) \in F \land (s\langle a \rangle, \emptyset) \in F$ is a sign that the process's engagement in a after s is decided nondeterministically. The divergences D are those sequences of events after which the process is deemed to be 'broken': ready to behave chaotically and able to break any co-operating process.

Roscoe (1988) introduces modifications of the 'improved failures model' for CSP (Brookes and Roscoe 1985) in order to cater for unbounded nondeterminism which is excluded by the axiom of the model $\mathcal{N}$:

$$(\forall X' \subseteq X \,.\, X' \text{ is finite} \Rightarrow (s, X') \in F) \Rightarrow (s, X) \in F$$

The first attempt is simply to omit that axiom, leaving the representation otherwise unchanged. This gives rise to the model $\mathcal{N}'$, where the nondeterminism order $\sqsubseteq$ defined by

$$(F, D) \sqsubseteq (F', D') \iff F \supseteq F' \land D \supseteq D'$$

is no longer a complete partial order, but the fixed point theory is rescued by the completeness of the definedness order $\leqslant$:

$$\begin{aligned}(F, D) &\leqslant (F', D') \iff \\ (F, D) &\sqsubseteq (F', D') \\ &\land F \setminus (D \times \mathbb{P}(\Sigma)) = F' \setminus (D \times \mathbb{P}(\Sigma)) \\ &\land \mu D \subseteq traces(F', D')\end{aligned} \qquad (*)$$

where μ and $traces$ are defined below. Because this model identifies the process which engages in any finite number of events before choosing nondeterministically whether to stop with the one which may, in addition, choose never to stop, several of the algebraic laws of $\mathcal{N}$ are lost. For instance, hiding is not commutative or associative.

It is in order to remove these infelicities that the second paper in (Roscoe 1988) introduces the model $\mathcal{U}$ which adds a third component of infinite traces ($I \subseteq \Sigma^\omega$). Again, the fixed point theory suffers (Roscoe shows that no complete partial order exists giving the 'correct' fixed points) but recursions involving CSP-definable functions can be shown to have least fixed points (Roscoe 1988, Barrett 1989). This is the model which we take as canonical.

Definition 1. $\mathcal{U} \subset \mathbb{P}(\Sigma^* \times \mathbb{P}(\Sigma)) \times \mathbb{P}(\Sigma^*) \times \mathbb{P}(\Sigma^\omega)$ is the set of triples (F, D, I) with $F \neq \emptyset$ satisfying

$$\begin{aligned}(st, \emptyset) \in F &\Rightarrow (s, \emptyset) \in F &(1)\\ (s, X) \in F \land Y \subset X &\Rightarrow (s, Y) \in F &(2)\\ (s, X) \in F \land (\forall a \in Y \,.\, (s\langle a \rangle, \emptyset) \notin F) &\Rightarrow (s, X \cup Y) \in F &(3)\\ s \in D &\Rightarrow st \in D &(4)\\ s \in D &\Rightarrow (s, X) \in F &(5)\end{aligned}$$

and

$$su \in I \Rightarrow (s, \emptyset) \in F \tag{6}$$
$$s \in D \Rightarrow su \in I \tag{7}$$
$$(s, \emptyset) \in F \Rightarrow \exists T . (\forall t \in T . (st, \{a \mid t\langle a \rangle \notin T\}) \in F) \wedge \forall u \in \overline{T} . su \in I \tag{8}$$

where $\overline{T}$ is defined below and (as throughout this chapter) a ranges over Σ, s, t over Σ^*, u over Σ^ω, X, Y over $\mathbb{P}(\Sigma)$, and T over non-empty prefix-closed subsets of Σ^*.

For non-empty sets S of sequences we define

$$\overline{S} = \{u \in \Sigma^\omega \mid \forall s \prec u . s \in S\}$$
$$\mu S = \{s \in S \mid \forall t \in S . t \not\prec s\}$$

where $\prec$ is the strict prefix ordering on sequences.

The first five of these axioms are those which must hold for a pair to be a member of $\mathcal{N}'$; together with the compactness axiom above, they form the axioms of the classic model $\mathcal{N}$.

Definition 2. *For a process $P = (F, D, I) \in \mathcal{U}$ the components of the value are selected by*

$$\begin{aligned} fails(P) &= F \\ divs(P) &= D \\ infs(P) &= I \\ traces(P) &= \{s \in \Sigma^* \mid (s, \emptyset) \in F\} \end{aligned}$$

Both the orders $\sqsubseteq$ and $\leqslant$ are extended to operate on $\mathcal{U}$ by inverse inclusion on the infinite traces; that is,

$$(F, D, I) \sqsubseteq_\mathcal{U} (F', D', I') \iff (F, D) \sqsubseteq_{\mathcal{N}'} (F', D') \wedge I \supseteq I'$$

and analogously for $\leqslant$. While neither partial order is (directed-set) complete, both remain consistently complete, and $\leqslant$ is still coarser than $\sqsubseteq$ in that $P \leqslant P' \Rightarrow P \sqsubseteq P'$. They share a common minimum element 'bottom': $\perp = \langle \Sigma^* \times \mathbb{P}(\Sigma), \Sigma^*, \Sigma^\omega \rangle$, the process which is broken before it does anything. Greatest lower bounds under $\sqsubseteq$, corresponding to nondeterministic choice, are given by the unions of each component.

3.2 Predeterministic processes

A process is *deterministic* if it is $\sqsubseteq$-maximal; it cannot refuse to engage in any event which it might perform after a given trace. The next lemma characterizes a weaker property: a *predeterministic* process's behaviour is deterministic unless it is divergent.

Lemma 3. *The following are equivalent for $Q = (F, D, I) \in \mathcal{U}$, where $T = traces(Q)$.*

1. $(s, X) \in F \Rightarrow (s \in D \vee X \cap \{a \mid s\langle a \rangle \in T\} = \emptyset)$

2. $\forall P \in \mathcal{U} \,.\, Q \sqsubseteq P \Rightarrow Q \leqslant P$

3. (F, D) *is a $\sqsubseteq$-maximum in $\mathcal{N}'$ among those with traces T and divergences D, and $I = \overline{T}$.*

Proof.

1⇒3 Axiom 3 implies the presence of a minimal set of failures connected with any trace set; (1) states that those are the only ones present, whence the maximality result. Axiom 6 implies $I \subseteq \overline{T}$; to show inclusion in the other direction consider $u \in \overline{T}$. If $\exists s \in D \,.\, s \prec u$, then Axiom 7 gives $u \in I$; otherwise Axiom 8 instantiated at $s = \langle \rangle$ gives the existence of a non-empty prefix-closed set of traces T' such that $\overline{T'} \subseteq I$. A simple induction on the length shows that $\forall t \prec u \,.\, t \in T'$: since (1) gives $t\langle a \rangle \prec u \Rightarrow (t, \{a\}) \notin F$, Axiom 2 and the defining condition on T' yield $t\langle a \rangle \in T'$.

3⇒1 If F contains any failure (s, X), $X \neq \emptyset$, which is not required by Axioms 3 or 5, that is, is not allowed by (1), then the pair $R = (F', D)$ where $F' = F \setminus \{(s, Y) \mid X \subseteq Y\}$ satisfies the axioms of $\mathcal{N}'$ and has the same trace set as Q, but $(F, D) \sqsubset R$, contradicting the maximality condition of (3).

1⇒2 Suppose $Q \sqsubseteq P = (F', D', I')$: it remains to show the other two lines in $(*)$. If $Q = \bot$ then trivially $Q \leqslant P$; otherwise $\langle \rangle \notin \mu D$, and a typical minimal divergence can be written $s \langle a \rangle$ with $s \notin D$. An inductive appeal to Axiom 3 together with (1) shows that $s\langle a \rangle \in traces(P)$. Similarly, the refusals on the non-divergent traces of Q do not allow any of the traces to be omitted, and are the minimal set allowed by Axiom 3, and thus equal the refusals on those traces in P.

2⇒1 As above, if F contains any failure (s, X), $X \neq \emptyset$, contradicting (1) then let $F' = F \setminus \{(s, Y) \mid X \subseteq Y\}$; $P = (F', D, I)$ is a process in $\mathcal{U}$ with $Q \sqsubset P$, but $s \notin D$ and $F \cap (\{s\} \times \mathbb{P}(\Sigma)) \neq F' \cap (\{s\} \times \mathbb{P}(\Sigma))$, so that $Q \not\leqslant P$.

□

Predeterministic processes

We shall represent the set of processes possessing this property by $\mathcal{P}$.

By Lemma 3(3) there is at most one predeterministic process with given trace and divergence sets T and D; in fact, the following definition gives a member of $\mathcal{P}$ for every pair of sets $D \subseteq T$ with T non-empty and prefix-closed and D suffix-closed.

Definition 4. *The predeterministic process* $[T, D] \in \mathcal{P}$ *is defined by*

$$divs\,[T, D] = D$$
$$fails\,[T, D] = \{(s, X) \mid s \in T \wedge \forall a \in X \,.\, s\langle a\rangle \notin T\} \cup \{(s, X) \mid s \in D\}$$
$$infs\,[T, D] = \overline{T}$$

Of course, for any $Q \in \mathcal{P}$, $Q = [traces(Q), divs(Q)]$.

By Lemma 3(2) $\sqsubseteq$ and $\leqslant$ agree on $\mathcal{P}$. The next lemma shows that this ordering can be simply calculated from the traces and divergences.

Lemma 5. $[T, D] \sqsubseteq [T', D'] \iff T \supseteq T' \wedge D \supseteq D' \wedge D \setminus \mu D \supseteq T \setminus T'$

Proof.

$\Rightarrow$ The first two conjuncts follow immediately from the definition of $\sqsubseteq$:

$$T = traces\,[T, D] \supseteq traces\,[T', D'] = T'$$
$$D = divs\,[T, D] \supseteq divs\,[T', D'] = D'$$

Let $s \in T \setminus T'$ and $t = \bigsqcup_{\preceq} \{s' \in T' \mid s' \preceq s\}$, then $t \prec s$; so define a by $t\langle a\rangle \preceq s$. Certainly $t\langle a\rangle \notin T'$ so that $(t, \{a\}) \in fails\,[T', D'] \subseteq fails\,[T, D]$. But now, since $t\langle a\rangle \in T$, we must have $t \in D$, so since $t \prec s$, $s \in D \setminus \mu D$.

$\Leftarrow$ $divs\,[T, D] = D \supseteq D' = divs\,[T', D']$. Also, since $infs\,[T, D]$ is constructed positively from T and $T \supset T'$, $infs\,[T, D] \supseteq infs\,[T', D']$. We are left to show $fails\,[T, D] \supseteq fails\,[T', D']$. Let $(s, X) \in fails\,[T', D']$. Suppose $s \in D'$; then $s \in D$ so $(s, X) \in fails\,[T, D]$. Suppose $s \notin D'$; then $\forall a \in X \,.\, s\langle a\rangle \notin T'$. Either $\forall a \in X \,.\, s\langle a\rangle \notin T$, in which case $(s, X) \in fails\,[T, D]$ or $\exists a \in X \,.\, s\langle a\rangle \in T$. Now, $s\langle a\rangle \in T \setminus T' \subseteq D \setminus \mu D$, whence $s \in D$ and $(s, X) \in fails\,[T, D]$. $\square$

Definition 6. *A predeterministic process Q is said to be an implementation of $P \in \mathcal{U}$ if $P \sqsubseteq Q$. The set of implementations of P is*

$$imps(P) \triangleq \{Q \in \mathcal{P} \mid P \sqsubseteq Q\}$$

Lemma 7. $imps(P) \neq \emptyset$ *and* $P = \bigsqcap_{\sqsubseteq} imps(P)$

Note that with careful development this can be used as an alternative to Axiom 8 (as indeed is done by Roscoe (1988)).

Proof. Let $P = (F, D, I)$ and $T = traces(P)$: by Axiom 8 there is a trace set $T_{\langle\rangle} \subseteq T$ such that $\overline{T_{\langle\rangle}} \subseteq I$ and it is readily checked that $P \sqsubseteq [T_{\langle\rangle}, \emptyset]$. This establishes the first conjunct. In general, $P \sqsubseteq [T', \emptyset]$ provided $T' \subseteq T$ and $\overline{T'} \subseteq I$, so that T' is not too large, and provided that if $s \in T'$ then $(s, \{a \mid s\langle a\rangle \notin T'\}) \in F$, so that T' is not so small that Axiom 3 forces in extra failures.

Since the set of implementations is non-empty and the partial order is consistently-complete, we know that $P' = (F', D', I') = \bigsqcap_{\sqsubseteq} imps(P)$ is a well-defined process. It is immediate that $P \sqsubseteq P'$.

To show the inclusion the other way, we appeal to a technical result proved below (Lemma 8): for any trace w of P, finite or infinite, there exists a set S_w such that $w \in S_w \cup \overline{S_w}$ and $P \sqsubseteq [S_w, \emptyset]$. For each element of each component of P we shall exhibit an implementation of P with that behaviour; as each component of P' is the union of that component of all implementations, this establishes the result.

Now for each $u \in I$, $u \in \overline{S_u} = infs[S_u, \emptyset]$.

For $s \in D$, let $D_s = \{st \mid t \in \Sigma^*\}$; clearly $[S_s \cup D_s, D_s] \in imps(P)$.

For the failures, if $s \in traces(P)$ write T_s for a witness to P satisfying Axiom 8 at s. Now for $(s, X) \in F$

$$Q \triangleq \left[\{t \in S_s \mid s \not\prec t\} \cup \{s\langle a\rangle t \mid a \notin X \wedge s\langle a\rangle \in traces(P) \wedge t \in T_{s\langle a\rangle}\}, \emptyset\right]$$

clearly has (s, X) as a failure. If Q is an implementation of P we are done. Trivially $divs(Q) \subseteq D$; for $u \in infs(Q)$ either for some a, $u = s\langle a\rangle v$ with $v \in T_{s\langle a\rangle}$ or $s \wedge u \preceq s$ and $u \in \overline{S_s}$, and in either case $u \in I$. Each failure $(t, Y) \in fails(Q)$ falls into one of three classes: $s \not\prec t$ and $(t, Y) \in fails[S_s, \emptyset]$; $s = t$ and $Y \subseteq X \cup \{a \mid s\langle a\rangle \notin traces(P)\}$ and so $(t, Y) \in F$ by Axioms 3 and 2, or $s \prec t$ so that for some a, $s\langle a\rangle \preceq t$ and $(t, Y) \in F$ by Axiom 8. □

All that remains is to prove the following technical result.

Lemma 8. If $P \in \mathcal{U}$ and $w \in traces(P) \cup infs(P)$, then there exists $S_w \subseteq traces(P)$ such that $w \in S_w \cup \overline{S_w}$ and $P \sqsubseteq [S_w, \emptyset]$.

Proof. Let $v \wedge w$ denote the longest common prefix of v and w. Then with T_s as in the proof of Lemma 7

$$S_w \triangleq \{st \mid s \preceq w \wedge t \in T_s \wedge s = st \wedge w\}$$

meets the requirements. First note that every prefix of w is an element of S_w, for if $s \preceq w$ then $s = s\langle\rangle \wedge w$ and $\langle\rangle \in T_s$. Therefore, $w \in S_w \cup \overline{S_w}$.

Restriction and the metric space

To show that S_w is the trace-set of an implementation, first consider the infinite traces: suppose $u \in \overline{S_w}$. If $u = w$ then $u \in I$ by hypothesis. Otherwise, let $s = u \curlywedge w$. Now by construction all extensions of s in S_w lie in T_s, so there is some $v \in \overline{T_s}$ such that $u = sv$ and, by definition of T_s, $sv \in I$ as required.

It is left to show that $\{(t, \{a \mid t\langle a\rangle \notin S_w\}) \mid t \in S_w\} \subseteq F$. For each $t \in S_w$, if $s = t \curlywedge w$ and $t = sr$ then

$$t\langle a\rangle \in S_w \Leftrightarrow r\langle a\rangle \in T_s \vee (s = t \wedge s\langle a\rangle \preceq w)$$
$$\Leftarrow r\langle a\rangle \in T_s$$

so that $\{a \mid t\langle a\rangle \notin S_w\} \subseteq \{a \mid r\langle a\rangle \notin T_s\}$. But, by Axiom 8, this latter set is a refusal of P after sr and so by Axiom 2 the first set is a refusal after t. □

3.3 Restriction and the metric space

Definition 9. *If $Q = [T, D] \in \mathcal{P}$ and $n \geqslant 0$, the n-step restriction of Q, $Q \downarrow n$, is defined by*

$$divs(Q \downarrow n) \; \widehat{=} \; D \cup \{st \mid s \in traces(Q) \wedge \#s = n\}$$
$$traces(Q \downarrow n) \; \widehat{=} \; T \cup divs(Q \downarrow n)$$

That is, $Q \downarrow n$ is a process which 'behaves like' Q for the first n events and then breaks.

It is simple to verify that $m \leqslant n \Rightarrow Q \downarrow m \sqsubseteq Q \downarrow n \sqsubseteq Q$, $(Q \downarrow n) \downarrow m = Q \downarrow m$ and that the operator is well defined.

Lemma 10. *Let the function $d : \mathcal{P} \times \mathcal{P} \to \mathbb{R}$ be given by*

$$d(Q, R) = \inf \left\{2^{-n} \mid Q \downarrow n = R \downarrow n\right\}$$

1. *d is a metric on $\mathcal{P}$.*

2. *d is an ultra-metric, so that $d(Q, Q') \leqslant \max \{d(Q, R), d(R, Q')\}$.*

3. *$(\mathcal{P}, d)$ is a complete metric space.*

Proof. The first two results are well known in $(\mathcal{N}, d)$. Since $\mathcal{P}$ is a subspace of $\mathcal{N}$, the results carry through immediately.

Suppose $(Q_r)_{r\in\omega}$ is a Cauchy sequence. Choose an increasing sequence of indices $(i_n)_{n\in\omega}$ such that $d(Q_{i_n}, Q_j) < 2^{-n}$ whenever $j > i_n$. Now let $[T_n, D_n] = Q_{i_n} \downarrow n$. It is easily verified that $(Q_r)_{r\in\omega}$ converges to the limit $[\bigcap_{n\in\omega} T_n, \bigcap_{n\in\omega} D_n]$. □

One consequence of d's being an ultra-metric is the following technical result, that open balls have no proper overlap.

Lemma 11.

1. Either $B_\varepsilon(Q) \cap B_\varepsilon(Q') = \emptyset$ or else $B_\varepsilon(Q) = B_\varepsilon(Q')$.

2. All open balls are clopen.

Proof.

1. Suppose $x \in B_\varepsilon(Q) \cap B_\varepsilon(Q')$; then $d(Q,x) < \varepsilon$ and $d(Q',x) < \varepsilon$. Therefore
$$d(Q,Q') \leq \max(d(Q,x), d(Q',x)) < \varepsilon$$
Let $y \in B_\varepsilon(Q)$ so that $d(y,Q) < \varepsilon$. Now
$$d(y,Q') \leq \max(d(y,Q), d(Q,Q')) < \varepsilon$$
so that $y \in B_\varepsilon(Q')$, whence, by symmetry, $B_\varepsilon(Q) = B_\varepsilon(Q')$.

2. Immediately from (1), we have
$$\mathcal{P} \setminus B_\varepsilon(Q) = \bigcup \{B_\varepsilon(Q') \mid Q' \notin B_\varepsilon(Q)\}$$

□

Now every subset of $\mathcal{P}$ determines an element of $\mathcal{U}$, its greatest lower bound, and by Lemma 7 every element of $\mathcal{U}$ determines a particular subset of $\mathcal{P}$. Thus if $P \in \mathcal{U}$ and $\mathcal{X} \subseteq \mathcal{P}$ we have
$$\bigsqcap_\sqsubseteq imps(P) = P$$
$$imps\left(\bigsqcap_\sqsubseteq \mathcal{X}\right) \supseteq \mathcal{X}$$

We want to identify elements of $\mathcal{U}$ with their set of implementations, so we are particularly interested in those $\mathcal{X}$ where equality holds in the second line.

Lemma 12. $[T, D] \in imps(\bigsqcap_{\lambda \in \Lambda} [T_\lambda, D_\lambda])$ iff

$$T \subseteq \bigcup_{\lambda \in \Lambda} T_\lambda \qquad D \subseteq \bigcup_{\lambda \in \Lambda} D_\lambda \qquad \overline{T} \subseteq \bigcup_{\lambda \in \Lambda} \overline{T_\lambda}$$

and

$$s \in T \wedge s\langle a \rangle \in \bigcup_{\lambda \in \Lambda} T_\lambda \setminus T$$
$$\Rightarrow \exists \lambda \in \Lambda \,.\, s \in T_\lambda \wedge (\forall a \,.\, s\langle a \rangle \notin T \Rightarrow s\langle a \rangle \notin T_\lambda)$$

Restriction and the metric space

Proof. The conditions on D and $\overline{T}$ are precisely those required to ensure inclusion of the divergences and infinite trace components. The first condition states that if $(s, \emptyset)$ is a failure of the candidate implementation then it must be a failure of the lower bound, and the fourth states the same for (s, X) when $X \neq \emptyset$; they are therefore both clearly necessary for inclusion of the failures. That they are sufficient is a consequence of Axiom 2. □

An important point in this lemma is that the behaviour of an implementation after different traces may arise from behaviours of different members of the nondeterministically composed set of generators. In particular the next result follows.

Lemma 13. Suppose $P \in \mathcal{U}$, $\mathcal{X} = imps(P)$, and that $R \in \mathcal{X}$. Then let $S \subseteq traces(R)$ be a set of traces pairwise incomparable by $\preceq$ such that every trace of R is comparable with at least one of them, and let $([T_t, D_t])_{t \in S}$ be a family of elements of $\mathcal{X}$ such that $t \in T_t$ for each t. We can deduce that $[T', D'] \in \mathcal{X}$ whenever D' is suffix-closed and

$$T' = \{s \mid \exists t \in S . s \prec t\} \cup \bigcup_{t \in S} \{s \in T_t \mid t \preceq s\}$$
$$D' \subseteq T' \cap divs(P)$$

Proof. Clearly the first three conditions of Lemma 12 are satisfied. For the fourth, R will serve as the $[T_\lambda, D_\lambda]$ for a trace s which is a proper prefix of an element of S, and Q_t for any $s \succeq t$. Since by construction every $s \in T'$ falls under one of these heads, the result follows immediately. □

We now extend the restriction operator to $\mathcal{U}$ by defining it to be distributive.

Definition 14. $P \downarrow n = \bigcap_{Q \in imps(P)} Q \downarrow n$

The important property of this definition is as follows.

Lemma 15. $imps(P \downarrow n) = \{R \mid \exists Q \in imps(P) . Q \downarrow n \sqsubseteq R\}$

whose proof is simplified by the following technical lemma.

Lemma 16. If $R \in imps(P \downarrow n)$, then there exists $Q \in imps(P)$ such that $R \downarrow n = Q \downarrow n$.

Proof. The proof is effected by the construction of Q as follows:

$$divs(Q) = \{st \mid s \in divs(R) \land \#s < n\}$$
$$traces(Q) = \{s \in traces(R) \mid \#s < n\} \cup divs(Q)$$
$$\cup \bigcup \{st \mid s \in traces(R) \land \#s = n \land t \in T_s\}$$

where T_s is a trace-set whose existence is implied by Axiom 8.

That $Q \downarrow n = R \downarrow n$ is a simple calculation. It is left to show that $Q \in imps(P)$. Each part of the necessary containment is easily deduced from the fact that the behaviours of R with length less than n are behaviours of P, and that the behaviours of length greater than or equal to n are the consequence of a particular set T_s. □

The proof of Lemma 15 is now easy.

Proof.

⊇ $P \sqsubseteq Q \wedge Q \downarrow n \sqsubseteq R \Rightarrow P \downarrow n \sqsubseteq Q \downarrow n \sqsubseteq R \Rightarrow R \in imps(P \downarrow n)$

⊆ Let $R \in imps(P \downarrow n)$ so that $P \downarrow n \sqsubseteq R$. By Lemma 16, we can construct $Q \in imps(P)$ with $Q \downarrow n = R \downarrow n \sqsubseteq R$, whence R is an element of the right-hand side. □

3.4 Topological properties of process implementations

In this section we investigate the relationship between semantic properties of processes and topological properties of their set of implementations in the metric space. It transpires that the property of having a closed set of implementations characterizes a significant subset of $\mathcal{U}$.

Lemma 17. *The following are equivalent:*

1. $imps(P)$ is closed
2. $P = \bigsqcup_{n<\omega} P \downarrow n$
3. $infs(P) = \overline{traces(P)}$

Proof.

3⇒2 Since $P \downarrow n \sqsubseteq P$ for all n, $\bigsqcup_{n<\omega} P \downarrow n$ exists; call it $\overline{P}$. Note $\overline{P} \sqsubseteq P$. Clearly, $divs(P) = divs(\overline{P})$ and $fails(P) = fails(\overline{P})$. By Axiom 6 and (3), $infs(\overline{P}) \subseteq infs(P)$. But $\overline{P} \sqsubseteq P$ so $infs(P) \subseteq infs(\overline{P})$.

2⇒1 Let $(Q_r) \in imps(P)^\omega$ be such that $Q_r \to Q$. Now for every n there exists M such that for all $m \geq M$, $d(Q_m, Q) < 2^{-n}$, so that $Q_m \downarrow n = Q \downarrow n$. Since $P \sqsubseteq Q_m$, $P \downarrow n \sqsubseteq Q_m \downarrow n = Q \downarrow n \sqsubseteq Q$, whence $P = \bigsqcup_{n<\omega} P \downarrow n \sqsubseteq Q$ and $Q \in imps(P)$.

$1 \Rightarrow 3$ Define predeterministic Q by

$$divs(Q) = divs(P)$$
$$traces(Q) = traces(P)$$

Clearly, for any finite number n, $Q \in imps(P \downarrow n)$, so there exists a $Q_n \in imps(P)$ with $Q_n \downarrow n = Q \downarrow n$, that is, $d(Q_n, Q) < 2^{-n}$. Now $Q_n \to Q$ and since $imps(P)$ is closed, $Q \in imps(P)$. Further, $infs(Q) = \overline{traces(P)}$, whence the set is contained in $infs(P)$. Axiom 6 gives containment in the other direction. □

Since all the infinite behaviour of a restricted process arises from divergence within a bounded number of events, there is a natural correspondence between the restriction of a process in $\mathcal{N}'$ and the restrictions of the processes in $\mathcal{U}$ with the same failures and divergences. As $P = \bigsqcup_{n<\omega} P \downarrow n$ is an identity in $\mathcal{N}'$, Lemma 17 leads us to identify a subset of $\mathcal{U}$ such that the pair of functions

$$\iota : \mathcal{N}' \to \{P \in \mathcal{U} \mid imps(P) \text{ is closed}\}$$
$$\pi : \{P \in \mathcal{U} \mid imps(P) \text{ is closed}\} \to \mathcal{N}'$$

given by

$$\pi(F, D, I) = (F, D)$$
$$\iota(F, D) = (F, D, \overline{\{s \mid (s, \emptyset) \in F\}})$$

are mutually inverse bijective ordermorphisms with respect to both $\sqsubseteq$ and $\leqslant$. This means that any monotonic function which preserves closedness can be treated by the established fixed-point theory of $\mathcal{N}'$.

Note also that, by criteria Lemma 3(3) and Lemma 17(3), $imps(Q)$ is closed for all $Q \in \mathcal{P}$.

In contrast, the processes with open sets of implementations are rather less interesting. We first establish a technical result.

Lemma 18. *If U is $\sqsubseteq$-closed upwards then U is open iff $\forall Q \in U. \exists n. Q \downarrow n \in U$.*

Proof.

$\Rightarrow$ If U is open then for $Q \in U$ there is a ball of radius ε about Q which is contained in U. Choose n such that $2^{-n} \leqslant \varepsilon$. $d(Q \downarrow n, Q) < 2^{-n} \leqslant \varepsilon$, so $Q \downarrow n \in U$.

⇐ Choose $Q \in U$ and let $Q \downarrow n \in U$. Suppose $d(Q, R) < 2^{-n}$, then $Q \downarrow n = R \downarrow n \sqsubseteq R$ so since U is $\sqsubseteq$-closed, $R \in U$. Thus $B_{2^{-n}}(Q) \subseteq U$ and U, as the union of such balls as Q ranges over U, is open. □

This leads us to make a definition which plays a role dual to that of n-step approximation in the characterization of closure (Lemma 17(2)).

Definition 19. *Given a process P, let $\mathcal{X}_n = \{Q \in \mathcal{P} \mid Q \downarrow n \in imps(P)\}$. If $\mathcal{X}_n$ is non-empty, then define $P \uparrow n = \sqcap \mathcal{X}_n$.*

Note that $\mathcal{X}_n$ is always open and whenever $P \uparrow n$ is defined, $P \sqsubseteq P \uparrow n$.

Lemma 20. *There exists an open, $\sqsubseteq$-closed upwards, and non-empty set U of predeterministic processes such that $P = \sqcap U$ iff there exists a least $m < \omega$ such that $P \uparrow m$ is defined and $P = \sqcap_{m \leqslant n < \omega} P \uparrow n$.*

Proof.

⇒ Since U is non-empty it contains some element Q and because U is open and $\sqsubseteq$-closed upwards there exists some n such that $Q \downarrow n \in U$. Now, $\mathcal{X}_n$ is non-empty and $P \uparrow n$ is defined. Therefore, there is a least integer such that $P \uparrow n$ is defined. If we can show that U is contained in the union of the $\mathcal{X}_n$, then

$$P \sqsubseteq \sqcap \bigcup_{n < \omega} \mathcal{X}_n \sqsubseteq \sqcap U = P$$

shows that $P = \sqcap \bigcup_{n < \omega} \mathcal{X}_n = \sqcap_{m \leqslant n < \omega} P \uparrow n$ where m is the least such that $\mathcal{X}_m \neq \emptyset$. To prove the containment

$$\bigcup_{n < \omega} \mathcal{X}_n = \{Q \in \mathcal{P} \mid \exists n . Q \downarrow n \in imps(P)\}$$
$$\supseteq \{Q \in U \mid \exists n . Q \downarrow n \in U\}$$
$$= U$$

⇐ To prove the backward implication, it is sufficient to show that $\bigcup_{n < \omega} \mathcal{X}_n$ is open. This in turn is shown by each $\mathcal{X}_n$'s being open. □

Note that while all the infinite behaviour of a process P with $imps(P)$ open is divergent, it does not follow that $infs(P) = \overline{traces(P)}$ and so that $imps(P)$ is closed, for there may be increasing $\preceq$-chains in $traces(P)$ with each element taken from the non-divergent behaviour of a different P_n of Lemma 20.

For example, consider $\bigsqcap_{n<\omega} Q_n$ where the Q_n are predeterministic and given by

$$divs(Q_n) = \emptyset$$
$$traces(Q_n) = \{\langle a\rangle^m \mid m \leqslant n\}$$

For each Q_n, $Q_n \downarrow (n+1) = Q_n$ and so $\{Q_n \mid n < \omega\}$ is open and $\sqsubseteq$-closed upwards. The closure of the set of implementations includes the infinite trace of as:

$$divs(Q_\omega) = \emptyset$$
$$traces(Q_\omega) = \{\langle a\rangle^m \mid m < \omega\}$$

The process $\bigsqcap_{n<\omega} n \to Q_n$ has an open $\sqsubseteq$-closed set of implementations but does not have an open set of implementations, for the implementation $n : \omega \to Q_n$ does not have any restriction which is an implementation.

3.5 The branching order of closed processes

The main result of this chapter is a way of finding a bound on the nondeterminism required to express closed processes (that is, those with a closed set of implementations).

The formal language of CSP in the second of the unbounded nondeterminism papers (Roscoe 1988) is implicitly parameterized not only by the alphabet Σ but also by a regular cardinal (strict) bound on the cardinality of the sets of processes which may be combined by the generalized nondeterministic-choice operator. Roscoe chooses to insist that this bounding cardinal ξ also be greater than the cardinality of Σ; relaxing this we must restrict the cardinality of the event-set argument to the hiding operator and of the inverse image of any event under admissible alphabet transformations to be less than ξ. We shall write CSP_ξ when we wish to be explicit about the syntactic bound.

A trivial result is that the index of nondeterminacy of the canonical transition system given by the operational semantics to a syntactic process in CSP_ξ is no greater than ξ, so that syntactic expressibility results bound the complexity of the machinery required to resolve the nondeterminism. Note also that requiring a bound on nondeterminism does not, paradoxically, mean that we are not discussing unbounded nondeterminism, for if we pick ξ greater than the cardinality of $\mathcal{U}$, then the constraints are trivially satisfied.

For a fixed alphabet, the languages CSP_ξ as ξ varies induce a chain of subsets of $\mathcal{U}$ under the denotational semantics. We shall establish a bound on the ξ required for any closed process to enter the chain in terms of a covering property of its implementations. Since the CSP theory is couched in terms of strict regular cardinal bounds, we choose to use a slightly modified (and more expressive) variant of the Lindelöf generalization.

Definition 21. *Given a topological space $(S, \mathcal{T})$ a set $X \subseteq S$ has Lindelöf bound ξ iff (ξ is a regular cardinal and) every cover $\mathcal{V} \subseteq \mathcal{T}$ of X contains a subcover $\mathcal{V}' \subseteq \mathcal{V}$ of X with the cardinality of $\mathcal{V}'$ strictly less than ξ.*

The standard definition of Lindelöf number substitutes 'less than or equal to' for 'strictly less than' (and drops the regularity requirement). Thus a compact set has Lindelöf bound ω, and a Lindelöf set has Lindelöf bound ω_1, the first uncountable ordinal. We are now in a position to state the theorem.

Theorem 22. *If $\mathcal{X}$ is a set of Lindelöf bound ξ and $P = \sqcap \mathcal{X}$ is closed then P can be expressed in CSP_ξ.*

The result is strictly better than requiring that the Lindelöf bound of the set of implementations be ξ. For instance, consider the binary choice

$$STOP \sqcap n : \omega \to STOP$$

which is manifestly expressible in CSP_ω. However, for any $X \subseteq \omega$ the process $n : X \to STOP$ is an implementation, and these implementations can be separated from each other by open sets since they differ on the very first step. Therefore the Lindelöf bound of the set of implementations is 2^ω, which is not such a good bound.

That the result requires some restriction, such as the implementations of P being closed, can be seen by considering the space of processes over a finite alphabet Σ of cardinality $m > 1$: for each n, the number of distinct n-step behaviours is clearly finite, so the set of open balls is countable and every subset is Lindelöf, but not every such process is expressible with countable nondeterminism. For a proof of this we must turn to the operational semantics: every process in CSP_{ω_1} gives rise (in the operational semantics) to a countably branching labelled transition system which can have only a countable number of accessible nodes; without loss of generality we can identify these nodes with elements of ω, and a transition system is determined up to isomorphism by the $m + 1$ transition relations for each label (in $\Sigma \cup \{\tau\}$), each of which is a subset of $\omega \times \omega$. Thus there are at most 2^ω non-isomorphic countable transition systems and, as the map from operational to denotational domains respects isomorphism, at most 2^ω elements of $\mathcal{U}$ in the image of CSP_{ω_1}. But now $\langle \Sigma^*, \Sigma^* \times \mathbb{P}\Sigma, W \rangle$ is a process for every $W \subseteq \Sigma^\omega$, and $\mathcal{U}$ has cardinality at least 2^{2^ω}. Therefore not every process whose implementations have Lindelöf bound ω_1 is expressible in CSP_{ω_1}.

The branching order of closed processes

Before we turn to the proof of the theorem, we must first introduce the concept of the behaviour of a process after trace s and study some of its properties.

Definition 23. *Provided that $s \in traces(P)$, define P / s (read P after s) by*

$$infs(P / s) = \{u \mid su \in infs(P)\}$$
$$divs(P / s) = \{t \mid st \in divs(P)\}$$
$$fails(P / s) = \{(t, X) \mid (st, X) \in fails(P)\}$$

The first property is that a process which can perform every possible infinite trace does not lose that ability, so that we have the following lemma.

Lemma 24. *If P is a closed process and s is a trace of P, then P / s is a closed process.*

Proof. The proof rests in observing that $\forall s \, . \, \overline{\{t \mid st \in \overline{T}\}} = \overline{\{t \mid st \in T\}}$. □

The other important property is the way in which the operator distributes through nondeterministic choice. The rule expresses the fact that, having performed trace s, any behaviour of the nondeterministic choice of processes must be the behaviour of one of the processes after the trace s, and vice versa.

Lemma 25. $(\sqcap \mathcal{X}) / s = \sqcap \{P / s \mid P \in \mathcal{X} \wedge s \in traces(P)\}$, *in the sense that if either side is defined then so is the other and equality holds.*

The proof is entirely trivial. It is useful to augment the operator to sets.

Definition 26. $\mathcal{X} / s = \{P / s \mid P \in \mathcal{X} \wedge s \in traces(P)\}$

so that the body of the lemma can be rewritten:

$$(\sqcap \mathcal{X}) / s = \sqcap (\mathcal{X} / s)$$

A corollary of the previous two lemmata is that $\sqcap (\mathcal{X} / s)$ is closed whenever $\sqcap \mathcal{X}$ is.

Although $/s$ is not a total operator, it does satisfy the continuity property in that the inverse image of an open set of predeterministic processes is open. The following lemma leads up to this result.

Lemma 27.

1. $[T, D] \,/\, s = [\{t \mid st \in T\}, \{t \mid st \in D\}]$ if $s \in T$
2. $([T, D] \,/\, s) \downarrow n = ([T, D] \downarrow (n + \#s)) \,/\, s$ if $s \in T$
3. $/s$ is continuous in the metric, where defined.

Proof.

1. Trivial.

2. Let $[T_1, D_1] = ([T, D] \,/\, s) \downarrow n$ and $[T_2, D_2] = ([T, D] \downarrow (n + \#s)) \,/\, s$. Consider $t \in D_1$. If $\#t < n$, then $t \in divs([T, D] \,/\, s)$ and so $st \in D$; now, $st \in divs([T, D] \downarrow (n + \#s))$ and $t \in D_2$. The argument follows in reverse for $t \in D_2$ with $\#t < n$ and, by a similar argument, the elements of T_1 and T_2 of length less than n are equal. Note that the elements of T_i of length greater than or equal to n are the same as those of D_i. Consider $t \in T_1$ with $\#t \geqslant n$. There exists $r \in traces([T, D] \,/\, s)$ with $\#r = n$ and $r \preceq t$; now, $sr \in T$ has length $\#s + n$ so that $st \in divs([T, D] \downarrow (n + \#s))$ and hence $t \in T_2$. The argument follows in reverse.

3. Suppose that $s \in traces(Q)$ and that $d(Q, Q') < 2^{-(n + \#s)}$; that is, $Q \downarrow (n + \#s) = Q' \downarrow (n + \#s)$, so that $s \in traces(Q')$ and moreover $(Q \downarrow (n + \#s)) \,/\, s = (Q' \downarrow (n + \#s)) \,/\, s$. Therefore, by (2), we have $(Q \,/\, s) \downarrow n = (Q' \,/\, s) \downarrow n$, giving $d(Q \,/\, s, Q' \,/\, s) < 2^{-n}$.

□

The import of this is in the following adaptation of a general topological result. The general result states that the continuous image of a set of Lindelöf bound ξ has Lindelöf bound ξ. Since $/s$ is not total, it is important that the set of processes on which it is not defined is open for us to obtain the same result.

Lemma 28. If $\mathcal{X}$ has Lindelöf bound ξ, then so has $\mathcal{X} \,/\, s$.

Proof. For $\mathcal{Y} \subseteq \mathcal{P}$ write $s \rightrightarrows \mathcal{Y}$ for $\{Q \in \mathcal{P} \mid s \in traces(Q) \wedge Q \,/\, s \in \mathcal{Y}\}$, a canonical inverse image of $\mathcal{Y}$ under $/s$ which is open whenever $\mathcal{Y}$ is. We take an open cover $\mathcal{S}$ of $\mathcal{X} \,/\, s$ and form a cover of $\mathcal{X}$ by taking the sets $\{s \rightrightarrows U \mid U \in \mathcal{S}\}$ together with the set $V = \{Q \in \mathcal{P} \mid s \notin traces\, Q\}$, which covers all those elements of $\mathcal{X}$ which are not in the inverse image of some element of $\mathcal{S}$. To see that V is open, note that if $d(Q, R) < 2^{-(\#s+1)}$, then s is a trace of Q if and only if it is a trace of R.

This cover has a subcover of cardinality less than ξ which consists of the set $\{s \rightrightarrows U \mid U \in \mathcal{V}\}$ for some set $\mathcal{V} \subseteq \mathcal{S}$ consisting of one element of $\mathcal{S}$ for each U together with, possibly, the set V. Clearly, $\mathcal{V}$ is a cover for $\mathcal{X} \,/\, s$ with cardinality less than ξ.

□

The branching order of closed processes

One final result relates the behaviour after one step to the behaviour of the whole process.

Definition 29. $inits(P) = \{a \in \Sigma \mid \langle a \rangle \in traces(P)\}$

Lemma 30. $P \neq \bot \Rightarrow P = \bigsqcap_{Q \in imps(P)} x : inits(Q) \to P / \langle x \rangle$

Proof. Recall that the semantics of $x : B \to P_x$ are

$$fails(x : B \to P_x) = \{(\langle\rangle, X) \mid X \cap B = \emptyset\}$$
$$\cup \{(\langle a \rangle s, X) \mid a \in B \land (s, X) \in fails(P_a)\}$$
$$divs(x : B \to P_x) = \{\langle a \rangle s \mid s \in divs(P_a)\}$$
$$infs(x : B \to P_x) = \{\langle a \rangle u \mid u \in infs(P_a)\}$$

Since $\bigcup_{Q \in imps(P)} inits(Q) = inits(P)$ it is immediate that the two sides of the equation agree on divergences (by assumption not including $\langle\rangle$), infinite traces, and failures on non-empty traces. All that remains is to check failures of the form $(\langle\rangle, X)$. If $(\langle\rangle, X) \in fails(P)$, then there exists $Q \in imps(P)$ with $(\langle\rangle, X) \in fails(Q)$, so $X \cap inits(Q) = \emptyset$ and $(\langle\rangle, X)$ is a failure of the right-hand side by virtue of the component of the choice due to Q. Conversely, a failure of the right-hand side is a failure of the component due to some Q, $X \cap inits(Q) = \emptyset$ and so $(\langle\rangle, X) \in fails(Q) \subseteq fails(P)$. □

We are now in a position to prove the main theorem.

Proof. The basic idea is to find a set of states and a set of transitions between these states which give the required function. The set of states is taken to be the set of subsets of $\mathcal{P}$ with Lindelöf bound ξ whose greatest lower bound is a closed process. This set is called Λ. We define a function $F : \mathcal{N}'^{\Lambda} \to \mathcal{N}'^{\Lambda}$ as follows. If $\bot \in \mathcal{X}$, then $F_{\mathcal{X}}(\underline{P}) = \bot$. Otherwise $\mathcal{V} = \{B_1(Q) \mid Q \in \mathcal{X}\}$ is a disjoint open cover of $\mathcal{X}$ and so has cardinality less than ξ. Since $Q, Q' \in U \in \mathcal{V} \Rightarrow inits(Q) = inits(Q')$ we can abuse the notation and write $inits(U)$ for the initial events of any member of U; note that by construction $\emptyset \notin \mathcal{V}$. Define

$$F_{\mathcal{X}}(\underline{P}) = \bigsqcap_{U \in \mathcal{V}} x : inits(U) \to P_{\mathcal{X}/\langle x \rangle}$$

which is well defined by previous lemmata.

Clearly, F is a contraction mapping on a product of the space of processes with closed implementations. This latter space is complete, and therefore so is any product. Hence F has a unique fixed point in this space. Furthermore, since the implementations of $\bot$ form a closed set, the unique fixed point is the least fixed point in the whole space $\mathcal{U}^{\Lambda}$. If this fixed point is $\underline{L}$, then we claim that $L_{\mathcal{X}} = \bigsqcap \mathcal{X}$. To prove this, we only need show that the vector $\underline{\Lambda}$ with $\Lambda_{\mathcal{X}} = \bigsqcap \mathcal{X}$ is a fixed point of F, that is, $F_{\mathcal{X}}(\underline{\Lambda}) = \bigsqcap \mathcal{X}$. In the case of $\bot \in \mathcal{X}$, this is immediate from the definition. In all other cases, we have

$$F_{\mathcal{X}}(\underline{\Lambda}) = \bigsqcap_{U \in \mathcal{V}} x : inits(U) \to \bigsqcap \mathcal{X} / \langle x \rangle$$

which by Lemma 30 and the idempotence of $\sqcap$ is clearly $\bigsqcap \mathcal{X}$. □

3.6 Conclusions

We have presented some elementary results about the topology of pre-deterministic processes and a more complicated result which gives a bound on the nondeterminacy required to express closed processes. The method which has been presented to determine this bound cannot be extended to non-closed processes because it essentially involves delaying nondeterministic choices until the point at which the nondeterminism can be detected. The sort of nondeterminism which is modelled by non-closed processes cannot be detected at any finite point in the execution of a process; it requires an infinite amount of time in order to unfold. It might, at first, appear that the dual method is all that is applicable: namely a transformation which brings all nondeterministic choices to first step of a process. However, we can improve on this by applying the above method after any trace s for which P / s is closed. Further investigations will concentrate on finding a bound for processes in which P / s is not closed for every trace s and proving that the given bound is the best that can be found for closed processes.

References

Barrett, G. (1989). A dominated convergence theorem for CSP. Submitted to *Formal Aspects of Computing*.

Brookes, S. D., and Roscoe, A. W. (1985). An Improved Failures Model for Communicating Processes. In *Seminar on Concurrency*, S. D. Brookes, A. W. Roscoe, and G. Winskel (Eds.), Lecture Notes in Computer Science, **197**, 281–305, Springer Verlag, Berlin.

Hoare, C. A. R. (1985). *Communicating Sequential Processes*. Prentice-Hall International, London.

Roscoe, A. W. (1988). *Two Papers on CSP*. Technical Monograph, PRG-67, Oxford University Programming Research Group.

4
Algebraic posets, algebraic cpo's and models of concurrency

MICHAEL W. MISLOVE*

Abstract

In this chapter we develop the theory of algebraic posets: partially ordered sets in which each element is the directed supremum of compact elements. We show that each such poset can be completed into an algebraic cpo in a natural way, so that the completion has exactly the same set of compact elements as does the underlying algebraic poset. We use the spectral theory of locally compact sober spaces to construct this completion, which is also known as the *sobrification* of the underlying algebraic poset in the Scott topology. We apply this theory to the problem of defining a denotational semantics for abstract languages supporting concurrency. By an abstract language, we mean one whose syntax is given in terms of uninterpreted atomic actions. To demonstrate our results, we show how to obtain denotational models for various fragments of CSP using our theory. The theory requires that the operations of the language satisfy certain conditions in order that they extend naturally to continuous operations in the denotational model, and it also indicates the problems which can occur if these conditions are not satisfied. We focus on the hiding operator of CSP to demonstrate this aspect, providing a remedy for this ill-behaved operator in one model, as well as showing how to capture the failures model of Brookes *et al.* (1984) in our setting.

4.1 Introduction

In this chapter we develop the theory of *algebraic posets*, which are partially ordered sets in which each element is the supremum of the directed set of compact elements below it, and we apply this theory to show how semantic models for languages supporting concurrency can be derived in

---

*This work was partially supported by the Office of Naval Research and the Louisiana State Board of Regents.

a very simple way. Our premise is that we are working with an *abstract* language for concurrency, that is, a language whose syntax is given in terms of an alphabet A of uninterpreted atomic actions. As a simple example, we show how the algebraic poset of all finite words over the alphabet A of atomic actions for such a language leads naturally to a traces model for the language, how operations from the language can be extended to this model in a natural way, and how the continuity of the extended operations on the semantic model can be deduced from properties of the operations of the language which they extend. Specifically, we derive a traces model for a fragment of CSP using our theory, and we expand the model to a failures model for CSP based on the one derived in (Brookes *et al.* 1984).

This is but one application of the more general theory we describe which shows how the theory of algebraic posets is related to the theory of algebraic cpo's. We use the spectral theory of locally compact sober spaces to show how any algebraic poset can be completed to an algebraic cpo with the same set of compact elements, and how the Scott closed subsets of the poset give a natural model for concurrency. We also show how each Scott continuous function on the poset extends naturally to a Scott continuous function on the family of Scott closed subsets, and how the Lawson continuity of the extension is determined from properties of the function on the algebraic poset which it extends.

We apply this theory to give simple criteria on the operations of a language such as we consider in order that our theory generates a semantic model for the language in which all operations are continuous. In addition, our theory predicts what will go awry with the extension to the semantic model of an operation from the language which does not satisfy the necessary criteria. The examples we present are fragments of the language CSP, and we focus in particular on the hiding operator of CSP. For the traces model, we show that the hiding operator extends to a Scott continuous operation on the model, but this extension is not Lawson continuous. We also provide a remedy for this, which is dictated by the criterion for Lawson continuity which the hiding operator fails to satisfy.

With the failures model, the problem with the hiding operator persists, but our theory again indicates a remedy to correct the problem. The remedy we adopt is different from the one given in (Brookes *et al.* 1984), but we also capture their model in our setting. Our results show clearly that it is the hiding operator alone which forces the failures model to have the structure that it does. Indeed, we present a much simpler model for our fragment of CSP with the hiding operator replaced by a family of operators which, while not being as powerful as the hiding operator, have the advantage of being continuous.

The chapter is organized as follows. In the following section we give the general theory of algebraic posets and their completions into algebraic cpo's. Once this abstract theory is in place, then Sections 4.3 and 4.4

4.2 Algebraic posets and algebraic cpo's

Recall that an element k in a partially ordered set P is *compact* if, for each directed subset $D \subseteq P$ for which $\bigvee D$ exists, $k \leqslant \bigvee D$ implies $k \leqslant d$ for some $d \in D$. The set of compact elements of P is denoted $K(P)$, and, for each $x \in P$, $K(x) = \{k \in K(P) \mid k \leqslant x\}$ denotes the set of compact elements below x. Our first goal is to establish a general theory of algebraic posets.

Definition 1. *An* algebraic poset *is a partially ordered set P satisfying the property that $K(x)$ is directed and $x = \bigvee K(x)$ for each $x \in P$. The partially ordered set P is a* complete partial order *(cpo), if each directed subset of P has a supremum. An* algebraic cpo *is an algebraic poset which is also a cpo.*

Here is a simple motivating example.

Example 2. *Let A be a set, and let A^* denote the family of all finite words from A, including the empty word ϵ. Order A^* by*

$$s \sqsubseteq t \iff (\exists u \in A^*)\, t = su$$

Then A^* is a partially ordered set in this order. For any alphabet A, it is clear that A^* is an algebraic poset, since $K(A^*) = A^*$.

The following result shows that every compact element of $\downarrow x$ is compact in P if P is an algebraic poset.

Proposition 3. *If P is an algebraic poset, and $x \in P$, then $K(\downarrow x) \subseteq K(P)$.*

Proof. Let $x \in P$ and let $k \in K(\downarrow x)$. Then $K(k) \subseteq \downarrow x$ is directed and $k = \bigvee K(k)$, so there is some $k' \in K(k)$ with $k \leqslant k'$. But then $k = k' \in K(D)$. □

The converse of this result is false. Indeed, the following example shows that the theory of algebraic posets is not entirely a local theory.

Example 4. Let $P = \{x_n \mid n \geq 0\} \cup \{a, b\}$, where $x_n < x_{n+1} < a, b$ for each $n \geq 0$. Then $K(P) = P$, so P is an algebraic poset. However, for example, $\downarrow a \cap K(P) = \downarrow a$, while $K(\downarrow a) = \{x_n \mid n \geq 0\}$.

We are interested in the Scott topology on an algebraic poset. The definition is the now standard one.

Definition 5. A subset $U \subseteq P$ of an algebraic poset P is Scott open if

1. $U = \uparrow U$, and
2. for all $D \subseteq P$ directed, if $\bigvee D$ exists and $\bigvee D \in U$, then $D \cap U \neq \emptyset$.

We denote the set of Scott open subsets of P by $\Sigma(P)$, and the family of Scott closed sets by $\Gamma(D)$.

Theorem 6. Let P be an algebraic poset. Then the following hold.

1. $\{\uparrow k \mid k \in K(P)\}$ is a basis for the Scott topology on P.
2. $\Sigma(P)$ is locally compact and T_0.
3. For each $x \in P$, $\overline{\{x\}} = \downarrow x$.
4. If P and P' are algebraic posets, then so is $P \times P'$, and the Scott topology on $P \times P'$ is the product of the Scott topologies on P and P'.
5. The function $f : P \to P'$ between algebraic posets is Scott continuous if and only if f preserves sups of directed sets, that is, if and only if, for every directed subset $D \subseteq P$ for which $\bigvee D$ exists, $\bigvee f(D)$ exists and $f(\bigvee D) = \bigvee f(D)$.
6. If P, P', and P'' are algebraic posets and $f : P \times P' \to P''$ is a function, then f is continuous if and only if, for each $x_0 \in P$ and for each $y_0 \in P'$, the functions $f_{x_0} : P' \to P''$ by $f_{x_0}(y) = f(x_0, y)$ and $f_{y_0} : P \to P''$ by $f_{y_0}(x) = f(x, y_0)$ are continuous.

Proof. It is clear that $\{\uparrow k \mid k \in K(P)\}$ is a basis for the Scott topology on the algebraic poset P. Clearly, each set $\uparrow k$ is compact, since Scott open sets are upper sets. The topology is T_0 since, given $x, y \in P$ with $x \neq y$, there is some $k \in K(P)$ which is below one of x and y, but not below the other, and so $\uparrow k$ is a Scott open set containing exactly one of x and y. This same fact implies that $\downarrow x$ is Scott closed, but this set is clearly a subset of $\overline{\{x\}}$ and so they are equal. This shows (1)–(3).

For (4), suppose that P and P' are algebraic posets. Then, clearly $K(P) \times K(P') \subseteq K(P \times P')$, and the reverse inclusion is also simple to verify. But then each element $(x, y) \in P \times P'$ satisfies the property that $(x, y) = \bigvee K((x, y))$ and $K((x, y))$ is directed. Thus $P \times P'$ is an algebraic poset. Then (1) implies that a basis for the Scott topology is $\{\uparrow(k, k') \mid (k, k') \in K(P) \times K(P')\}$, and this implies that the product of the Scott topologies is the Scott topology of $P \times P'$. Hence (4) holds.

Suppose $f : P \to P'$ is Scott continuous. We first show that f is monotone. Indeed, given $x \leq y \in P$, let $k \in K(f(x))$ in P'. Then $f^{-1}(\uparrow k)$ is Scott open in P and contains x, so $y \in f^{-1}(\uparrow k)$ as well. Thus, $k \leq f(y)$, and since $k \in K(f(x))$ is arbitrary, we conclude that $f(x) = \bigvee K(f(x)) \leq f(y)$.

Now, given $D \subseteq P$ directed for which $\bigvee D$ exists, the monotonicity of f implies that $f(D)$ is directed and $f(\bigvee D)$ is an upper bound for $f(D)$. Conversely, let $y \in P'$ be any upper bound of $f(D)$, and let $k \in K(f(\bigvee D))$. Then $\uparrow k$ is a Scott open set in P' and contains $f(\bigvee D)$, so $f^{-1}(\uparrow k)$ is Scott open in P and contains $\bigvee D$. Thus, $D \cap f^{-1}(\uparrow k) \neq \emptyset$. Now, if $d \in D \cap f^{-1}(\uparrow k)$, then $f(d) \in \uparrow k$, which means $k \leq f(d) \leq y$. Thus, $k \leq y$ for each $k \in K(f(\bigvee D))$, and so $f(\bigvee D) \leq y$. Hence $f(\bigvee D) = \bigvee f(D)$, so f preserves directed suprema.

Conversely, suppose $f : P \to P'$ preserves directed suprema. Now, clearly f is monotone, since $x \leq y \in P$ implies $\{x, y\}$ is directed with supremum y. To show that f is Scott continuous, it is sufficient to show that $f^{-1}(\uparrow k)$ is Scott open in P, for each $k \in K(P')$. If $k \in K(P')$, then $f^{-1}(\uparrow k)$ is an upper set since f is monotone. And, if $D \subseteq P$ satisfies $\bigvee D \in f^{-1}(\uparrow k)$, then $\bigvee f(D) = f(\bigvee D) \in \uparrow k$, since f preserves directed suprema. But, $f(D)$ is directed since f is monotone, and so $k \in K(P')$ implies there is some $d \in D$ with $f(d) \in \uparrow k$. Hence, $d \in f^{-1}(\uparrow k)$, so that $D \cap f^{-1}(\uparrow k) \neq \emptyset$. It follows that $f^{-1}(\uparrow k)$ is Scott open in P. Thus (5) holds.

For (6), it is clear that any continuous function $f : P \times P' \to P''$ satisfies the property that the functions f_{x_0} and f_{y_0} are continuous for each $(x_0, y_0) \in P \times P'$. Conversely, suppose f_{x_0} and f_{y_0} are continuous for each $(x_0, y_0) \in P \times P'$. Then, it is routine to show that f must be monotone. If $D \subseteq P \times P'$ is a directed set and $(x, y) = \bigvee D$, then the monotonicity of f implies that $f(d) \leq f(\bigvee D)$ for all $d \in D$, and so $f(\bigvee D)$ is an upper bound of $f(D)$.

To show $f(\bigvee D)$ is the least upper bound of $f(D)$, let z be any upper bound of $f(D)$ in P''. If we write the elements of D as pairs (d, d'), then we have

$$f(\bigvee D) = f(x, y) = f(\bigvee_{(d,d') \in D} (d, y))$$

$$\begin{aligned}
&= \bigvee_{(d,d')\in D} f(d,y) \qquad (f_y \text{ is continuous}) \\
&= \bigvee_{(d,d')\in D} f(d, \bigvee_{(e,e')\in D} e') \\
&= \bigvee_{(d,d')\in D} \bigvee_{(e,e')\in D} f(d, e')
\end{aligned}$$

The last equality follows from the continuity of f_d. Since D is directed, for each $(d, d'), (e, e') \in D$, we can find $(c, c') \in D$ with $(d, d'), (e, e') \leqslant (c, c')$, and since f is monotone, we have $f(d, e') \leqslant f(c, c') \leqslant z$, so

$$f(\bigvee D) = \bigvee_{(d,d')\in D} \bigvee_{(e,e')\in D} f(d, e') \leqslant z$$

Hence $f(\bigvee D) \leqslant z$ for all upper bounds z of $f(D)$, which proves the desired result.

□

Note. The proof given of part (6) of Theorem 6 involves only a minor modification of the proof of Lemma II-2.9 of (Gierz et al. 1980), which is the analogous result for continuous lattices.

Our goal is to associate with each algebraic poset P an algebraic cpo which has the same set of compact elements as P. To do this, we use spectral theory, which studies the relationship between a topological space and its lattice of open (or closed) subsets. The heart of spectral theory is the map

$$\eta_X : X \to \mathcal{O}(X)$$
$$\eta_X(x) = X \setminus \overline{\{x\}}$$

where $\mathcal{O}(X)$ denotes the lattice of open subsets of the topological space X. For reasons which will become clearer below, we wish to use the closed subsets, so we want the map

$$\chi_X : X \to \Gamma(X)$$
$$\chi_X(x) = \overline{\{x\}}$$

The lattice $\Gamma(X)$ of closed subsets of the topological space X is a complete Heyting algebra; that is, it satisfies the infinite distributivity law

$$C \cup (\bigcap_{i \in I} C_i) = \bigcap_{i \in I} (C \cup C_i)$$

for any family $\{C_i \mid i \in I\} \subseteq \Gamma(X)$. Moreover, the map $\chi_X : X \to \Gamma(X)$ defined by $\chi_X(x) = \overline{\{x\}}$ is a one–one map of X into $\Gamma(X)$ if X is T_0. In fact, each closed set from X of the form $\overline{\{x\}}$ is a $\cup$-prime in $\Gamma(X)$: if A and B are Scott closed sets and $\overline{\{x\}} \subseteq A \cup B$, then $\overline{\{x\}} \subseteq A$ or $\overline{\{x\}} \subseteq B$ (such sets are also called *irreducible*). If we denote by $Spec_\cup(\Gamma(X))$ the family of $\cup$-primes of $\Gamma(X)$, then $\chi_X(X) \subseteq Spec_\cup(\Gamma(X))$.

Definition 7. *The topological space X is* sober *if, for each $\cup$-prime A in the lattice $\Gamma(X)$ of closed subsets of X, there is a unique $x \in X$ with $A = \overline{\{x\}}$. In other words, the space X is sober if and only if the map $x \mapsto \overline{\{x\}}: X \to \Gamma(X)$ is a bijection from X onto the set of $\cup$-primes of $\Gamma(X)$.*

Since each $\cup$-prime of $\Gamma(X)$ is the closure of a unique point, it is clear that a sober space is T_0, but the converse may fail (see Corollary 10 below). For a sober space X, the map $\chi_X : X \to Spec_\cup(\Gamma(X))$ is a bijection, and we now show how to topologize the latter so that the map is a homeomorphism. Indeed, each $C \in \Gamma(X)$ is a closed subset of X, and

$$\begin{aligned} \downarrow C \cap Spec_\cup(\Gamma(X)) &= \{\chi_X(x) \mid \chi_X(x) \subseteq C\} \\ &= \{\overline{\{x\}} \mid x \in C\} \\ &= \chi_X(C) \end{aligned}$$

If we define the topology on $Spec_\cup(\Gamma(X))$ which has the family $\{\downarrow C \cap Spec_\cup(\Gamma(X)) \mid C \in \Gamma(X)\}$ as a basis of closed sets, then the mapping χ_X becomes a homeomorphism of X onto $Spec_\cup(\Gamma(X))$. But, even if X is only T_0, the mapping χ_X is still a homeomorphism onto its image. This topology is called the *hull–kernel topology*, a terminology which arises from ring theory.

Theorem 8. *(Hofmann and Lawson 1978) If X is a T_0 space, then the mapping $\chi_X : X \to Spec_\cup(\Gamma(X))$ defined by $\chi_X(x) = \overline{\{x\}}$ is a homeomorphism of X onto its image in $Spec_\cup(\Gamma(X))$ endowed with the hull–kernel topology τ.*

The space $(Spec_\cup(\Gamma(X)), \tau)$ is a sober space, and each continuous map $f : X \to Y$ from X into the sober space Y has a unique extension to $(Spec_\cup(\Gamma(X)), \tau)$. Moreover, $(Spec_\cup(\Gamma(X)), \tau)$ is the largest T_0 space having the same lattice of closed sets as does X.

The space $(Spec_\cup(\Gamma(X)), \tau)$ is called the *sobrification* of X. Our interest in this construction is evidenced by the following result.

Theorem 9. *If P is an algebraic poset, then $Spec_\cup(\Gamma(P))$ is an algebraic cpo under set inclusion, $K(Spec_\cup(\Gamma(P))) = \chi_P(K(P))$, and the hull–kernel topology on $Spec_\cup(\Gamma(P))$ is the Scott topology.*

Proof. It is routine to verify that, in any complete lattice L, the supremum of a directed family of $\vee$-primes is a $\vee$-prime, and so $Spec_\cup(\Gamma(P))$ is a cpo under set inclusion, since this is the inherited order from $\Gamma(P)$.

Let $k \in K(P)$, and let $\{C_i \mid i \in I\}$ be a directed family of closed subsets of P such that $\chi_P(k) = \downarrow k \subseteq \bigvee C_i$. Now, $\bigvee C_i = \overline{\cup_i C_i}$, so $k \in \overline{\cup_i C_i}$. But, $\uparrow k$ is a Scott open set containing k, and so $\uparrow k \cap (\cup_i C_i) \neq \emptyset$. Since $C_i = \downarrow C_i$ for each $i \in I$, it follows that $k \in \cup_i C_i$, and so there is some index i with $k \in C_i$. But then $\downarrow k \subseteq \downarrow C_i = C_i$. This shows that $\chi_P(k) = \downarrow k \in K(Spec_\cup(\Gamma(P))$ for each $k \in K(P)$.

If $C \in Spec_\cup(\Gamma(P))$, then $C = \bigvee\{\downarrow k \mid k \in K(C)\}$, since P is an algebraic poset. If $k, k' \in K(C)$, then $P \setminus \uparrow k$ and $P \setminus \uparrow k'$ are closed subsets of P. If $(\uparrow k \cap C) \cap (\uparrow k' \cap C) = \emptyset$, then $C \subseteq P \setminus (\uparrow k \cap \uparrow k') = P \setminus \uparrow k \cup P \setminus \uparrow k'$, which contradicts the fact that C is $\cup$-prime. Thus, $(\uparrow k \cap C) \cap (\uparrow k' \cap C) \neq \emptyset$. If $x \in (\uparrow k \cap C) \cap (\uparrow k' \cap C)$, then $k, k' \in K(x)$, so there is some $k'' \in K(x)$ with $k, k' \leqslant k''$. Then $k'' \in C$ since C is a lower set, which implies $K(C)$ is directed, and so the family $\{\downarrow k \mid k \in K(C)\}$ is directed as well. Thus, $C = \bigvee\{\downarrow k \mid k \in C \cap K(P)\}$ is the supremum of a directed family of compact elements from $Spec_\cup(\Gamma(P))$. In particular, if $C \in K(Spec_\cup(\Gamma(P)))$, then $C = \bigvee\{\downarrow k \mid k \in C \cap K(P)\}$, and so there is some $k \in C \cap K(P)$ with $C = \downarrow k = \overline{\{k\}}$. Thus $\{\downarrow k \mid k \in K(P)\} = K(Spec_\cup(\Gamma(P)))$, and it follows that $Spec_\cup(\Gamma(P))$ is an algebraic cpo.

Since $K(Spec_\cup(\Gamma(P))) = \{\downarrow k \mid k \in K(P)\}$, the basic Scott open sets are of the form

$$\begin{aligned}
U_k &= \{C \in Spec_\cup(\Gamma(P)) \mid \downarrow k \subseteq C\} \\
&= \{C \in Spec_\cup(\Gamma(P)) \mid k \in C\} \\
&= \{C \in Spec_\cup(\Gamma(P)) \mid \uparrow k \cap C \neq \emptyset\} \\
&= Spec_\cup(\Gamma(P)) \setminus \{C \mid C \subseteq P \setminus \uparrow k\} \\
&= Spec_\cup(\Gamma(P)) \setminus \downarrow(P \setminus \uparrow k))
\end{aligned}$$

which is open in the hull–kernel topology, by definition.

Conversely, if $A \in \Gamma(P)$ is a closed set, then the set $Spec_\cup(\Gamma(P)) \setminus \{C \mid C \subseteq A\}$ is open in the Scott topology. Indeed, given $C \in Spec_\cup(\Gamma(P))$ with $C \not\subseteq A$, there is some $x \in C \setminus A$, and so there is some $k \in K(x) \setminus A$. Then $\downarrow k \subseteq C$ since C is a lower set, and so $C \in U_k$. Moreover, if $C' \in U_k$ then $k \in C'$, and so $C' \not\subseteq A$. Hence

$$C \in U_k \subseteq Spec_\cup(\Gamma(P)) \setminus \{C \mid C \subseteq A\}$$

This proves that the hull–kernel topology is the same as the Scott topology. □

Corollary 10. *If P is an algebraic poset, then P is a sober space in the Scott topology if and only if P is a cpo.*

Proof. Suppose that P is an algebraic cpo, and let $A \in Spec_\cup \Gamma(P)$. Then A is a lower set in P, and the same argument we used in the proof of Theorem 9 shows that $K(A)$ is directed and $A = \bigvee \{\downarrow k \mid k \in K(A)\}$, and so $A = \downarrow \bigvee K(A) = \overline{\{\bigvee K(A)\}}$. Thus P is sober.

Conversely, if P is sober, then Theorem 8 implies that the map $\chi_P : P \to Spec_\cup(\Gamma(P))$ is a homeomorphism, and this map is clearly an order isomorphism. Since Theorem 9 implies that $Spec_\cup(\Gamma(P))$ is an algebraic cpo, the same holds true for P. $\square$

Given a Scott continuous map $f : D \to E$ between algebraic posets, the map $\Gamma(f) = f^{-1} : \Gamma(E) \to \Gamma(D)$ preserves all the set operations; that is, $\Gamma(f)$ preserves all intersections and all finite unions. The following result is then important.

Theorem 11. (Gierz et al. 1980, Theorems 0-3.4 and IV-1.26) *If L and M are complete lattices and $f : L \to M$ preserves all infima, then the map $g : M \to L$ defined by $g(m) = \wedge f^{-1}(\uparrow m)$ preserves all suprema. If, in addition, f preserves finite suprema, then g preserves $\vee$-primes; that is, $g(Spec_\vee(M)) \subseteq Spec_\vee(L)$.*

The map g is called the *lower adjoint* of the map f, which is called the *upper adjoint* of g. Now, $\Gamma(f)$ is an upper adjoint, and Theorem 11 implies that $\Gamma(f)$ has a lower adjoint $F : \Gamma(D) \to \Gamma(E)$ defined by

$$F(C) = \bigcap \Gamma(f)^{-1}(\uparrow C) = \bigcap \{C' \mid C' \supseteq f(C)\} = \overline{f(C)}$$

It is also important to note that, since $F : \Gamma(D) \to \Gamma(E)$ is a lower adjoint, it preserves all suprema. Furthermore, since $\Gamma(f)$ preserves finite unions, the map F preserves $\cup$-primes; that is, F has a restriction

$$\overline{F} : Spec_\cup(\Gamma(D)) \to Spec_\cup(\Gamma(E))$$

Now, for $x \in D$, we calculate

$$\begin{aligned} \overline{F} \circ \chi_D(x) &= \overline{F}(\overline{\{x\}}) \\ &= \bigcap \{C' \mid C' \supseteq f(\overline{\{x\}})\} \\ &= \overline{\{f(x)\}} \\ &= \chi_E \circ f(x) \end{aligned}$$

Moreover, the very definition of $\overline{F}$ implies that it is continuous with respect to the hull–kernel topologies. Thus $F : \Gamma(D) \to \Gamma(E)$ is the natural extension of the map induced by $f : D \to E$.

We want to cast this result in terms of adjoint functors. To do this, we form the categories

$\mathcal{AP}$: the category of algebraic posets and Scott continuous maps, and

$\mathcal{AC}$: the category of algebraic cpo's and Scott continuous maps.

Theorem 12. *The functor* $\mathcal{S} : \mathcal{AP} \to \mathcal{AC}$ *which sends an algebraic poset P to the algebraic cpo* $\mathcal{S}(P) = Spec_\cup(\Gamma(P))$ *and the Scott continuous map $f : P \to P'$ to the Scott continuous map* $\mathcal{S}(f) = \overline{F} : \mathcal{S}(P) \to \mathcal{S}(P')$ *is left adjoint to the forgetful functor from $\mathcal{AC}$ to $\mathcal{AP}$.*

Proof. The unit of the adjunction is the mapping $\chi_P : P \to \mathcal{S}(P)$, which is a homeomorphism onto its image, and hence is Scott continuous. Given a Scott continuous map $f : P \to P'$ between algebraic posets, we have the map $\mathcal{S}(f) : \mathcal{S}(P) \to \mathcal{S}(P')$, which extends f, by our comments above. So, it only remains to check what happens when P' is an algebraic cpo. But in that case $\chi_{P'} : P' \to \mathcal{S}(P')$ is a homeomorphism of P' onto $\mathcal{S}(P')$ which satisfies

$$\chi_{P'} \circ f = \mathcal{S}(f) \circ \chi_P$$

So, if we define $\hat{f} : \mathcal{S}(P) \to P'$ to be the composition $\hat{f} = \chi_{P'}^{-1} \circ \mathcal{S}(f)$, then clearly $\hat{f} \circ \chi_P = f$. That $\hat{f} : \mathcal{S}(P) \to P'$ is the unique mapping which makes the appropriate diagram commute follows from the fact that $\chi_P(P)$ is dense in $\mathcal{S}(P)$ in the Scott topology, and $\hat{f}$ is continuous with respect to this topology. □

Remark. The fact that $\mathcal{S}$ is a left adjoint means that each Scott continuous map $f : P \to P'$ between algebraic posets extends uniquely to a Scott continuous mapping between the algebraic cpo's $\mathcal{S}(P)$ and $\mathcal{S}(P')$. For example, if we consider the algebraic poset A^*, then it is routine to verify that $\mathcal{S}(P) \simeq A^* \cup A^\infty$, the algebraic cpo consisting of all finite or infinite words over A in the prefix order. Moreover, a Scott continuous self-map $f : A^* \to A^*$ is nothing more than a monotone self-map on A^*, since $A^* = K(A^*)$. Thus, each such mapping extends to a Scott continuous mapping $\mathcal{S}(f) : A^* \cup A^\infty \to A^* \cup A^\infty$ in a unique way.

We also need to investigate the structure of the lattice $\Gamma(P)$ for an algebraic poset P. In fact, $\Gamma(P)$ is a bialgebraic lattice; that is, $(\Gamma(P), \subseteq)$ and $(\Gamma(P), \supseteq)$ are both algebraic lattices.

Theorem 13. *If P is an algebraic poset, then $\Gamma(P)$ is a bialgebraic lattice:*

1. $K(\Gamma(P), \subseteq) = \{\downarrow F \mid F \subseteq K(P) \text{ finite}\}$
2. $K(\Gamma(P), \supseteq) = \{P \setminus \uparrow F \mid F \subseteq K(P) \text{ finite}\}$

Proof. Let $F \subseteq K(P)$ be a finite set, and let $\{C_i\} \subseteq \Gamma(P)$ be a directed family of Scott closed sets with $\downarrow F \subseteq \bigvee C_i$. Then, $\downarrow F \subseteq \overline{\bigcup_i C_i}$. Since each $k \in F$ is compact, this means that $k \in C_{i_k}$ for some i_k, for each $k \in F$. But then $F \subseteq C_i$ for some index i, since F is finite and $\{C_i\}$ is directed. Thus $\downarrow F \subseteq \downarrow C_i = C_i$, and we have shown that $\downarrow F \in K(\Gamma(P), \subseteq)$ for each finite subset $F \subseteq K(P)$. If $C \in \Gamma(P)$, then $C = \bigvee\{\downarrow F \mid F \subseteq C \cap K(P) \text{ finite}\}$, and this family is clearly directed. Hence $\downarrow C \cap K(\Gamma(P), \subseteq)$ contains a directed family whose supremum is C, for each $C \in \Gamma(P)$. As in the proof of Theorem 9, it follows that $K(\Gamma(P), \subseteq) = \{\downarrow F \mid F \subseteq K(P) \text{ finite}\}$, and so $(\Gamma(P), \subseteq)$ is an algebraic lattice.

For part (2), let $F \subseteq K(P)$, and let $\{C_i\} \subseteq \Gamma(P)$ be a filtered family such that $\bigcap_i C_i \subseteq P \setminus {\uparrow}F$. Note that $P \setminus {\uparrow}F$ is Scott closed only because ${\uparrow}F \subseteq P$ is Scott open. If $k \in F$, then there is some C_{i_k} with $k \notin C_{i_k}$. Then ${\uparrow}k \cap C_{i_k} = \emptyset$ since C_{i_k} is a lower set. Hence $C_{i_k} \subseteq P \setminus {\uparrow}k$. Since F is finite, there is some index i with $C_i \subseteq P \setminus {\uparrow}k$ for each $k \in F$, and so $C_i \subseteq \bigcap_{k \in F} P \setminus {\uparrow}k = P \setminus {\uparrow}F$. This shows that $P \setminus {\uparrow}F \in K(\Gamma(P), \supseteq)$. Moreover, if $C \in \Gamma(P)$, then

$$C = \bigcap\{P \setminus {\uparrow}F \mid C \subseteq P \setminus {\uparrow}F \wedge F \subseteq K(P) \text{ finite}\}$$

since C is Scott closed, and the family of such sets $P \setminus {\uparrow}F$ is clearly filtered. We have shown that each Scott closed set is the filtered intersection of a family of compact elements from $(\Gamma(P), \supseteq)$, and so this lattice is algebraic, and $K(\Gamma(P), \supseteq) = \{P \setminus {\uparrow}F \mid F \subseteq K(P) \text{ finite}\}$. □

Given a Scott continuous map $f : P \to P'$ between algebraic posets, the lower adjoint $F : \Gamma(P) \to \Gamma(P')$ described above preserves all suprema, which are closures of unions. In particular, $F : (\Gamma(P), \subseteq) \to (\Gamma(P'), \subseteq)$ is Scott continuous.

Corollary 14. *Let $f : P \to P'$ be a Scott continuous map between algebraic posets. If $F : (\Gamma(P), \subseteq) \to (\Gamma(P'), \subseteq)$ is the lower adjoint to the map $\Gamma(f) : \Gamma(P') \to \Gamma(P)$, then F preserves all suprema, and hence is Scott continuous.*

We also want to consider the dual Scott topology on $\Gamma(P)$, that is, the Scott topology of $(\Gamma(P), \supseteq)$. The results of Hofmann and Lawson have something to tell us here, as well, but we must first define some terminology.

Definition 15. *Let P be an algebraic poset. The lower topology on P has the family $\{\emptyset, P\} \cup \{{\uparrow}x \mid x \in P\}$ as a subbasis for the closed sets. The Lawson topology (also called the λ-topology) is the common refinement of the Scott and lower topologies.*

On an algebraic lattice, the Lawson topology is a compact Hausdorff topology (Gierz et al. 1980, Chapter III), and on the bialgebraic lattice $\Gamma(P)$, this is the same as the common refinement of the Scott and dual Scott topologies.

Theorem 16. *Let P be an algebraic poset. For each $F \subseteq K(P)$ finite, let $U_F = \{C \in \Gamma(P) \mid F \subseteq C\}$ and let $V_F = \{C \in \Gamma(P) \mid C \subseteq P \setminus {\uparrow}F\}$. Then the family*
$$\{U_F \setminus V_G \mid F, G \subseteq K(P) \text{ finite}\}$$
forms a basis for the Lawson topology on $\Gamma(P)$. In particular, the family
$$\{U_{\{k\}} \setminus V_{\{k'\}} \mid k, k' \in K(P)\}$$
forms a subbasis for the Lawson topology.

Proof. Proposition VII–2.2 of (Gierz et al. 1980) implies that the Lawson topology on $\Gamma(P)$ is the common refinement of the Scott and dual Scott topologies. Theorem 13 then implies that the sets of the form U_F and V_F are the upper and dual upper sets generated by the compact element U_F and the cocompact element V_F, respectively. This implies the first statement, and the second is a trivial corollary of the first. □

If X is a topological space, then a subset of X is *saturated* if it is the intersection of open sets. In Hausdorff spaces, all subsets are saturated, but this is not true in general. For an algebraic poset in the Scott topology, it is easy to show that a set is saturated if and only if it is an upper set. A function $f : X \to Y$ is *proper* if $f^{-1}(C)$ is compact for each compact saturated subset C of Y. The following result is then of relevance.

Theorem 17. *(Hofmann and Lawson 1978) Let $f : X \to Y$ be a continuous function between locally compact sober spaces. Then the lower adjoint $F : \Gamma(X) \to \Gamma(Y)$ to $\Gamma(f) : \Gamma(Y) \to \Gamma(X)$ is Lawson continuous if and only if f is proper.*

Corollary 18. *If $f : P \to P'$ is a Scott continuous map between algebraic posets, then the map $F : \Gamma(P) \to \Gamma(P')$ is Lawson continuous if and only if $f^{-1}({\uparrow}k) = {\uparrow}X$ for some finite subset $X \subseteq K(P)$, for each $k \in K(P')$. In particular, if $f : P \to P'$ is pre-image finite, then $F : \Gamma(P) \to \Gamma(P')$ is Lawson continuous.*

Proof. According to Theorem 17, given a Scott continuous map $f : P \to P'$ between algebraic posets, the map $F : \Gamma(P) \to \Gamma(P')$ is Lawson continuous if and only if the restriction of F, $\overline{F} : Spec_\cup(\Gamma(P)) \to Spec_\cup(\Gamma(P'))$ is proper, that is, if and only if $\overline{F}^{-1}(C)$ is compact for every saturated compact subset C of $Spec_\cup(P')$. Now, if $\overline{F}$ is proper, then, for each $k \in K(P')$, the set $W_k \equiv U_k \cap Spec_\cup(P')$ is a compact open subset of $Spec_\cup(P')$, and so $\overline{F}^{-1}(U_k)$ must be compact. Since $\overline{F}$ is Scott continuous and W_k is open, this set is Scott open. So, there is some subset $Y \subseteq K(P)$ such that $\overline{F}^{-1}(W_k) = \cup\{W_y \mid y \in Y\}$. Since $\overline{F}^{-1}(W_k)$ is compact, this open cover has a finite subcover; that is, there is some finite subset $X \subseteq Y$ with $\overline{F}^{-1}(W_k) = \cup\{W_x \mid x \in X\}$. Translating this to P and P' via χ_P and $\chi_{P'}$, we conclude that, if $\overline{F}$ is proper, then for each $k \in K(P')$, there is a finite subset $X \subseteq K(P)$ such that

$$\begin{aligned} f^{-1}(\uparrow k) &= \chi_P^{-1}(\overline{F}^{-1}(W_k)) \\ &= \chi_P^{-1}(\bigcup\{W_x \mid x \in X\}) \\ &= \uparrow X \end{aligned}$$

which shows the only if implication.

Conversely, suppose that $f^{-1}(\uparrow k)$ is compact for each $k \in K(P')$. Let $C \subseteq Spec_\cup(P')$ be saturated and compact. Then C is an upper set with

$$C = \bigcap\{\cup_{x \in X} W_x \mid C \subseteq \cup_{x \in X} W_x \wedge X \subseteq K(P') \text{ is finite}\}$$

Now, for any such set $X \subseteq K(P')$, we have

$$\overline{F}^{-1}(\bigcup_{x \in X} W_x) = \bigcup_{x \in X} \overline{F}^{-1}(W_x)$$

and this set is compact and open, since it is the union of finitely many compact open sets $\overline{F}^{-1}(W_x)$, by hypothesis. Hence,

$$\begin{aligned} \overline{F}^{-1}(C) &= \overline{F}^{-1}(\bigcap\{\bigcup_{x \in X} W_x \mid C \subseteq \bigcup_{x \in X} W_x \wedge X \subseteq K(P') \text{ is finite}\}) \\ &= \bigcap\{\overline{F}^{-1}(\bigcup_{x \in X} W_x) \mid C \subseteq \bigcup_{x \in X} W_x \wedge X \subseteq K(P') \text{ is finite}\} \end{aligned}$$

is the intersection of a filtered family of compact open sets. By Proposition 2.19 of (Hofmann and Mislove, 1979), this intersection is compact. Thus $\overline{F}$ is proper, and so F is Lawson continuous.

The last comment is an obvious corollary to the first part of the result. □

We need one further result about extending maps between algebraic posets to their lattices of Scott closed subsets. In the definition of an abstract language, we find operations such as sequential composition and nondeterministic choice which correspond to unary, binary, ternary, etc. operations on the language. In a denotational model for the language, each of these operations must have a continuous analogue, and so we are led to consider how to extend, say, binary operations $+ : P \times P \to P$, which are defined on an algebraic poset P, to the lattice $\Gamma(P)$. Given such an operation, it is natural to consider the induced map $\Gamma(+) : \Gamma(P \times P) \to \Gamma(P)$, but Γ, being a left adjoint, preserves coproducts, not products. That is, $\Gamma(P \times P) \not\cong \Gamma(P) \times \Gamma(P)$ (this non-isomorphism is obvious if we state it as the fact that there are more closed subsets on the product $P \times P$ than simply products of closed subsets from P). Nonetheless, we can still obtain the result we need.

Theorem 19. Let P be an algebraic poset, and let $+ : P \times \cdots \times P \to P$ be a Scott continuous n-ary operation on P. Then there is a Scott continuous n-ary operation
$$\oplus : \Gamma(P) \times \cdots \times \Gamma(P) \to \Gamma(P)$$
which extends $+$. Moreover, $\oplus$ is Lawson continuous if and only if, for each $k \in K(P)$, there is some finite subset $F \subseteq K(P) \times \cdots \times K(P)$ so that $+^{-1}(\uparrow k) = \bigcup \{\uparrow(k_1, ..., k_n) \mid (k_1, ..., k_n) \in F\}$. In particular, if $+$ is pre-image finite, then the induced operation $\oplus$ is Lawson continuous.

Proof. Since $+ : P \times \cdots \times P \to P$ is Scott continuous, there is a continuous mapping $\uplus : \Gamma(P \times \cdots \times P) \to \Gamma(P)$ which extends $+$, by Theorem 12. Now, consider the subset $\Gamma(P) \times \cdots \times \Gamma(P)$ of $\Gamma(P \times \cdots \times P)$. It is routine to verify that $\Gamma(P) \times \cdots \times \Gamma(P)$ is closed in $\Gamma(P \times \cdots \times P)$ under all intersections and all increasing unions (that is, the intersection of a family of products of closed sets is a product of closed sets, and the closure of the union of a family of products of closed sets is a product of closed sets). Moreover, the unit of the adjunction between $\mathcal{AP}$ and $\mathcal{AC}$ sends $P \times \cdots \times P$ into $\Gamma(P) \times \cdots \times \Gamma(P)$. Thus the inclusion map $\iota_n : \Gamma(P) \times \cdots \times \Gamma(P) \to \Gamma(P \times \cdots \times P)$ preserves all intersections and all directed unions, and so it is Scott continuous. Hence, the map $\oplus = \uplus \circ \iota_n : \Gamma(P) \times \cdots \times \Gamma(P) \to \Gamma(P)$ is the desired Scott continuous extension of $+$.

The comment about Lawson continuity of the extension follows from Corollary 18, once it is noted that the Lawson topology on $\Gamma(P) \times \cdots \times \Gamma(P)$ is the product of the Lawson topologies on each factor; this is shown, for example, in (Gierz et al. 1980, Proposition III–2.2). □

4.3 A trace semantics for CSP

In this section we apply the results from Section 4.2 to derive a traces model for the language CSP (Brookes et al. 1984). This model is not new, but we use it to develop additional insights into CSP. In the next section, we apply the same techniques to develop an alternative failures model for CSP. We begin by recalling the definition of a subset of CSP.

Let A be a non-empty set of atomic actions which a process can perform; this set may be finite or infinite. Processes are understood in terms of the actions from A which they execute, and so we use strings of such actions as trace histories of the processes under study. The algebraic structure of CSP processes we consider is suggested by the following BNF-like description:

$$P ::= SKIP \mid STOP \mid a \to P \mid P; Q \mid P \sqcap Q \mid P \parallel Q \mid P \mid\mid\mid Q$$
$$\mid P \setminus a \mid \textbf{repeat } P$$

Here is a brief description of the processes defined by this syntax. $SKIP$ is a process whose sole action is immediate normal termination. $STOP$ is a process which cannot engage in any event; it is deadlocked. If $a \in A$, then $a \to P$ is a process which first does the action a, and then acts like P. More generally, $P; Q$ is a process which first acts like P, and then like Q. The process $P \sqcap Q$ is the (internal) nondeterministic choice between P and Q; it acts like P or like Q, and the choice is not influenced by the environment. The process $P \parallel Q$ is the synchronized parallel composition of P and Q; this process must synchronize each action of P and Q, and so the actions of either must be duplicated in the other in exactly the same order. This amounts to lockstep synchronization of P and Q. The process $P \mid\mid\mid Q$, on the other hand, consists of an arbitrary interleaving of the actions of P and Q. There is no synchronization here; the processes are each allowed to perform their actions completely independently. The process $P \setminus a$ acts like P, but with all occurrences of the action a hidden from the environment. The process **repeat** P is the simple recursive construct. We utilize this simple form of recursion to avoid the complications which treating environments entails.

The syntax given above means that the simplest processes P end with $STOP$ or $SKIP$, and all other processes are constructed from such processes as these. So, if we want a process which performs the action a and then normally terminates, we should take $a \to SKIP$, while if we want one which performs a and then deadlocks, we should take $a \to STOP$. In any case, the term a alone is not a process. Also, the process **repeat** $SKIP$ is the same as $STOP$, so even if we had not taken $STOP$ as a primitive process, it would still have appeared in our language. One other property of $STOP$ is that it is a left zero for ;, that is, $STOP; P = STOP$ for any process P.

Before we define a traces model for this fragment of CSP, we note the following. If we are observing processes via their visible actions, then it is certainly reasonable to assume that we can observe normal termination of a process. So, we add a symbol $\checkmark$ to our alphabet A of observable actions. The trace semantics we give for CSP consists of recording the traces of all the actions which a process P can perform. Since $\checkmark$ is one such action, we expect that a trace of a process P consists of a string $s \in A^* \cup A^*\checkmark$, where a string from the latter set indicates the normal termination of the process. So, we define a map

$$\Phi : CSP \to \Gamma(A^* \cup A^*\checkmark)$$
$$\Phi(P) = \{s \mid s \text{ is a trace of } P\}$$

For $\Phi(P)$ to be a Scott closed subset of $A^* \cup A^*\checkmark$, we must assume the following.

1. If $s \in \Phi(P)$ and $t \leqslant s$, then $t \in \Phi(P)$.
2. If $D \subseteq \Phi(P)$ is directed, then $\bigvee D \in \Phi(P)$.

The first condition just says that any prefix of a trace of the process P must also be a trace of P. The second assumes that the traces of a process form a Scott closed set, but since $A^* \cup A^*\checkmark$ has no infinite directed subsets, this condition is vacuously satisfied.

To define a trace semantics for CSP using Φ, we must define operations on $\Gamma(A^* \cup A^*\checkmark)$ to reflect the operations of our language. But, before we define the necessary operations on $\Gamma(A^* \cup A^*\checkmark)$, we first describe the λ-topology on $\Gamma(A^* \cup A^*\checkmark)$ in more suggestive terms.

Since $K(A^* \cup A^*\checkmark) = A^* \cup A^*\checkmark$, Theorem 16 implies that a subbasis for the λ-topology on $\Gamma(A^*)$ is the family

$$\{ \{C \in \Gamma(A^* \cup A^*\checkmark) \mid s \in C\} \mid s \in A^* \cup A^*\checkmark \} \cup$$
$$\{ \{C \in \Gamma(A^* \cup A^*\checkmark) \mid s \notin C\} \mid s \in A^* \cup A^*\checkmark \}$$

Definition 20. Let $s \in A^* \cup A^*\checkmark$. The *liveness test determined by* s is the set

$$U_{\{s\}} = \{C \in \Gamma(A^* \cup A^*\checkmark) \mid s \in C\}$$

in $\Gamma(A^* \cup A^*\checkmark)$. The *safety test determined by* s is the set

$$V_{\{s\}} = \{C \in \Gamma(A^* \cup A^*\checkmark) \mid s \notin C\}$$

in $\Gamma(A^* \cup A^*\checkmark)$.

So, one description of the λ-topology on $\Gamma(A^* \cup A^*\checkmark)$ is that it is the topology generated by the liveness and safety tests.

Now, in order that the mapping Φ be a semantic function, it is necessary that Φ preserve the operations of the language. For example, we must have $\Phi(P \;|||\; Q) = \Phi(P) \;|||\; \Phi(Q)$. But this means that we must define operations like $|||$ on $\Gamma(A^* \cup A^*\sqrt{})$ and then prove that Φ takes each operation CSP to the corresponding operation on $\Gamma(A^* \cup A^*\sqrt{})$. This is precisely where the theory we developed in Section 4.2 comes into play. Since $A^* \cup A^*\sqrt{}$ is an algebraic poset, we can define continuous operations on $\Gamma(A^* \cup A^*\sqrt{})$ by simply defining those operations on $A^* \cup A^*\sqrt{}$ and then use Theorem 12 or Corollary 18 to induce the desired continuous extensions to $\Gamma(A^* \cup A^*\sqrt{})$. Moreover, since $A^* \cup A^*\sqrt{}$ has no infinite directed subsets which have suprema, the required continuous operations on $A^* \cup A^*\sqrt{}$ need only be monotone for the extension to be liveness continuous, or monotone and proper for the extension to be safety and liveness continuous. The following list gives the definitions for all operations except the hiding operator \:

$\to\colon A \times (A^* \cup A^*\sqrt{}) \to A^* \cup A^*\sqrt{}$ is defined by $(a, s) \mapsto as$.

$;\colon (A^* \cup A^*\sqrt{}) \times (A^* \cup A^*\sqrt{}) \to A^* \cup A^*\sqrt{}$ is defined by

$$(s, t) \mapsto \begin{cases} s, & \text{if } s \notin A^*\sqrt{}, \\ s't, & \text{if } s = s'\sqrt{} \in A^*\sqrt{} \end{cases}$$

$\sqcap\colon (A^* \cup A^*\sqrt{}) \times (A^* \cup A^*\sqrt{}) \to \Gamma(A^* \cup A^*\sqrt{})$ is defined by $(s,t) \mapsto {\downarrow}s \cup {\downarrow}t$.

$\|\colon (A^* \cup A^*\sqrt{}) \times (A^* \cup A^*\sqrt{}) \to A^* \cup A^*\sqrt{}$ is defined by $(s,t) \mapsto s \wedge t$, where $s \wedge t$ is the largest prefix common to both s and t.

$|||\colon (A^* \cup A^*\sqrt{}) \times (A^* \cup A^*\sqrt{}) \to \Gamma(A^* \cup A^*\sqrt{})$ is defined by

$$s \;|||\; t = \bigcup\{{\downarrow}u \mid u \text{ is an interleaving of } s \text{ and } t\}$$

where an interleaving of s and t allows the symbol $\sqrt{}$ to appear only at the end (if at all).

Proposition 21. *Each of the operations defined above is monotone and proper, and so induces a safety and liveness continuous operation on $\Gamma(A^* \cup A^*\sqrt{})$.*

Proof. It is routine to verify that $\to$ is monotone and one–one, and so it induces the desired safety and liveness continuous operation on $\Gamma(A^* \cup A^*\sqrt{})$ by Corollary 18.

The case of ; is more interesting. Indeed, the definition we have given for ; above is monotone. If $u \in A^* \cup A^*\sqrt{}$, then consider the set

$$U = \{(s,t) \in (A^* \cup A^*\sqrt{}) \times (A^* \cup A^*\sqrt{}) \mid s;t \geqslant u\}$$

Since ; is monotone, $U = \uparrow U_0$, where

$$U_0 = \{(s,t) \in (A^* \cup A^*\sqrt{}) \times (A^* \cup A^*\sqrt{}) \mid s;t = u\}$$

The string u has only finitely many factorizations $u = xy$, where $x, y \in A^* \cup A^*\sqrt{}$. For each of these with $x < u$,

$$\{(s,t) \in (A^* \cup A^*\sqrt{}) \times (A^* \cup A^*\sqrt{}) \mid s;t = u \wedge s = x\sqrt{}\} = \{(x\sqrt{}, y)\}$$

has only one element. If $u \in A^*\sqrt{}$, then

$$U_0 = \cup \{(x\sqrt{}, y) \mid xy = u\}$$

and so $U = \uparrow U_0$ is the upper set of finitely many compact elements from $A^* \cup A^*\sqrt{}$.

On the other hand, if $u \in A^*$, then

$$U_0 \setminus (\cup\{(x\sqrt{}, y) \mid xy = u\}) = \{(u, t) \mid t \in A^* \cup A^*\sqrt{}\} = \uparrow\{(u, \epsilon)\}$$

and so $U = \uparrow U_0$ is again the upper set of a finite set of compact elements. Thus Corollary 18 applies to give a safety and liveness continuous extension of ; to $\Gamma(A^* \cup A^*\sqrt{})$.

The operation $\sqcap : (A^* \cup A^*\sqrt{}) \times (A^* \cup A^*\sqrt{}) \to \Gamma(A^* \cup A^*\sqrt{})$ defined by $(s,t) \mapsto \downarrow s \cup \downarrow t$ maps $(A^* \cup A^*\sqrt{}) \times (A^* \cup A^*\sqrt{})$ into $K(\Gamma(A^* \cup A^*\sqrt{}))$, and it is easy to see that this operation is monotone. Moreover, given any finite subset $F \subseteq A^* \cup A^*\sqrt{}$, the set

$$\sqcap^{-1}(\downarrow F) = \{(s_1, ..., s_n) \in (A^* \cup A^*\sqrt{})^n \mid \{s_1, ..., s_n\} = F\}$$

is a finite set, and so

$$\sqcap^{-1}(U_F) = \uparrow\{(s_1, ..., s_n) \in (A^* \cup A^*\sqrt{})^n \mid \{s_1, ..., s_n\} = F\}$$

is finitely generated. This shows there is a safety and liveness continuous extension of $\sqcap$, since the functor $\mathcal{S} : \mathcal{AP} \to \mathcal{AC}$ is a left adjoint by Theorem 12.

The mapping $(s,t) \mapsto s \wedge v : (A^* \cup A^*\sqrt{}) \times (A^* \cup A^*\sqrt{}) \to (A^* \cup A^*\sqrt{})$ defines the semilattice structure on $A^* \cup A^*\sqrt{}$, and so it is monotone. If $u \in A^* \cup A^*\sqrt{}$ and $(s,t) \in (A^* \cup A^*\sqrt{}) \times (A^* \cup A^*\sqrt{})$ with $s \wedge t = u$, then clearly $u \leq s$ and $u \leq t$, so $(s,t) \in \uparrow\{(u,u)\}$; that is, $\wedge^{-1}(u) = \uparrow\{(u,u)\}$.

An argument similar to that for the operation ; shows ||| has a safety and liveness extension. Indeed, our definition of the interleaving operation (that is, with the restriction that $\sqrt{}$ appears only at the end) is certainly monotone from $(A^* \cup A^*\sqrt{}) \times (A^* \cup A^*\sqrt{})$ to $\Gamma(A^* \cup A^*\sqrt{})$. Moreover, given any string $u \in A^* \cup A^*\sqrt{}$ there are only finitely many strings s and t which can be interwoven to create u, and this implies that the interleaving operation is proper. □

A trace semantics for CSP 93

The hiding operator does not fall within the scope of Corollary 18, since it is not proper (as we demonstrate below). But we can deduce a partial continuity of this operator.

Proposition 22. For each $a \in A$, the hiding operator $\setminus a : \Gamma(A^* \cup A^*\sqrt{}) \to \Gamma(A^* \cup A^*\sqrt{})$ defined by $X \setminus a = \{s \setminus a \mid s \in X\}$ is continuous with respect to liveness tests on $\Gamma(A^* \cup A^*\sqrt{})$.

Proof. If we define $\setminus a : A^* \cup A^*\sqrt{} \to A^* \cup A^*\sqrt{}$ by $s \setminus a = t$ is the word s with all occurrences of a deleted, then it is routine to show that this operator is monotone. Thus it induces a liveness continuous operator $\setminus a : \Gamma(A^* \cup A^*\sqrt{}) \to \Gamma(A^* \cup A^*\sqrt{})$. □

We can now give the full definition of the semantic function Φ. Indeed, the following clauses suffice:

$$\begin{aligned}
\Phi(SKIP) &= \{\epsilon, \sqrt{}\} \\
\Phi(STOP) &= \{\epsilon\} \\
\Phi(a \to P) &= \{as \mid s \in \Phi(P)\} \\
\Phi(P;Q) &= \Phi(P); \Phi(Q) \\
\Phi(P \sqcap Q) &= \Phi(P) \sqcap \Phi(Q) \\
\Phi(P \parallel Q) &= \Phi(P) \parallel \Phi(Q) \\
\Phi(P \mid\!\mid\!\mid Q) &= \Phi(P) \mid\!\mid\!\mid \Phi(Q) \\
\Phi(P \setminus a) &= \Phi(P) \setminus a \\
\Phi(\textbf{repeat } P) &= \bigcup_n \Phi(P)^n
\end{aligned}$$

where $\Phi(P)^0 = \{\epsilon\}$ and $\Phi(P)^{n+1} = \Phi(P); \Phi(P)^n$ for each $n \geq 0$.

Note that our definition of ; works as we expect. Namely, the set of traces of the process $P;Q$ consists of traces $s \in \Phi(P) \cap A^*$ together with traces st where $s\sqrt{} \in \Phi(P)$ and $t \in \Phi(Q)$. If the trace $s \in \Phi(P) \cap A^*$ has no extension in $\Phi(P)$, then it is not composed with any trace t of Q. Also, our definition of $\Phi(\textbf{repeat } P)$ works correctly. Indeed, for any process P, $\Phi(P)^0 = \{\epsilon\}$, and so

$$\begin{aligned}
\Phi(P)^1 &= \Phi(P); \Phi(P)^0 \\
&= \{s;t \mid s \in \Phi(P) \wedge t \in \Phi(P)^0\} \\
&= \Phi(P) \cap (A^* \cup A^\infty) \cup \{s\epsilon \mid s\sqrt{} \in \Phi(P)\} \\
&= \Phi(P) \cap (A^* \cup A^\infty) \cup \{s \mid s\sqrt{} \in \Phi(P)\} \\
&\subseteq A^* \cup A^\infty
\end{aligned}$$

It then follows that $\Phi(\text{repeat } P) \subseteq A^* \cup A^\infty$, since this set is Scott closed in $A^* \cup A^*\sqrt{} \cup A^\infty$. So, for example, $\Phi(\text{repeat } SKIP) = \{\epsilon\} = \Phi(STOP)$. More generally, $\Phi((\text{repeat } P); Q) = \Phi(\text{repeat } P); \Phi(Q) = \Phi(\text{repeat } P)$ for any processes P and Q, since $\Phi(\text{repeat } P) \subseteq A^* \cup A^\infty$. The idea of using $\sqrt{}$ in this way is inspired by a similar analysis presented in (Oles 1987).

The work we did above to induce the appropriate operators on the semantic model $\Gamma(A^* \cup A^*\sqrt{})$ makes these definitions a matter of simply listing them. We are guaranteed that they are all continuous and Φ preserves all the operators, by definition. The following result shows the relationship of the topology on $\Gamma(A^* \cup A^*\sqrt{})$ to the meanings $\Phi(P)$ for processes P.

Proposition 23. *Let P be a CSP process. Then we have the following.*

1. *$\Phi(P)$ is the union of the liveness tests which it passes. Thus $\Phi(P)$ is the smallest element of $\Gamma(A^* \cup A^*\sqrt{})$ which passes all of P's liveness tests.*

2. *$\Phi(P)$ is the intersection of the safety tests which contain $\Phi(P)$. Thus $\Phi(P)$ is the largest element of $\Gamma(A^* \cup A^*\sqrt{})$ which passes all of P's safety tests.*

We can also use the inherent order on $\Gamma(A^* \cup A^*\sqrt{})$ to define an order on CSP. Namely, we define the *traces order* on CSP by

$$P \sqsubseteq Q \iff \Phi(Q) \subseteq \Phi(P)$$

This orders CSP so that the more nondeterministic a process P is, the lower P is in the order. It is sometimes called the *Smyth order*.

The fact that hiding is not continuous with respect to the safety tests can be seen with the following example. Let $a, b \in A$ with $a \neq b$. Then $\{s \in A^* \mid s \setminus a = b\}$ contains the family $ab, a^2b, \ldots, a^n b, \ldots$, which is an infinite family of minimal elements from A^*. Thus $(\setminus a)^{-1}(U_{\{b\}})$ is not the upper set of a finite subset of $K(A^* \cup A^*\sqrt{})$. Corollary 18 then implies that this operator does not have a safety continuous extension. More succinctly, it is not safe to use the hiding operator. To remedy this problem, we introduce a new operator to the language CSP to replace the hiding operator.

Definition 24. *Define the operator*

$$\setminus_1 : A^* \times A \to A^*$$
$$s \setminus_1 a = t$$

where t is the word s with the first occurrence (from the left) of the letter a deleted.

Proposition 25. *For each $n > 0$ and each $a \in A$, the operator $\backslash_n a = \backslash_1 a \circ \backslash_1 a \circ \cdots \circ \backslash_1 a$ is monotone and pre-image finite, and so it induces a λ-continuous operator $\backslash_n a : \Gamma(A^* \cup A^* \sqrt{}) \to \Gamma(A^* \cup A^* \sqrt{})$. Moreover, for each $X \in \Gamma(A^* \cup A^* \sqrt{})$, we have*

$$X \backslash a = \bigcap \{X \backslash_n a \mid n > 0\}$$

Proof. It is routine to check that $\backslash_1 a$ is monotone, and it is pre-image finite since there are only finitely many ways to add just one a to any word s. Moreover, the same is true of the n-fold composition of this operator with itself, for each $n > 0$ and each $a \in A$. It is also trivial to check the asserted equality. □

So, we can replace the ill-behaved hiding operator $\backslash$ by the family of operators $\{\backslash_n \mid n > 0\}$, and thus achieve a language CSP' in which all of the operators are continuous with respect to both liveness and safety tests. We then extend the semantic function Φ given above to include the new operator(s) by the simple definition

$$\Phi(P \backslash_1 a) = \Phi(P) \backslash_1 a$$

and then define

$$\Phi(P \backslash_n a) = (\cdots((\Phi(P) \backslash_1 a) \backslash_1 a) \cdots) \backslash_1 a$$

The resulting operators are all liveness and safety continuous.

We can see by the definition of the operators $\backslash_n$ what the real problem with the hiding operator is. Namely, we can regard the hiding operator as the infinite composition of $\backslash_1 a$ with itself; that is, $\backslash a = \texttt{repeat} \; \backslash_1 a$. Then hiding is just like an unguarded recursion and, as such, cannot be expected to be continuous with respect to the safety tests. Replacing this operator with the 'guarded recursions' $\backslash_n$ thus cures the problem.

It is important to know when a process P from CSP might *diverge*, by which we mean it might engage in an infinite chatter between subprocesses and thus never respond to its environment. In CSP, the hiding operator is the cause of this problem, since the process $P \backslash a$ may engage in just such a chatter (represented by the action a which has been hidden). But this is not a problem for CSP', since, in this language, the event a is hidden only a finite number of times, and so such a chatter will eventually be seen if it occurs. In fact, it is precisely the continuity of all of the operators in the language with respect to the safety tests which reflect this property. Thus divergence has been eliminated in CSP'. On the other hand, we have also eliminated one of the most useful operators of CSP, and so we can justifiably raise the question of whether CSP' is as powerful as CSP.

4.4 A failures model for CSP

The fragment of CSP for which we gave a traces model in Section 4.3 does not contain the external nondeterministic choice operation which is available within CSP. The reason is that the trace semantics does not provide a mathematical model which can distinguish this operation from internal choice. If P and Q are processes, then the external nondeterministic choice of P and Q, denoted $P \square Q$, should act like P or like Q, with the environment able to influence which choice is made on the first step. So, if the environment offers an action which P can participate in, but which Q cannot, then P should be chosen, and if the action offered is one which Q but not P can participate in, then Q should be chosen. If both can participate, then $P \square Q$ should be just like $P \sqcap Q$. But there is no way to model this in a traces model. Indeed, it is easy to construct examples of processes P and Q for which the traces of $P \square Q$ and of $P \sqcap Q$ are the same, but the processes are different (Hoare 1985). This explains the need to expand the semantics of CSP to the failures model. First, we give an algebraic description of the syntax of our expanded language.

The language CSP'' consists of processes described by

$$P ::= SKIP \mid STOP \mid a \to P \mid P; Q \mid P \sqcap Q \mid P \square Q$$
$$P \parallel Q \mid P \parallel\parallel Q \mid P \setminus_n a \mid \textbf{repeat } P$$

The operations from CSP' are the same as before, and we have just described the new operation $\square$, external nondeterminism.

To develop a failures model for CSP'' similar to the one presented in (Brookes et al. 1984), we first expand the set $A^* \cup A^*\checkmark$ so that we encode more information about a process P than merely the actions in which it participates. Following (Brookes et al. 1984), we add to the information about a process a second component, which consists of all the actions which it might refuse to participate in after participating in the sequence of actions which comprise its trace to that point. That is, a *refusal set* for the process P after the sequence of actions s is a subset $X \subseteq A$ consisting of events the process P might refuse after executing s. If $\mathcal{P}(X)$ denotes the family of all subsets of the set X, then the failures model for CSP'' consists of sets of pairs $(s, X) \in (A^* \cup A^*\checkmark) \times \mathcal{P}(A)$, where s is a possible trace of the process P, and X a refusal set for P after executing s. To realize the family of failures sets as the closed subsets of an algebraic poset, we must first define a partial order on $(A^* \cup A^*\checkmark) \times \mathcal{P}(A)$.

Definition 26. *Consider the set* $(A^* \cup A^*\checkmark) \times \mathcal{P}(A)$. *Define a partial order on* $(A^* \cup A^*\checkmark) \times \mathcal{P}(A)$ *by*

$$(s, X) \sqsubseteq (t, Y) \iff (s = t \wedge X \subseteq Y) \vee (s < t \wedge X = \emptyset)$$

Proposition 27. *In the order given above, $(A^* \cup A^*\checkmark) \times \mathcal{P}(A)$ is an algebraic poset whose set of compact elements is $(A^* \cup A^*\checkmark) \times \mathcal{P}_{<\omega}(A)$, where $\mathcal{P}_{<\omega}(A)$ denotes the family of finite subsets of A.*

Proof. It is routine to verify that the order given above defines a partial order on $(A^* \cup A^*\checkmark) \times \mathcal{P}(A)$, and that $(A^* \cup A^*\checkmark) \times \mathcal{P}_{<\omega}(A) = K((A^* \cup A^*\checkmark) \times \mathcal{P}(A))$. Thus $(A^* \cup A^*\checkmark) \times \mathcal{P}(A)$ is an algebraic poset in this order. $\square$

We can therefore define the family of Scott closed sets $\Gamma((A^* \cup A^*\checkmark) \times \mathcal{P}(A))$, and use the theory of algebraic posets that we developed to achieve a model for CSP''. It should be noted that we need the 'full power' of the theory at this point, since, unlike the algebraic poset $A^* \cup A^*\checkmark$, the algebraic poset $(A^* \cup A^*\checkmark) \times \mathcal{P}(A)$ consists of more than compact elements. That is, the algebraic poset $(A^* \cup A^*\checkmark) \times \mathcal{P}(A)$ is not an algebraic cpo, but its set of compact elements is a proper subset.

We can define a *failures model* for CSP'' via the map

$$\Psi : CSP'' \to \Gamma((A^* \cup A^*\checkmark) \times \mathcal{P}(A))$$

given by

$$\Psi(P) = \{(s, X) \mid s \text{ is a trace of } P \text{ and } X \text{ is a refusal set for } P \text{ after } s\}$$

As with our traces model from Section 4.3, in order that the map Ψ be a semantic mapping, we must define continuous operations on $\Gamma((A^* \cup A^*\checkmark) \times \mathcal{P}(A))$ for each of the operations of CSP'', and show that Ψ preserves these operations. Here are the definitions of most of the operations:

$\to : A \times ((A^* \cup A^*\checkmark) \times \mathcal{P}(A)) \to (A^* \cup A^*\checkmark) \times \mathcal{P}(A)$ is defined by

$$a \to (s, X) = (as, X)$$

$;: ((A^* \cup A^*\checkmark) \times \mathcal{P}(A)) \times ((A^* \cup A^*\checkmark) \times \mathcal{P}(A)) \to (A^* \cup A^*\checkmark) \times \mathcal{P}(A)$ is defined by

$$(s, X); (t, Y) = \begin{cases} (s, X), & \text{if } s \in A^* \\ (s_0 t, Y) & \text{if } s = s_0\checkmark \end{cases}$$

$\sqcap : ((A^* \cup A^*\checkmark) \times \mathcal{P}(A)) \times ((A^* \cup A^*\checkmark) \times \mathcal{P}(A)) \to \Gamma((A^* \cup A^*\checkmark) \times \mathcal{P}(A))$ is defined by

$$(s, X) \sqcap (t, Y) \mapsto {\downarrow}(s, X) \cup {\downarrow}(t, Y)$$

$||| : ((A^* \cup A^*\checkmark) \times \mathcal{P}(A) \times ((A^* \cup A^*\checkmark) \times \mathcal{P}(A)) \to \Gamma((A^* \cup A^*\checkmark) \times \mathcal{P}(A))$
is defined by

$$(s, X) \,|||\, (t, Y) = \cup\{\downarrow(u, X \cap Y) \mid u = s \,|||\, t\}$$

where $s \,|||\, t$ is the set of all interleavings of s and t in which $\checkmark$ appears only at the end (if at all).

For each $a \in A$ and each $n > 0$, the operator $\backslash_n a : (A^* \cup A^*\checkmark) \times \mathcal{P}(A) \to (A^* \cup A^*\checkmark) \times \mathcal{P}(A)$ is defined by

$$(s, X) \backslash_n a = (s \backslash_n a, X)$$

where $s \backslash_n a$ is the string s with the first n occurrences of a deleted.

Proposition 28. *Each operation defined above is monotone and proper, and hence induces a λ-continuous operation on $\Gamma((A^* \cup A^*\checkmark) \times \mathcal{P}(A))$.*

Proof. Arguments similar to those given in the proof of Proposition 21 show that the restriction of each of the above operations to $K((A^* \cup A^*\checkmark) \times \mathcal{P}(A))$ is monotone and proper, and so each of the operations given above induces a λ-continuous operation on $\Gamma((A^* \cup A^*\checkmark) \times \mathcal{P}(A))$. Moreover, if we verify that each original operation is also Scott continuous on $(A^* \cup A^*\checkmark) \times \mathcal{P}(A)$, then we know that the guaranteed extension extends the original definition. This argument is easy for the operations $\to$, ;, $\sqcap$, and $\backslash_n a$, for each $a \in A$ and each $n > 0$. For $|||$, the argument relies on the fact that $\cap$ is Scott continuous on $\mathcal{P}(A)$, which is also obvious (again, $A^* \cup A^*\checkmark$ has no non-compact elements, so Scott continuity in this component is simply monotonicity). □

Neither the new operation $\square$ nor the old operation $\|$ is definable on the new model in the same way the operations above are defined. The reason in both cases is the same. In introducing the refusals set component of a pair (s, X), we are incorporating information about the future behaviour of the process, and this information is not discernible from the order on $(A^* \cup A^*\checkmark) \times \mathcal{P}(A)$. For example, $P \square Q$ is supposed to be influenced by the environment *on the first step*, and, thereafter it is to be like $P \sqcap Q$. Such an operation cannot be defined on the basis of elements $(s, X), (t, Y) \in (A^* \cup A^*\checkmark) \times \mathcal{P}(A)$ alone, but instead relies on the full history of the processes P and Q.

We define the operations $\|$ and $\square$ directly on the proposed model $\Gamma((A^* \cup A^*\checkmark) \times \mathcal{P}(A))$. Here are the relevant definitions.

Definition 29. $\square : \Gamma((A^* \cup A^*\checkmark) \times \mathcal{P}(A)) \times \Gamma((A^* \cup A^*\checkmark) \times \mathcal{P}(A)) \to \Gamma((A^* \cup A^*\checkmark) \times \mathcal{P}(A))$ *is defined by*

$$C \square D = \big((\{\epsilon\} \times \mathcal{P}(A)) \cap C \cap D\big) \cup \big((C \cup D) \setminus (\{\epsilon\} \times \mathcal{P}(A))\big)$$

A failures model for CSP

$$\| : \Gamma((A^* \cup A^*\sqrt{}) \times \mathcal{P}(A)) \times \Gamma((A^* \cup A^*\sqrt{}) \times \mathcal{P}(A)) \to \Gamma((A^* \cup A^*\sqrt{}) \times \mathcal{P}(A))$$

is defined by

$$C \| D = \{(s, X \cup Y) \mid (s, X) \in C \land (s, Y) \in D\}$$

Proposition 30. *The operations $\square$ and $\|$ are λ-continuous on $\Gamma((A^* \cup A^*\sqrt{}) \times \mathcal{P}(A))$.*

Proof. The definition of the partial order on $(A^* \cup A^*\sqrt{}) \times \mathcal{P}(A)$ ensures that $C \square D$ is Scott closed if C and D are. $\Gamma((A^* \cup A^*\sqrt{}) \times \mathcal{P}(A))$ is a compact topological lattice in the λ-topology (Gierz et al. 1980, Corollary VII-2.4), and the operations are $\cup$ and $\cap$. From this it follows readily that $\square$ is λ-continuous.

Since $\Gamma((A^* \cup A^*\sqrt{}) \times \mathcal{P}(A))$ is a bialgebraic lattice whose λ-topology is the common refinement of the Scott and dual Scott topologies, we only need show that an operation is Scott and dual Scott continuous to show it is λ-continuous. Moreover, we can show that this holds on $\Gamma((A^* \cup A^*\sqrt{}) \times \mathcal{P}(A)) \times \Gamma((A^* \cup A^*\sqrt{}) \times \mathcal{P}(A))$ in each variable separately. Now, it is clear that, for a fixed $C \in \Gamma((A^* \cup A^*\sqrt{}) \times \mathcal{P}(A))$, the mapping

$$D \mapsto \{(s, X \cup Y) \mid (s, X) \in C \land (s, Y) \in D\}$$

is monotone and preserves increasing suprema (that is, closures of increasing unions). Likewise, this mapping also preserves decreasing intersections, as is readily verified. Hence the operation $\|$ is Scott and dual Scott continuous in the second variable, and a similar argument shows the same result for the first variable. $\square$

Note. One way to check whether the definitions for $\square$ or $\|$ could be induced from mappings defined on $(A^* \cup A^*\sqrt{}) \times \mathcal{P}(A)$ using the theory of Section 4.2 is to consider the restriction of each operation to the principal lower sets in $\Gamma((A^* \cup A^*\sqrt{}) \times \mathcal{P}(A))$, and to see if the extension of that restriction guaranteed by the theory agrees with the intended operation. In both cases, this fails to be true. For example, the operation $\|$ restricts to the function

$$(\downarrow(s, X), \downarrow(t, Y)) \mapsto \downarrow(s \land t, \emptyset)$$

whose extension to $\Gamma((A^* \cup A^*\sqrt{}) \times \mathcal{P}(A))$ using spectral theory would always give the result $\emptyset$ in the second component.

The semantic map $\Psi : CSP'' \to \Gamma((A^* \cup A^*\sqrt{}) \times \mathcal{P}(A))$ can now be defined. The relevant clauses are as follows:

$$\Psi(STOP) = \{(\epsilon, X) \mid X \in \mathcal{P}(A)\}$$
$$\Psi(SKIP) = \{(\epsilon, X) \mid \sqrt{} \notin X \in \mathcal{P}(A)\} \cup \{(\sqrt{}, X) \mid X \in \mathcal{P}(A)\}$$
$$\Psi(a \to P) = \{(\epsilon, X) \mid a \notin X \in \mathcal{P}(A)\} \cup \{(as, X) \mid (s, X) \in \Psi(P)\}$$
$$\Psi(P \sqcap Q) = \Psi(P) \cup \Psi(Q)$$
$$\Psi(P \square Q) = \Psi(P) \square \Psi(Q)$$
$$\Psi(P \;|||\; Q) = \Psi(P) \;|||\; \Psi(Q)$$
$$\Psi(P \parallel Q) = \Psi(P) \parallel \Psi(Q)$$
$$\Psi(P; Q) = \Psi(P); \Psi(Q)$$
$$\Psi(P \setminus_n a) = \Psi(P) \setminus_n a$$
$$\Psi(\textbf{repeat } P) = \overline{\bigcup_n \Psi(P)^n}$$

where $\Psi(P)^0 = \{\epsilon\}$ and $\Psi(P)^n = \Psi(P); \Psi(P)^n$ for each $n \geq 1$.

Once again, the work we did above made these definitions a simple matter. Moreover, we can assert the same result for the 'liveness' and 'safety' tests in this model as we did for the traces model. Namely, for each $(s, X) \in (A^* \cup A^*\sqrt{}) \times \mathcal{P}_{<\omega}(A)$, we can define a *liveness test* to be the set

$$U_{(s,X)} = \{C \in \Gamma((A^* \cup A^*\sqrt{}) \times \mathcal{P}(A)) \mid (s, X) \in C\}$$

and we can define a *safety test* to be the set

$$V_{(s,X)} = \{C \in \Gamma((A^* \cup A^*\sqrt{}) \times \mathcal{P}(A)) \mid (s, X) \notin C\}$$

These sets form a subbasis for the λ-topology on $\Gamma((A^* \cup A^*\sqrt{}) \times \mathcal{P}(A))$, and so the following result again is valid.

Proposition 31. *Let P be a CSP'' process. Then we have the following.*

1. *$\Psi(P)$ is the union of the liveness tests which it passes. Thus $\Psi(P)$ is the smallest element of $\Gamma((A^* \cup A^*\sqrt{}) \times \mathcal{P}(A))$ which passes all of P's liveness tests.*

2. *$\Psi(P)$ is the intersection of the safety tests which it passes. Thus $\Psi(P)$ is the largest element of $\Gamma((A^* \cup A^*\sqrt{}) \times \mathcal{P}(A))$ which passes all of P's safety tests.*

Furthermore, we can define the *failures order* on CSP'' by

$$P \sqsubseteq_F Q \iff \Psi(Q) \subseteq \Psi(P)$$

This order refines the traces order we defined in Section 4.3. The simplest way to see this is to note that the mapping projection $\pi : (A^* \cup A^*\sqrt{}) \times \mathcal{P}(A) \to A^* \cup A^*\sqrt{}$ induces a continuous projection $\Pi : \Gamma((A^* \cup A^*\sqrt{}) \times \mathcal{P}(A)) \to \Gamma(A^* \cup A^*\sqrt{})$ which satisfies the property that $\Phi(P) = \Pi \circ \Psi(P)$ for all P.

4.5 Making hiding continuous

So far, we have not mentioned the hiding operator in relation to the failures model. The reason is that this operator is once again problematical. Actually, that is to be expected, since our comment above means the traces model we defined in Section 4.3 can be realized as an image of our failures model via the projection Π. So, we shall not be able to eliminate the continuity problems of this operator in this larger model $\Gamma((A^* \cup A^*\sqrt{}) \times \mathcal{P}(A))$.

Here is one possible definition for the hiding operator:

$$\backslash : ((A^* \cup A^*\sqrt{}) \times \mathcal{P}(A)) \times A \to (A^* \cup A^*\sqrt{}) \times \mathcal{P}(A)$$
$$(s, X) \backslash a = (s \backslash a, X)$$

It is routine to show that this definition is Scott continuous on $(A^* \cup A^*\sqrt{}) \times \mathcal{P}(A)$ for each $a \in A$, and so there is a Scott continuous extension to $\Gamma((A^* \cup A^*\sqrt{}) \times \mathcal{P}(A))$. Indeed, for each $a \in A$, the extension is given by

$$C \backslash a = \{(s \backslash a, X) \mid (s, X) \in C\}$$

While this defines a Scott continuous mapping for each $a \in A$, the extension to $\Gamma((A^* \cup A^*\sqrt{}) \times \mathcal{P}(A))$ is not λ-continuous, since this mapping is not proper. The following example points out the problem.

Example 32. Let $C_n = \downarrow(a^n b, A)$ for each $n > 0$, and let $C = \downarrow(\{a^n \mid n > 0\} \times \emptyset)$. Then $C = \lim_n C_n$ in $\Gamma((A^* \cup A^*\sqrt{}) \times \mathcal{P}(A))$, but $C_n \backslash a = \downarrow(b, A)$ for each $n > 0$, while $C \backslash a = \{\epsilon\} \times \emptyset$.

There is nothing special about b in the above example; we can realize the set C as the limit of a family defined similarly to the C_ns which end the string a^n with any element $s \in A^* \cup A^*\sqrt{}$ which we wish. Hence, to make the hiding operator continuous, we must add every such string to the set $C \backslash a$. But, what then of the C_ns? $C_n \backslash a$ would no longer converge to $C \backslash a$ unless we also add such strings to the C_ns, eventually! On the whole of $\Gamma((A^* \cup A^*\sqrt{}) \times \mathcal{P}(A))$ there is no way to decide how to do this (continuously). This led Brookes et al. (1984) to consider the following family of sets.

Definition 33. *The family $\mathcal{F}$ of all non-empty sets $F \in \Gamma((A^* \cup A^*\sqrt{}) \times \mathcal{P}(A))$ satisfying*

$$(s, X) \in F \wedge (sc, \emptyset) \notin F \Rightarrow (s, X \cup \{c\}) \in F \qquad (*)$$

is the failures model *for CSP*.

A routine verification shows the following.

Proposition 34. *The family $\mathcal{F}$ is precisely the family of sets which form the failures model of Brookes et al. (1984).*

Proposition 35. *The family $\mathcal{F}$ is closed in $\Gamma((A^* \cup A^*\sqrt{}) \times \mathcal{P}(A))$ under all suprema and under all filtered infima. Moreover, each non-empty Scott closed subset of $(A^* \cup A^*\sqrt{}) \times \mathcal{P}(A)$ is the intersection of the elements of $\mathcal{F}$ containing it.*

Proof. Suprema in $\Gamma((A^* \cup A^*\sqrt{}) \times \mathcal{P}(A))$ are closures of unions and filtered infima are descending intersections, and it is routine to verify that $\mathcal{F}$ is closed under these operations. Furthermore, given any element $G \in \Gamma((A^* \cup A^*\sqrt{}) \times \mathcal{P}(A))$, if $(s, X) \notin G$, then $G \subseteq ((A^* \cup A^*\sqrt{}) \times \mathcal{P}(A)) \setminus \uparrow(s, Y)$ for some finite subset $Y \subseteq X$. Then $((A^* \cup A^*\sqrt{}) \times \mathcal{P}(A)) \setminus \uparrow(s, Y) \in \mathcal{F}$ and $(s, X) \in \uparrow(s, Y)$. □

Corollary 36. *The family $\mathcal{F}$ is a consistently complete algebraic cpo under reverse set inclusion. The topology $\mathcal{F}$ inherits from $(\Gamma((A^* \cup A^*\sqrt{}) \times \mathcal{P}(A)), \lambda)$ is the λ-topology on $\mathcal{F}$.*

Proof. Since $\mathcal{F}$ is closed in $\Gamma((A^* \cup A^*\sqrt{}) \times \mathcal{P}(A))$ under all suprema and $\Gamma((A^* \cup A^*\sqrt{}) \times \mathcal{P}(A))$ is a consistently complete cpo, the same is true of $\mathcal{F}$. Also, $\chi : (\Gamma((A^* \cup A^*\sqrt{}) \times \mathcal{P}(A)), \supseteq) \to (\Gamma((A^* \cup A^*\sqrt{}) \times \mathcal{P}(A)), \supseteq)$ by $\chi(X) = \bigvee(\downarrow X \cap \mathcal{F})$ is a closure operator which preserves directed suprema in the opposite order (that is, preserves filtered intersections). Since $(\Gamma((A^* \cup A^*\sqrt{}) \times \mathcal{P}(A)), \supseteq)$ is an algebraic cpo, the same is true of $\chi(\Gamma((A^* \cup A^*\sqrt{}) \times \mathcal{P}(A)), \supseteq)$ and $K(\chi(\Gamma((A^* \cup A^*\sqrt{}) \times \mathcal{P}(A)), \supseteq)) = \chi(K(\Gamma((A^* \cup A^*\sqrt{}) \times \mathcal{P}(A)), \supseteq)))$ (Gierz et al. 1980, Proposition I-4.9). But $\chi(\Gamma((A^* \cup A^*\sqrt{}) \times \mathcal{P}(A))) = \mathcal{F} \cup \{\emptyset, \{(\epsilon, \emptyset)\}\}$, and $\mathcal{F}$ is clearly an algebraic subsemilattice of this lattice in the opposite order.

Since $\mathcal{F}$ is consistently complete, the λ-topology on $\mathcal{F}$ is compact and Hausdorff (Gierz et al. 1980, Theorem III–1.10), and the inclusion map from $\mathcal{F}$ into $\Gamma((A^* \cup A^*\sqrt{}) \times \mathcal{P}(A), \supseteq)$ is continuous. So, the λ-topology on $\mathcal{F}$ is the inherited λ-topology from the lattice $\Gamma((A^* \cup A^*\sqrt{}) \times \mathcal{P}(A), \supseteq)$. Since $\Gamma((A^* \cup A^*\sqrt{}) \times \mathcal{P}(A))$ is bialgebraic and distributive, this is the common refinement of the Scott and dual Scott topologies. □

To define a *failures semantics* for CSP using the family $\mathcal{F}$, we must show that the map Ψ defined above satisfies $\Psi(P) \in \mathcal{F}$ for all processes $P \in CSP''$, and we must define continuous operations required for CSP on $\mathcal{F}$ as we did for the failures model we defined above. This is more involved than appealing to our results in Section 4.2, however, since the family $\mathcal{F}$ is not all of $\Gamma((A^* \cup A^*\sqrt{}) \times \mathcal{P}(A))$, but rather a subset of that family. But we can still use the operations we defined above at the level of $(A^* \cup A^*\sqrt{}) \times \mathcal{P}(A)$ if we show they take elements of $\mathcal{F}$ to elements of $\mathcal{F}$. Indeed, the result that the λ-topology on $\mathcal{F}$ is the inherited λ-topology assures us that the operations are continuous on $\mathcal{F}$, once we have shown that they leave $\mathcal{F}$ invariant.

Making hiding continuous

Proposition 37. *Each of the operations defined above leaves $\mathcal{F}$ invariant, and hence induces a λ-continuous operation on $\mathcal{F}$.*

Proof. We must show that each of the induced operations leaves $\mathcal{F}$ invariant; that is, we must show that each takes sets satisfying (*) of Definition 33 to a set also satisfying (*). If $F \in \mathcal{F}$ and $a \in A$, then $a \to F = \{(as, X) \mid (s, X) \in F\}$. Now, since F satisfies (*), if $(s, X) \in F$ and $(sc, \emptyset) \notin F$, then $(s, X \cup \{c\}) \in F$. So, if $(t, Y) \in a \to F$, then $t = as$ and $(s, Y) \in F$. If $c \in A$ with $(tc, \emptyset) \notin a \to F$, then $(sc, \emptyset) \notin F$, and so $(s, X \cup \{c\}) \in F$. Thus, $(t, Y \cup \{c\}) \in a \to F$, so $\to$ leaves $\mathcal{F}$ invariant.

If $F, G \in \mathcal{F}$, then

$$F; G = \overline{\{(s, X); (t, Y) \mid (s, X) \in F \wedge (t, Y) \in G\}}$$

Since the sets satisfying (*) are closed under all suprema in $\Gamma((A^* \cup A^*\sqrt{}) \times \mathcal{P}(A))$, the closure of a set satisfying (*) also satisfies (*). So, we only need to show that (*) holds for the generating elements $(s, X); (t, Y)$. Suppose that $(s, X); (t, Y) = (u, V)$ is such a generating element, and that $c \in A$ satisfies $(uc, \emptyset) \notin F; G$. Either $s \notin A^*\sqrt{}$ or $u = s_0 t$. In the first case, we conclude that $(sc, \emptyset) \notin F$, so $(s, X \cup \{c\}) \in F$. Then $(s, X \cup \{c\}); (t, Y) = (s, X \cup \{c\}) = (u, V \cup \{c\}) \in F; G$. On the other hand, if $u = s_0 t$, then $(t, Y) \in G$ and, since $(uc, \emptyset) \notin F; G$, it follows that $(tc, \emptyset) \notin G$. Since G satisfies (*), we conclude that $(t, Y \cup \{c\}) \in G$, and so $(s, X); (t, Y \cup \{c\}) \in F; G$, as needed. Thus ; also leaves $\mathcal{F}$ invariant.

It is obvious that $\sqcap$ leaves $\mathcal{F}$ invariant, and so we consider the operation $|||$. If $F, G \in \mathcal{F}$, then

$$F \,|||\, G = \overline{\{(s, X) \,|||\, (t, Y) \mid (s, X) \in F \wedge (t, Y) \in G\}}$$

Suppose that $(u, V) = (s, X) \,|||\, (t, Y)$ and $c \in A$ with $(uc, \emptyset) \notin F \,|||\, G$. Then u is an interleaving of the letters in the string s with those from t, and $V = X \cap Y$. Since $(uc, \emptyset) \notin F \,|||\, G$, we can conclude that $(sc, \emptyset) \notin F$ and $(tc, \emptyset) \notin Y$, from which it follows that $(s, X \cup \{c\}) \in F$ and $(t, Y \cup \{c\}) \in G$. But then $(s, X \cup \{c\}) \,|||\, (t, Y \cup \{c\}) \in F \,|||\, G$, and so $(u, (X \cup \{c\}) \cap (Y \cup \{c\})) \in F \,|||\, G$. Since sets satisfying (*) are closed, the result follows.

Suppose that $F, G \in \mathcal{F}$. Then

$$F \,\square\, G = \overline{\{(s, X) \,\square\, (t, Y) \mid (s, X) \in F \wedge (t, Y) \in G\}}$$

Again, showing that the property is satisfied for the generating pairs shows that it also holds for the closure, and so we consider only the case of $(s, X) \in F$ and $(t, Y) \in G$. Since F and G satisfy (*), the only case to consider for $(s, X) \,\square\, (t, Y)$ is for some $c \in A$ with $(c, \emptyset) = (\epsilon c, \emptyset) \notin (s, X) \,\square\, (t, Y)$. Then $(c, \emptyset) \notin F$ or $(c, \emptyset) \notin G$ must hold. Now, if $s = \epsilon$, then $(s, X) \in F \cap G$

by the definition of $\square$, and then $(s, X \cup \{c\}) \in F \cap G$ holds, which implies $(s, X \cup \{c\}) \in F \square G$. On the other hand, if $s \neq \epsilon$, then $(s, X) \in F \cup G$. If $(s, X) \in F$, then $(c, \emptyset) \in F$ and $F \in \mathcal{F}$ imply $(s, X \cup \{c\}) \in F$, so that $(s, X \cup \{c\}) \in F \square G$. The argument in case $(s, X) \in G$ is similar. $\square$

We introduced the family $\mathcal{F}$ in order to realize a semantic model for CSP including the external choice operation, in which the hiding operator is continuous. We saw above that we can define a Scott continuous hiding operator on $\Gamma((A^* \cup A^*\sqrt{}) \times \mathcal{P}(A))$, but that operator is not λ-continuous. Hence the operator we defined above preserves directed suprema (that is, closures of increasing unions), but it does not preserve filtered infima (that is, descending intersections). That is, the hiding operator we defined above is continuous with respect to the usual inclusion order on $\Gamma((A^* \cup A^*\sqrt{}) \times \mathcal{P}(A))$, but not with respect to the reverse inclusion order. This latter order, which is sometimes called the *Smyth order*, is more appealing for modelling nondeterminism, since the higher a process is in the order, the more deterministic the process is. We now show how to define the hiding operator on the family $\mathcal{F}$ so that it is continuous with respect to reverse inclusion; however, this operator is not continuous with respect to inclusion. In fact, we do not know of a model where both continuity properties for the hiding operator hold.

The condition (*) holds the key to defining the hiding operator on $\mathcal{F}$. We saw above that the problem with the full lattice $\Gamma((A^* \cup A^*\sqrt{}) \times \mathcal{P}(A))$ is the difficulty in deciding which traces and refusal sets to add to the simply defined set $C \setminus a = \{(s \setminus a, X) \mid (s, X) \in C\}$. Indeed, if we want to define $C \setminus a$ for some $C \in \mathcal{F}$, and we want to know which traces to add as the first components of the result, we can see whether $(sa, \emptyset) \in C$. Because of the condition (*), the only case in which we need to add traces to $C \setminus a$ is for those $(s, X) \in C$ for which $(s, X \cup \{a\}) \notin C$. This leads us to the following.

Definition 38. For each $a \in A$, the operator $\setminus a : \mathcal{F} \to \mathcal{F}$ is defined by

$$C \setminus a = \{(s \setminus a, X) \mid (s, X \cup \{a\}) \in C\} \cup$$
$$\{((s \setminus a)t, X) \mid |(\forall n > 0)(sa^n, \emptyset) \in C, t \in A^* \cup A^*\sqrt{}, X \in \mathcal{P}(A)\}$$

We now show that $\setminus a : \mathcal{F} \to \mathcal{F}$ preserves descending intersections. We establish this via a series of technical lemmata. Throughout this discussion, we define

$$B = (A^* \cup A^*\sqrt{}) \times \mathcal{P}(A)$$

to denote the algebraic poset $(A^* \cup A^*\sqrt{}) \times \mathcal{P}(A)$.

Lemma 39. *If $C \in \mathcal{F}$, then $C \setminus a \in \mathcal{F}$.*

Proof. We must first show that $C \setminus a \in \Gamma(B)$. Let $(t, Y) \in C \setminus a$ and let $(r, Z) \sqsubseteq (t, Y)$. Now, there is some $s \in A^* \cup A^*\sqrt{}$ with either $s \setminus a = t$ and $(s, Y \cup \{a\}) \in C$, or else $(sa^n, \emptyset) \in C$ for all n and $s \setminus a \leqslant t$. If $r = t$, then, regardless of which condition holds for s, it is easy to show that $(r, Z) \in C \setminus a$.

So, suppose that $r < t$; then $Z = \emptyset$. In the case $s \setminus a = t$ and $(s, Y \cup \{a\}) \in C$, there is some $s_0 \leqslant s$ with $s_0 \setminus a = r$. Now, either $(s_0 a^n, \emptyset) \in C$ for all n, or else there is some n with $(s_0 a^n, \emptyset) \in C$, but $(s a^{n+1}, \emptyset) \notin C$. In the former case, the definition of $C \setminus a$ implies that $(s_0 \setminus a, W) \in C \setminus a$ for all sets W, and so $(r, Z) = (s_0 \setminus a, \emptyset) \in C \setminus a$. In the latter case, $(s_0 a^n, \{a\}) \in C$ as C satisfies (*), and so $(r, \emptyset) = (s_0 a^n, \emptyset) \in C \setminus a$ since $(s_0 a^n, \{a\}) \in C$. This shows that $C \setminus a$ is a lower set in B.

To show that $C \setminus a$ is Scott closed, let $(t_i, Y_i) \in C \setminus a$ be an increasing family with $(t, Y) = \bigvee(t_i, Y_i)$. Since $t \in A^* \cup A^*\sqrt{}$, it follows that there is some index j with $t_i = t$ for $i > j$, and so we can assume that this holds for all indices i. Then, $Y = \bigcup_i Y_i$. As in the proof that $C \setminus a$ is a lower set, there is some s_i with $s_i \setminus a = t$ and $(s_i, Y_i \cup \{a\}) \in C$ for each i, or else there is some s with $s \setminus a = t$ and $(sa^n, \emptyset) \in C$ for all n. If the latter holds, then clearly $(t, Y) \in C \setminus a$. So, we assume that the latter does not hold; this means that, for all s, if $s \setminus a \leqslant t$, then there is some n with $(sa^n, \emptyset) \notin C$. It then follows that there are only finitely many $s \in A^* \cup A^*\sqrt{}$ with $s \setminus a \leqslant t$. For each index i, there is some s_i with $s_i \setminus a = t$ and $(s_i, Y_i \cup \{a\}) \in C$, and since there are only finitely many such s_i, we can assume there is some fixed s with $s_i = s$ for all indices i. If $F \subset Y$ is finite, then there is some index i with $F \subseteq Y_i$, and so $(s, F \cup \{a\}) \in C$. Then, since $C \in \Gamma(B)$, it follows that $(s, Y \cup \{a\}) \in C$, and so $(t, Y) \in C \setminus a$. This completes the proof that $C \setminus a$ is closed.

Finally, we must show $C \setminus a \in \mathcal{F}$. Suppose that $(t, Y) \in C \setminus a$ and $(tc, \emptyset) \notin C \setminus a$. Then there is some s with either $s \setminus a = t$ and $(s, Y \cup \{a\}) \in C$, or else $s \setminus a \leqslant t$ and $(sa^n, \emptyset) \in C$ for all n. In the latter case, it is clear that $(t, Y \cup \{c\}) \in C \setminus a$. So, suppose that the latter case does not occur. Again, this means that there are only finitely many s with $s \setminus a \leqslant t$. If $c = a$, then, using any s with $(s, Y \cup \{a\}) \in C$, it follows that $(t, Y \cup \{a\}) \in C \setminus a$. So, assume that $c \neq a$. Since $(tc, \emptyset) \notin C \setminus a$, it follows that, for any s with $s \setminus a = t$, we have $(sc, \{a\}) \notin C$. If there is some such s with $(sca^n, \emptyset) \in C$ for all n, then $(sc \setminus a, W) \in C \setminus a$ for any set W, which implies that $(tc, \emptyset) \in C \setminus a$, a contradiction. Hence, for each such s, there is some n with $(sca^{n+1}, \emptyset) \notin C$. For any such s, we have $(sca^n, \{a\}) \in C$, and so $(tc, \emptyset) = (sca^n \setminus a, \emptyset) \in C \setminus a$, again a contradiction. Thus, we must have $(sca, \emptyset) \notin C$ for any s, which satisfies $(s, Y \cup \{a\}) \in C$, and the fact that $C \in \mathcal{F}$ implies that $(s, Y \cup \{a, c\}) \in C$, from which it follows that $(t, Y \cup \{c\}) \in C \setminus a$. This concludes the proof of the lemma. □

Lemma 40.

1. $B \setminus \uparrow(s, X) \in K(\mathcal{F})$ for each $(s, X) \in B$

2. If $(s, X) \in B$, then

$$B \setminus \uparrow(s, X) \bigvee\nolimits_{\mathcal{F}} B \setminus \uparrow(sc, \emptyset) \;=\; B \setminus (\uparrow(s, X \setminus \{c\}) \cup \uparrow(sc, \emptyset))$$

3. If $(s, X), (t, Y) \in B$ with $X \neq \emptyset$, and if $t \neq sc$ or $Y \neq \emptyset$, then

$$B \setminus \uparrow(s, X) \bigvee\nolimits_{\mathcal{F}} B \setminus \uparrow(t, Y) \;=\; B \setminus (\uparrow(s, X) \cup (t, Y))$$

Proof. We noted in the proof of Corollary 36 that $K(\mathcal{F}) = \chi(K(\Gamma(B), \supseteq))$, which is the family $\{B \setminus \uparrow F \mid F \subseteq K(B) \text{ finite}\}$, by Theorem 13. Thus, to verify (1), we only need to show that $B \setminus \uparrow(s, X) \in \mathcal{F}$ for each $(s, X) \in B$. Now, certainly each such set is a closed non-empty subset of B. Suppose that $(r, Z) \in B \setminus \uparrow(s, X)$ and $(rc, \emptyset) \notin B \setminus \uparrow(s, X)$. Then $(s, X) \sqsubseteq (rc, \emptyset)$, and so $s \leqslant rc$ and $X = \emptyset$. If $s < rc$, then $s \leqslant r$, and since $X = \emptyset \subseteq Z$, it then follows that $(s, X) \sqsubseteq (r, Z)$, a contradiction. Hence we must have $s = rc$. Then $r < s$, and so $\{r\} \times \mathcal{P}(A) \subseteq B \setminus \uparrow(s, X)$. It follows that $(r, Z \cup \{c\}) \in B \setminus \uparrow(s, X)$, as required.

For (2), we first note that $B \setminus \uparrow(s, X) \bigvee\nolimits_{\mathcal{F}} B \setminus \uparrow(sc, \emptyset) \subseteq B \setminus (\uparrow(s, X) \cup \uparrow(sc, \emptyset))$, and that the latter is not an element of $\mathcal{F}$, since it does not satisfy condition (*) of Definition 33. Since it is routine to show that $B \setminus (\uparrow(s, X \setminus \{c\}) \cup \uparrow(sc, \emptyset)) \in \mathcal{F}$, we only show that any element $C \in \mathcal{F}$ with $C \subseteq B \setminus \uparrow(s, X) \bigvee\nolimits_{\mathcal{F}} B \setminus \uparrow(sc, \emptyset)$ satisfies $C \subseteq B \setminus (\uparrow(s, X \setminus \{c\}) \cup \uparrow(sc, \emptyset))$. Suppose $C \in \mathcal{F}$ with $C \subseteq B \setminus \uparrow(s, X) \bigvee\nolimits_{\mathcal{F}} B \setminus \uparrow(sc, \emptyset)$. Suppose that $(s, Z) \in C$; we must show that $X \setminus \{c\} \not\subseteq Z$. Now, since $C \subseteq B \setminus \uparrow(s, X) \bigvee\nolimits_{\mathcal{F}} B \setminus \uparrow(sc, \emptyset)$, we know that $(sc, \emptyset) \notin C$, and since $C \in \mathcal{F}$, it follows that $(s, Z \cup \{c\}) \in C$. Hence $(s, Z \cup \{c\}) \notin \uparrow(s, X)$, so $X \not\subseteq Z \cup \{c\}$, and so $X \setminus \{c\} \not\subseteq Z$. Therefore $(s, Z) \notin \uparrow(s, X \setminus \{c\})$, and so $(s, Z) \in B \setminus (\uparrow(s, X \setminus \{c\}) \cup \uparrow(sc, \emptyset))$. It follows that $C \sqsubseteq B \setminus (\uparrow(s, X \setminus \{c\}) \cup \uparrow(sc, \emptyset))$, and (2) is established.

For (3), we only need to show that $B \setminus (\uparrow(s, X) \cup (t, Y)) \in \mathcal{F}$ under the stated hypotheses, which means showing that this set satisfies (*) of Definition 33, and this is easy to do. □

Note. We can use this lemma to determine the compact elements of $\mathcal{F}$. Indeed, as we commented in the proof, $K(\mathcal{F}) = \chi(K(\Gamma(B), \supseteq)) = \{\chi(B \setminus \uparrow F) \mid F \subseteq K(B) \text{ finite}\}$. We have just shown that χ leaves certain of these sets fixed, and we have calculated the images of some of the others under χ. A simple inductive argument can now be used to calculate $\chi(B \setminus \uparrow F)$ for any finite subset $F \subseteq K(B)$.

Lemma 41. For each $a \in A$, the function $\backslash a : \mathcal{F} \to \mathcal{F}$ preserves descending intersections and finite unions.

Proof. It is clear from the definition of $\backslash a$ that the map is monotone, and it is a simple exercise to show that $\backslash a$ preserves finite unions. Showing that $\backslash a$ preserves descending intersections is the same thing as showing that the map is Scott continuous on $(\mathcal{F}, \supseteq)$. Since $(\mathcal{F}, \supseteq)$ is an algebraic semilattice, it is enough to show that $\backslash a^{-1}(U)$ is Scott open for each Scott open subset U of $\mathcal{F}$, and it is enough to show this for $U = \uparrow k$, where k is a compact element of $\mathcal{F}$. According to Lemma 40, the compact elements of $\mathcal{F}$ are the sets of the form $B \setminus \uparrow F$, where $F \subseteq K(B)$ is finite. So, it is enough to show that $\backslash a^{-1}(\uparrow(B \setminus \uparrow F))$ is Scott open in $\mathcal{F}$ for appropriate finite subsets $F \subset K(B)$.

Now, Lemma 40 implies that every compact element of $\mathcal{F}$ has the form $\uparrow(B \setminus \uparrow F)$ where $F \subseteq K(B)$ is finite. If $F = \{(s_1, X_1), \ldots, (s_n, X_n)\}$, and if $s_i = s_j c$ for some i and j, then either $X_i \neq \emptyset$ or $c \notin X_j$.

Consider first the case when F is a singleton set. Then we need to show that $\backslash a^{-1}(B \setminus \uparrow(s, X))$ is Scott open in $\mathcal{F}$. If $C \in \backslash a^{-1}(B \setminus \uparrow(s, X))$, then for each t with $t \setminus a \leqslant s$, there is some m_t with $(ta^{m_t}, \emptyset) \notin C$. It follows that there are only finitely many $t \in A^* \cup A^* \sqrt{\ }$ with $t \setminus a = s$ and $(t, \emptyset) \in C$. Moreover, for each such t with $t \setminus a = s$, there is some subset $F_t \subseteq X$ with $(t, F_t) \notin C$. So, if $T = \{t \mid t \setminus a \leqslant s$ and $(t, \emptyset) \in C\}$, then

$$C \subseteq B \setminus (\bigcup_{t \in T} \uparrow(ta^{m_t}, \emptyset) \cup \bigcup_{t \in T_0} \uparrow(t, F_t))$$

where $T_0 = \{t \in T \mid t \setminus a = s\}$. If $t \in T_0$ and $m_t = 1$, then $C \in \mathcal{F}$ implies that $C \subseteq B \setminus \uparrow(t, F_t \setminus \{a\})$, and so we can assume $F_t \subseteq A \setminus \{a\}$ for $t \in T_0$. The set

$$D = B \setminus (\bigcup_{t \in T} \uparrow(ta^{m_t}, \emptyset) \cup \bigcup_{t \in T_0} \uparrow(t, F_t))$$

is then a Scott closed subset of B, and it is easy to show that D satisfies the property (*) of Definition 33. That is, $D \in \mathcal{F}$. Moreover, the construction of D implies that $D \setminus a \subseteq B \setminus \uparrow(s, X)$.

In the case that $F = \{(s_i, X_i) \mid i = 1, \ldots, n\}$ has more than one element, it is easy to show that

$$\backslash a^{-1}(B \setminus \uparrow F) = \cup(\uparrow \bigvee_i D_i)$$

where D_i is a set of the form of D above for $B \setminus \uparrow(s_i, X_i)$, and the union is taken over all possible choices of the D_is. For each choice of D_is, the set $\bigvee_i D_i \in K(\mathcal{F})$, and so $\uparrow \bigvee_i D_i$ is Scott open in $\mathcal{F}$. This proves that the inverse image of the upper set of a compact element in $\mathcal{F}$ is Scott open in $\mathcal{F}$, which completes the proof of the lemma. □

We also show that $\backslash a$ does not preserve closures of increasing unions.

Example 42. Fix $b \in A$ with $a \neq b$. For each n, let

$$C_n = B \setminus (\bigcup_{m<n} \uparrow(a^m b, \emptyset) \cup \uparrow(a^n, \emptyset))$$

and let

$$C = B \setminus (\bigcup_{m} \uparrow(a^m b, \emptyset))$$

Then each of these sets is closed in B, and $C = \bigcup_n C_n$, as is readily verified. Now, for each n, it is easy to show that $C_n \setminus a \subseteq B \setminus \uparrow(b, \emptyset)$. However, $(a^n, \emptyset) \in C$ for each n, and so $B \subseteq C \setminus a$. That is, $\backslash a$ does not preserve closures of increasing unions.

It is easy to show that the definitions of $\Psi(STOP)$, $\Psi(SKIP)$, and $\Psi(a \to P)$ all lie within $\mathcal{F}$ for each process P such that $\Psi(P) \in \mathcal{F}$. We can then define a failures model to include the operator $\backslash a$ for each $a \in A$, by simply adding the definition

$$\Psi(P \setminus a) = \Psi(P) \setminus a$$

for each $a \in A$, and we have a Scott continuous semantics whose range is the family $\mathcal{F}$. In fact, all the operators of the language save the hiding operator are λ-continuous on $\mathcal{F}$, which means they are both Scott and dual Scott continuous. This is the failures model presented in (Brookes et al. 1984). Thus we can summarize the results of this section as follows.

Theorem 43. Let $\mathcal{F} \subseteq \Gamma((A^* \cup A^*\checkmark) \times \mathcal{P}(A))$ be the family of subsets $F \subseteq (A^* \cup A^*\checkmark) \times \mathcal{P}(A)$ satisfying

$$(s, X) \in F \wedge (sc, \emptyset) \notin F \Rightarrow (s, X \cup \{c\}) \in F$$

for all $c \in A$. Then each of the operations $\to, ;, \sqcap, \sqcup, \|, \|\|, \backslash_n a$, and **repeat** are Lawson continuous on $\Gamma((A^* \cup A^*\checkmark) \times \mathcal{P}(A))$ and leave $\mathcal{F}$ invariant.

1. The operator $\backslash a : (A^* \cup A^*\checkmark) \times \mathcal{P}(A) \to (A^* \cup A^*\checkmark) \times \mathcal{P}(A)$ by $(s, X) \setminus a = (s \setminus a, X)$ induces an operator on $\Gamma(A^* \cup A^*\checkmark) \times \mathcal{P}(A))$ which is continuous with respect to liveness tests (that is, usual set inclusion), but which is not continuous with respect to safety test (that is, reverse inclusion). These operators also leave $\mathcal{F}$ invariant, and so the same result holds true for their restrictions to $\mathcal{F}$.

2. The operator $\backslash a : \mathcal{F} \to \mathcal{F}$ defined by

$$C \backslash a = \{(s \backslash a, X) \mid (s, X \cup \{a\}) \in C\} \\ \cup \{((s \backslash a)t, X) \mid (\forall n > 0)(sa^n, \emptyset) \in C, \\ t \in A^* \cup A^*\sqrt{}, X \in \mathcal{P}(A)\}$$

is safety continuous (that is, continuous with respect to reverse set inclusion), but is not liveness continuous (that is, with respect to usual set inclusion).

4.6 Summary

We have presented a general theory of algebraic posets and their relationship to algebraic cpo's, and applied that theory to the problem of deriving denotational models for fragments of the language CSP. The models we have given demonstrate two results. First, it is possible to provide denotational models for abstract languages in a uniform way. Much can be gained from this method, since the arguments about continuity become general in nature, rather than being *ad hoc* to a particular model or a particular language. Second, where the approach fails, it does so for a clearly discernible reason. The example of the hiding operator in the traces model shows that the theory can guide how to 'cure' the problem of a discontinuous operator by providing a clear indication of how to replace it with a continuous one. And, in the process, the theory provides insight into the bad behavior of this operator.

The theory also provides a natural model for the fragment of CSP including the external choice operation with the hiding operator replaced by the family $\backslash_n a$ of continuous operators. This model is easy to derive, and all the operations from the language give rise to continuous operations on the model. Finally, we saw how we could realize the failures model of Brookes et al. (1984) within this larger model. In fact, it is now clear why the failures model is constructed as it is: it is all to make hiding continuous. This author remains amazed at the ingenuity of this model, and that it was derived without a general theory such as the one we have presented. But we can now place the model within a larger uniform mathematical framework, and this allows a better appreciation of the subtleties of that model.

Clearly, the same theory could be extended to capture a version of the *improved failures model* of Brookes and Roscoe (1985). In fact, this same theory could be applied to obtain the spectrum of models presented in (Olderog and Hoare, 1986). Moreover, the author believes that this theory can be suitably modified to capture the models of *Timed CSP* recently presented by Reed (1990). However, this will require some non-trivial modifications of the underlying theory, since the models will have to allow for a component which is comprised of the unit interval. This will take us away from algebraic cpo's into the realm of continuous cpo's, and there the structure of the lattice of Scott closed sets is not quite so nice as in the algebraic case.

Another direction for future research would be to derive an operational semantics to go along with the denotational semantics that this theory provides. For abstract languages where all the operations are continuous, we believe there should be a natural method for devising such a semantic model. In fact, in joint work with others, the author is exploring this question for an abstract language which utilizes the rendezvous protocol for synchronization, and preliminary results indicate that a fully abstract semantics is achievable.

Acknowledgements

The author wishes to thank Dr Frank Oles for many valuable conversations, especially relating to the use of the symbol $\sqrt{}$ to signal normal termination. The ideas which appear in the treatment of the composition operator ; (in particular, how it can be made a monotone function using $\sqrt{}$) arose from these conversations, and the results belong as much to Dr Oles as to the author.

The author has also benefited from numerous conversations with Professor Boumediene Belkhouche on theoretical computer science in general, and concurrency in particular. Many of the ideas in this work stem from those conversations.

Preliminary versions of this work were presented on a number of occasions at a number of places, and the author would like to express his appreciation to the audiences on those occasions whose comments and questions helped with the development of this material. Special thanks go to Professor Rüdiger Göbel and Dr Manfred Droste of the Universität Essen for a very enjoyable visit during which aspects of this work were discussed, to Professor Karl Hofmann and other faculty and graduate students at the Technische Hochschule Darmstadt, and to the members of the Theoretical Computer Science Seminar at the University of Pennsylvania.

Thanks also go to Professor Jimmie D. Lawson, who volunteered to present the paper based on this work at the Oxford Symposium when the author was unable to attend the meeting, and whose incisive comments clarified for the author some subtle aspects of spectral theory which this paper relies on. Finally, the author thanks Han Zhang for carefully reading through a preliminary version of this chapter and pointing out several errors.

References

Brookes, S. D., Hoare, C. A. R., and Roscoe, A. W. (1984) A Theory of Communicating Processes. *Journal of the Association of Computing Machinery* **31**, 560–599.

Brookes, S. D. and Roscoe, A. W. (1985) An improved failures model for communicating processes, *Lecture Notes in Computer Science* **197**, 281–305.

Gierz, G., Hofmann, K. H., Keimel, K., Lawson, J. D., Mislove, M. W. and Scott, D. S. (1980) A Compendium of Continuous Lattices, Springer-Verlag, Berlin, Heidelberg and New York.

Hoare, C. A. R. (1985) Communicating Sequential Processes, Prentice-Hall International, Englewood Cliffs.

Hofmann, K. H. and Lawson, J. D. (1978) The spectral theory of distributive continuous lattices, *Transactions of the American Mathematical Society* **246**, 285–310.

Hofmann, K. H. and Mislove, M. W. (1979) Local compactness and continuous lattices, *Lecture Notes in Mathematics* **871**, 199–230.

Jung, A. (1988) Cartesian closed categories of domains, Ph. D. Dissertation, Darmstadt.

Olderog, E.-R. and Hoare, C. A. R. (1986) Specification-oriented semantics for communicating sequential processes, *Acta Informatica* **23**, 9–66.

Oles, F. J. (1987) The semantics of concurrency without powerdomains, *Proceedings of the 14th Annual ACM Symposium on Principles of Programming Languages*.

Reed, G. M. (1990) A hierarchy of domains for timed *CSP*, *Proceedings of the Fifth Conference on the Mathematical Foundations of Programming Semantics*, in: *Lecture Notes in Computer Science* **442**, to appear.

5
Concurrency semantics based on metric domain equations

J. W. DE BAKKER AND J. J. M. M. RUTTEN

Abstract

We show how domain equations may be solved in the category of complete metric spaces. For five example languages we demonstrate how to exploit domain equations in the design of their operational and denotational semantics. Two languages are schematic or uniform. Three have interpreted elementary actions involving individual variables and inducing state transformations. For the latter group we discuss three denotational models reflecting a variety of language notions considered. A central theme is the distinction, within the non-uniform setting, of linear time versus branching time models. Throughout, fruitful use is made of the technique of obtaining semantic mappings, operators, etc. as fixed points of higher-order functions. A brief discussion of the relationship between bisimulation and one of the domains considered concludes the paper.

5.1 Introduction

Concurrency semantics is concerned with the mathematical modelling of parallel behaviour. A parallel computation induces some form of simultaneous or interleaved execution of the elementary actions from the constituent (parallel) components. Accordingly, it is to be expected that the mathematical description of such a computation involves a detailed modelling of its intermediate steps — rather than just its input–output behaviour, as is mostly sufficient in a sequential setting. The collection of intermediate steps may be said to constitute the *history* of the computation. Two histories p_1, p_2 are close together if their first difference is exhibited only after many steps. This observation is at the basis of the metric approach to concurrency semantics. We introduce distances d such that

$$d(p_1, p_2) = 2^{-n} \qquad (1)$$

where $n = \sup \{k \mid p_1[k] = p_2[k]\}$, with $p[k]$ a truncation of p after k steps. It is our aim in this chapter to make this idea precise, and to illustrate how it may be exploited in the design of semantic models for a variety of concurrency phenomena.

Section 5.2 introduces a rigorous setting for the metric space techniques to be applied subsequently. The category $\mathcal{C}$ of *complete metric spaces* is introduced, and it is shown how metric spaces (P,d), or P for short, can be specified as solutions of *domain equations* $P = F(P)$ for a variety of functors $F : \mathcal{C} \to \mathcal{C}$. In the formation of these F, several composition operators such as $\times$ (cartesian product), $\cup$ (disjoint union), $\to$ (function space), $\mathbb{P}$ (powerset of), etc. are used. The main result of this section is the following. Provided a rather natural condition is satisfied for the recursive occurrences of P in the expression $F(P)$ (which condition ensures a kind of *contractivity* of F in P), the equation $P = F(P)$ can be solved and its solution is unique. The first application of metric spaces in order to obtain domains as solutions of such equations was described in (de Bakker and Zucker 1982), a paper in turn inspired by Nivat's general metric approach to semantics (for example, (Nivat 1979)). The ideas of (de Bakker and Zucker 1982) were generalized (to cover equations of the form $P = \cdots (P \to F_1(P)) \cdots$ also, a case missing in (de Bakker and Zucker 1982)) and put in a category-theoretic framework in (America and Rutten 1989a). Since the latter reference provides full mathematical details, including complete proofs, we restrict the treatment in Section 5.2 to a more concise one, not repeating these proofs, but with sufficient information to make the present chapter self-contained. Independently of (America and Rutten 1989a), the question of how to extend the ideas of (de Bakker and Zucker 1982) was also investigated by Majster-Cederbaum (1988, 1989, 199?); in these references the issues of the existence and uniqueness of solutions of the equation $P = F(P)$ are also investigated in a category-theoretic framework.

Section 5.3 constitutes the main body of our chapter. For five example languages L_i, $i = 0,\ldots,4$, we introduce operational ($\mathcal{O}_i$) and denotational ($\mathcal{D}_i$) semantic models, where O_i is a mapping $L_i \to R_i$, and D_i a mapping $L_i \to P_i$ (here we neglect one refinement to be discussed later), $i = 0,\ldots,4$. Determined by the range of programming concepts in the language L_i, we shall design a corresponding range of operational domains R_i and denotational domains P_i, $i = 0,\ldots,4$, each time as the solution of a (pair of) domain equation(s) geared to the construction of an appropriate model capturing the notions concerned. Of the languages L_0 to L_4, two are what we like to call *uniform* (the elementary actions are just symbols) (de Bakker et al. 1986, 1987, 1988). The other three are *non-uniform*: the elementary actions refer to individual variables, and we encounter states, assignments, etc. The models for L_2 to L_4 mention states and state transformations, or, put in mathematical terms, the corresponding functor F

Introduction

now has occurrences of the function space constructor. There are somewhat subtle (and not yet fully understood) differences between P_2, P_3, and P_4. Using a terminology mostly reserved for the uniform case, that is, the contrast between *linear time* (models with sets of sequences) versus *branching time* (models with trees or tree-like entities) (de Bakker et al. 1984), we might say that the domains P_2 and P_3 are (non-uniform and) linear time, whereas P_4 is (non-uniform and) branching time. Understanding the difference between P_2 and P_3 requires further study. The introduction and associated analysis of P_2 to P_4 appears here for the first time. In earlier work, we always used P_4 (or trivial variants), and for some time we did not see how to design a satisfactory non-uniform model with the linear time flavour. The domain P_2 was then proposed as a candidate to enable us to design a *fully abstract* $\mathcal{D}_2$ (with respect to the $\mathcal{O}_2$ to be given in Section 5.3). In the meantime it has been shown by Horita et al. (1990) that a certain extension P_2' of P_2 (P_2' ignores details present in P_2) indeed allows us to define a fully abstract denotational $\mathcal{D}_2'$ (with respect to $\mathcal{O}_2$ as to be given). For L_3, we do not know whether a similar result holds. For L_4, we do know that $\mathcal{D}_4$ is not fully abstract with respect to $\mathcal{O}_4$.

In general, the material in Section 5.3 is organized in such a way that it brings out the unifying effect of the metric approach. At least the following definitions and proof techniques all follow the same pattern (for $i = 0, \ldots, 4$):

- introduction of the transition system T_i (as in Plotkin's structured operational semantics) and the definition of the associated $\mathcal{O}_i$ as the fixed point of a contracting Ψ_i;

- introduction of the domains R_i, P_i, and definition of the various semantic operators (such as $\circ$, $\|$), for the P_i setting, in terms of fixed points of contracting $\Omega_\circ$, $\Omega_\|$;

- introducing the denotational semantics $\mathcal{D}_i$ as the fixed point of a contracting Φ_i;

- relating $\mathcal{O}_i$ and $\mathcal{D}_i$ through abstraction mappings abs_i, themselves obtained as fixed points of contracting Δ_i;

- establishing that $\mathcal{O}_i = abs_i \circ \mathcal{D}_i$, by introducing an intermediate semantics $\mathcal{I}_i : L_i \to P_i$ (with denotational codomain P_i, but obtained from the transition system T_i), deriving that $\mathcal{I}_i = \mathcal{D}_i$ (as in (Kok and Rutten 1988, de Bakker and Meyer 1988) and then proving that $abs_i \circ \mathcal{I}_i = \mathcal{O}_i$, once more by a fixed point argument.

In case the reader is not satisfied by the elementary character of L_0 to L_4, we emphasize that these languages have been selected for didactic reasons. Elsewhere we have demonstrated how the metric techniques described in the present chapter may be exploited in the treatment of substantially more complicated language notions. For the case of object-oriented programming languages, we refer to (America et al. 1989, America and de Bakker 1988, America and Rutten 1989b, Rutten 1990a); for a treatment of parallel logic programming semantics, we mention (de Bakker 1988, de Bakker and Kok 1988, 1990). Earlier introductory or overview presentations of metric concurrency semantics were given in (de Bakker and Meyer 1988, de Bakker 1989).

The last section of the chapter is devoted to a slightly more special topic. It is well known that the notion of *bisimulation* (Park 1981) is a central tool in concurrency semantics, and the question arises whether it may be related to results about domains in the style of P_0 to P_4. For a simple case (P_0 only), we prove the following theorem. Let s_1, s_2 be two states (here used as abstractions of the statements as introduced in Section 5.3) from a set S. We have that s_1 is bisimilar to s_2 (with respect to a given labelled transition system T) if and only if $\mathcal{M}[\![s_1]\!] = \mathcal{M}[\![s_2]\!]$, where $\mathcal{M}: S \to P_0$ is obtained from T in a manner which is the same as the way in which $\mathcal{I}$ (from Section 5.3) is obtained from T_0. Let us also draw attention to the fact that this result depends critically on the branching structure for P_0.

We conclude this introduction with two remarks about possible extensions of the reported results. In (Rutten 1989), a beginning has been made with the exploration of a technique which 'automatically' infers a denotational semantics $\mathcal{D}$ from a given transition system T (of course obeying the compositionality requirement on $\mathcal{D}$). A bonus of this automatic inference is, in particular, the possibility of avoiding *ad hoc* equivalence proofs for $\mathcal{O} = abs \circ \mathcal{D}$. A second important topic which we want to address in future work is the design of a fully abstract model for a language with process creation.

5.2 Metric spaces and domain equations

As mathematical domains for our operational and denotational semantics we shall use complete metric spaces satisfying a so-called *reflexive domain equation* of the following form:

$$P \cong F(P)$$

(The symbol $\cong$ should be read 'is isometric to' and is defined below.) Here $F(P)$ is an expression built from P and a number of standard constructions on metric spaces (also to be formally introduced shortly). A few examples are

$$P \cong A \cup (B \times P) \qquad (2)$$
$$P \cong A \cup \mathbb{P}_{co}(B \times P) \qquad (3)$$
$$P \cong A \cup (B \to P) \qquad (4)$$

where A and B are given fixed complete metric spaces. De Bakker and Zucker (1982) have first described how to solve these equations in a metric setting. Roughly, their approach amounts to the following. In order to solve $P \cong F(P)$ they define a sequence of complete metric spaces $(P_n)_n$ by $P_0 = A$ and $P_{n+1} = F(P_n)$, for $n > 0$, such that $P_0 \subseteq P_1 \subseteq \cdots$. Then they take the *metric completion* of the union of these spaces P_n, say $\bar{P}$, and show $\bar{P} \cong F(\bar{P})$. In this way they are able to solve equations (2), (3) and (4).

There is one type of equation for which this approach does not work, namely

$$P \cong A \cup (P \to^1 G(P)) \qquad (5)$$

in which P occurs at the *left* side of a function space arrow and $G(P)$ is an expression possibly containing P. This is due to the fact that it is not always the case that $P_n \subseteq F(P_n)$.

In (America and Rutten 1989a) the above approach is generalized in order to overcome this problem. The family of complete metric spaces is made into a *category* $\mathcal{C}$ by providing some additional structure. (For an extensive introduction to category theory we refer the reader to (Mac Lane 1971).) Then the expression F is interpreted as a *functor* $F : \mathcal{C} \to \mathcal{C}$ which is (in a sense) *contracting*. It is proved that a generalized version of Banach's theorem (see below) holds, that is, that contracting functors have a fixed point (up to isometry). Such a fixed point, satisfying $P \cong F(P)$, is a solution of the domain equation.

We shall now give a quick overview of these results, omitting many details and all proofs. For a full treatment we refer the reader to (America and Rutten 1989a). We start by listing the basic definitions and facts of metric topology that we shall need.

We assume the following notions to be known (the reader might consult (Dugundji 1966) or (Enkelking 1977)): metric space, ultra-metric space, complete (ultra-) metric space, continuous function, closed set, compact set. (In our definition the distance between two elements of a metric space is always bounded by 1.)

An arbitrary set A can be supplied with a metric d_A, called the *discrete* metric, defined by

$$d_A(x, y) = \begin{cases} 0 & \text{if } x = y \\ 1 & \text{if } x \neq y \end{cases}$$

Now (A, d_A) is a metric, even an ultra-metric, space.

Let (M_1, d_1) and (M_2, d_2) be two complete metric spaces. A function $f : M_1 \to M_2$ is called *non-expansive* if for all $x, y \in M_1$

$$d_2(f(x), f(y)) \le d_1(x, y)$$

The set of all non-expansive functions from M_1 to M_2 is denoted by $M_1 \to^1 M_2$. A function $f : M_1 \to M_2$ is called *contracting* (or a *contraction*) if there exists $\epsilon \in [0, 1)$ such that for all $x, y \in M_1$

$$d_2(f(x), f(y)) \le \epsilon \cdot d_1(x, y)$$

(Non-expansive functions and contractions are continuous.)

The following fact is known as Banach's theorem. Let (M, d) be a complete metric space and $f : M \to M$ a contraction. Then f has a unique fixed point, that is, there exists a unique solution $x \in M$ such that $f(x) = x$.

We call M_1 and M_2 *isometric* (notation: $M_1 \cong M_2$) if there exists a bijective mapping $f : M_1 \to M_2$ such that, for all $x, y \in M_1$,

$$d_2(f(x), f(y)) = d_1(x, y)$$

Definition 1. *Let $(M, d), (M_1, d_1), \ldots, (M_n, d_n)$ be metric spaces.*

1. *We define a metric d_F on the set $M_1 \to M_2$ of all functions from M_1 to M_2 as follows. For every $f_1, f_2 \in M_1 \to M_2$ we put*

$$d_F(f_1, f_2) = \sup_{x \in M_1} \{d_2(f_1(x), f_2(x))\}$$

This supremum always exists since the codomain of our metrics is always $[0, 1]$. The set $M_1 \to^1 M_2$ is a subset of $M_1 \to M_2$, and a metric on $M_1 \to^1 M_2$ can be obtained by taking the restriction of the corresponding d_F.

2. *With $M_1 \bar{\cup} \cdots \bar{\cup} M_n$ we denote the disjoint union of $M_1, \ldots, M_n$, which can be defined as $\{1\} \times M_1 \cup \cdots \cup \{n\} \times M_n$. We define a metric d_U on $M_1 \bar{\cup} \cdots \bar{\cup} M_n$ as follows. For every $x, y \in M_1 \bar{\cup} \cdots \bar{\cup} M_n$,*

$$d_U(x, y) = \begin{cases} d_j(x, y) & \text{if } x, y \in \{j\} \times M_j, 1 \le j \le n \\ 1 & \text{otherwise} \end{cases}$$

If no confusion is possible we shall often write $\cup$ rather than $\bar{\cup}$.

3. *We define a metric d_P on the cartesian product $M_1 \times \cdots \times M_n$ by the following clause. For every $(x_1, \ldots, x_n), (y_1, \ldots, y_n) \in M_1 \times \cdots \times M_n$,*

$$d_P((x_1, \ldots, x_n), (y_1, \ldots, y_n)) = \max_i \{d_i(x_i, y_i)\}$$

4. Let $\mathbb{P}_{cl}(M) = \{X \mid X \subseteq M \land X \text{ is closed}\}$. We define a metric d_H on $\mathbb{P}_{cl}(M)$, called the Hausdorff distance, as follows. For every $X, Y \in \mathbb{P}_{cl}(M)$,

$$d_H(X, Y) = \max\{\sup_{x \in X}\{d(x, Y)\}, \sup_{y \in Y}\{d(y, X)\}\}$$

where $d(x, Z) = \inf_{z \in Z}\{d(x, z)\}$ for every $Z \subseteq M$, $x \in M$. (We use the convention that $\sup \emptyset = 0$ and $\inf \emptyset = 1$.) The spaces

$$\mathbb{P}_{co}(M) = \{X \mid X \subseteq M \land X \text{ is compact}\}$$
$$\mathbb{P}_{nc}(M) = \{X \mid X \subseteq M \land X \text{ is non-empty and compact}\}$$

are supplied with a metric by taking the restriction of d_H.

5. For any real number ϵ with $\epsilon \in [0, 1]$ we define

$$id_\epsilon((M, d)) = (M, d')$$

where $d'(x, y) = \epsilon \cdot d(x, y)$, for every x and y in M.

Proposition 2. Let (M, d), $(M_1, d_1), \ldots, (M_n, d_n)$, d_F, d_U, d_P, and d_H be as in Definition 1 and suppose that (M, d), $(M_1, d_1), \ldots, (M_n, d_n)$ are complete. We have that

$$(M_1 \to M_2, d_F) \quad (M_1 \to^1 M_2, d_F) \tag{a}$$
$$(M_1 \,\dot\cup\, \cdots \,\dot\cup\, M_n, d_U) \tag{b}$$
$$(M_1 \times \cdots \times M_n, d_P) \tag{c}$$
$$(\mathbb{P}_{cl}(M), d_H) \quad (\mathbb{P}_{co}(M), d_H) \quad (\mathbb{P}_{nc}(M), d_H) \tag{d}$$
$$id_\epsilon((M, d)) \tag{e}$$

are complete metric spaces. If (M, d) and (M_i, d_i) are all ultra-metric spaces, then so are these composed spaces. (Strictly speaking, for the completeness of $M_1 \to M_2$ and $M_1 \to^1 M_2$ we do not need the completeness of M_1. The same holds for the ultra-metric property.)

Whenever in the sequel we write $M_1 \to M_2$, $M_1 \to^1 M_2$, $M_1 \,\dot\cup\, \cdots \,\dot\cup\, M_n$, $M_1 \times \cdots \times M_n$, $\mathbb{P}_{cl}(M)$, $\mathbb{P}_{co}(M)$, $\mathbb{P}_{nc}(M)$, or $id_\epsilon(M)$, we mean the metric space with the metric defined above.

The proofs of Proposition 2(a), (b), (c), and (e) are straightforward. Part (d) is more involved. It can be proved with the help of the following characterization of the completeness of $(\mathbb{P}_{cl}(M), d_H)$.

Proposition 3. Let $(\mathbb{P}_{cl}(M), d_H)$ be as in Definition 1. Let $(X_i)_i$ be a Cauchy sequence in $\mathbb{P}_{cl}(M)$. We have

$$\lim_{i \to \infty} X_i = \{ \lim_{i \to \infty} x_i \mid x_i \in X_i, (x_i)_i \text{ a Cauchy sequence in } M \}$$

Proofs of Propositions 2(d) and 3 can be found in, for instance, (Dugundji 1966) and (Enkelking 1977). The proofs are also repeated in (de Bakker and Zucker 1982). The completeness of the Hausdorff space containing compact sets is proved in (Michael 1951).

We proceed by introducing a category of complete metric spaces and some basic definitions, after which a categorical fixed point theorem will be formulated.

Definition 4. (Category of complete metric spaces) Let $\mathcal{C}$ denote the category that has complete metric spaces for its objects. The arrows ι in $\mathcal{C}$ are defined as follows. Let M_1, M_2 be complete metric spaces. Then $M_1 \to^\iota M_2$ denotes a pair of maps $M_1 \rightleftarrows^i_j M_2$, satisfying the following properties:

1. i is an isometric embedding;

2. j is non-distance-increasing (NDI);

3. $j \circ i = id_{M_1}$.

(We sometimes write $\langle i, j \rangle$ for ι.) Composition of the arrows is defined in the obvious way.

We can consider M_1 as an approximation to M_2. In a sense, the set M_2 contains more information than M_1, because M_1 can be isometrically embedded into M_2. Elements in M_2 are approximated by elements in M_1. For an element $m_2 \in M_2$ its (best) approximation in M_1 is given by $j(m_2)$. Clause 3 states that M_2 is a consistent extension of M_1.

Definition 5. For every arrow $M_1 \to^\iota M_2$ in $\mathcal{C}$ with $\iota = \langle i, j \rangle$ we define

$$\delta(\iota) = d_{M_2 \to M_1}(i \circ j, id_{M_2}) \quad (= \sup_{m_2 \in M_2} \{d_{M_2}(i \circ j(m_2), m_2)\})$$

This number can be regarded as a measure of the quality with which M_2 is approximated by M_1: the smaller $\delta(\iota)$, the better M_2 is approximated by M_1.

Increasing sequences of metric spaces are generalized in the following definition.

Definition 6. (Converging tower)

1. We call a sequence $(D_n, \iota_n)_n$ of complete metric spaces and arrows a tower whenever we have that $\forall n \in \mathbb{N} \cdot D_n \to^{\iota_n} D_{n+1} \in \mathcal{C}$.

2. The sequence $(D_n, \iota_n)_n$ is called a converging tower when furthermore the following condition is satisfied:

$$\forall \epsilon > 0 \cdot \exists N \in \mathbb{N} \cdot \forall m > n \geq N \cdot \delta(\iota_{nm}) < \epsilon$$

where $\iota_{nm} = \iota_{m-1} \circ \cdots \circ \iota_n : D_n \to D_m$.

A special case of a converging tower is a tower $(D_n, \iota_n)_n$ satisfying, for some ϵ with $0 \leq \epsilon < 1$,

$$\forall n \in \mathbb{N} \cdot \delta(\iota_{n+1}) \leq \epsilon \cdot \delta(\iota_n)$$

Note that

$$\delta(\iota_{nm}) \leq \delta(\iota_n) + \cdots + \delta(\iota_{m-1})$$
$$\leq \epsilon^n \cdot \delta(\iota_0) + \cdots + \epsilon^{m-1} \cdot \delta(\iota_0)$$
$$\leq \frac{\epsilon^n}{1-\epsilon} \cdot \delta(\iota_0)$$

We shall now generalize the technique of forming the metric completion of the union of an increasing sequence of metric spaces by proving that, in $\mathcal{C}$, every converging tower has an initial cone. The construction of such an initial cone for a given tower is called the direct limit construction. Before we treat this direct limit construction, we first give the definition of a cone and an initial cone.

Definition 7. (Cone) Let $(D_n, \iota_n)_n$ be a tower. Let D be a complete metric space and $(\gamma_n)_n$ a sequence of arrows. We call $(D, (\gamma_n)_n)$ a cone for $(D_n, \iota_n)_n$ whenever the following condition holds:

$$\forall n \in \mathbb{N} \cdot D_n \to^{\gamma_n} D \in \mathcal{C} \wedge \gamma_n = \gamma_{n+1} \circ \iota_n$$

Definition 8. (Initial cone) A cone $(D, (\gamma_n)_n)$ for a tower $(D_n, \iota_n)_n$ is called initial whenever for every other cone $(D', (\gamma'_n)_n)$ for $(D_n, \iota_n)_n$ there exists a unique arrow $\iota : D \to D'$ in $\mathcal{C}$ such that

$$\forall n \in \mathbb{N} \cdot \iota \circ \gamma_n = \gamma'_n$$

Definition 9. (Direct limit construction) Let $(D_n, \iota_n)_n$, with $\iota_n = \langle i_n, j_n \rangle$, be a converging tower. The direct limit of $(D_n, \iota_n)_n$ is a cone $(D, (\gamma_n)_n)$, with $\gamma_n = \langle g_n, h_n \rangle$, that is defined as follows:

$$D = \{(x_n)_n \mid \forall n \geqslant 0 \cdot x_n \in D_n \wedge j_n(x_{n+1}) = x_n\}$$

is equipped with a metric $d : D \times D \to [0, 1]$ defined by

$$d((x_n)_n, (y_n)_n) = \sup\{d_{D_n}(x_n, y_n)\}$$

for all $(x_n)_n$ and $(y_n)_n \in D$. The function $g_n : D_n \to D$ is defined by $g_n(x) = (x_k)_k$, where

$$x_k = \begin{cases} j_{kn}(x) & \text{if } k < n \\ x & \text{if } k = n \\ i_{nk}(x) & \text{if } k > n \end{cases}$$

$h_n : D \to D_n$ is defined by $h_n((x_k)_k) = x_n$.

Lemma 10. *The direct limit of a converging tower (as defined in Definition 9) is an initial cone for that tower.*

As a category-theoretic equivalent of a contracting function on a metric space, we have the following notion of a contracting functor on $\mathcal{C}$.

Definition 11. (Contracting functor) We call a functor $F : \mathcal{C} \to \mathcal{C}$ contracting whenever the following holds. There exists an ϵ, with $0 \leqslant \epsilon < 1$, such that, for all $D \to^\iota E \in \mathcal{C}$,

$$\delta(F(\iota)) \leqslant \epsilon \cdot \delta(\iota)$$

A contracting function on a complete metric space is continuous, so it preserves Cauchy sequences and their limits. Similarly, a contracting functor preserves converging towers and their initial cones.

Lemma 12. *Let $F : \mathcal{C} \to \mathcal{C}$ be a contracting functor, and let $(D_n, \iota_n)_n$ be a converging tower with an initial cone $(D, (\gamma_n)_n)$. Then $(F(D_n), F(\iota_n))_n$ is again a converging tower with $(F(D), (F(\gamma_n))_n)$ as an initial cone.*

Theorem 13. (Fixed point theorem) *Let F be a contracting functor $F : \mathcal{C} \to \mathcal{C}$ and let $D_0 \to^{\iota_0} F(D_0) \in \mathcal{C}$. Let the tower $(D_n, \iota_n)_n$ be defined by $D_{n+1} = F(D_n)$ and $\iota_{n+1} = F(\iota_n)$ for all $n \geqslant 0$. This tower is converging, so it has a direct limit $(D, (\gamma_n)_n)$. We have $D \cong F(D)$.*

In (America and Rutten 1989a) it is shown that contracting functors that are moreover contracting on all *hom-sets* (the sets of arrows in $\mathcal{C}$ between any two given complete metric spaces) have *unique* fixed points (up to isometry). It is also possible to impose certain restrictions upon the category $\mathcal{C}$ such that every contracting functor on $\mathcal{C}$ has a unique fixed point.

Let us now indicate how this theorem can be used to solve Equations (2)–(5) above. We define

$$F_1(P) = A \cup id_{1/2}(B \times P) \tag{6}$$
$$F_2(P) = A \cup \mathbb{P}_{co}(B \times id_{1/2}(P)) \tag{7}$$
$$F_3(P) = A \cup (B \to id_{1/2}(P)) \tag{8}$$

If the expression $G(P)$ in Equation (5) is equal to P, for example, then we define F_4 by

$$F_4(P) = A \cup id_{1/2}(P \to^1 P) \tag{9}$$

Note that the definitions of these functors specify, for each metric space (P, d_P), the metric on $F(P)$ implicitly (see Definition 1). These metrics all satisfy Equation (1) given in the introduction (Section 5.1) for a suitably defined truncation function.

Now it is easily verified that F_1, F_2, F_3, and F_4 are contracting functors on $\mathcal{C}$. Intuitively, this is a consequence of the fact that in the definitions above each occurrence of P is preceded by a factor $id_{1/2}$. Thus these functors have a fixed point, according to Theorem 13, which is a solution for the corresponding equation. (In the sequel we shall usually omit the factor $id_{1/2}$ in the reflexive domain equations, assuming that the reader will be able to fill in the details.)

In (America and Rutten 1989a) it is shown that functors like F_1 through F_4 are also contracting on hom-sets, which guarantees that they have *unique* fixed points (up to isometry).

The results above hold for complete *ultra-metric* spaces too, which can easily be verified.

In the next section, we shall encounter pairs of reflexive equations of the form

$$P \cong F(P,Q) \quad Q \cong G(P,Q)$$

where F and G are functors on $\mathcal{C} \times \mathcal{C}$. Equations like this can be solved by a straightforward generalization of the above theory.

5.3 Concurrency semantics

Introduction

In this section we demonstrate how (solutions of) metric domain equations can be exploited in the design of semantics for languages with some form of concurrency. Altogether we shall be concerned with five languages, and for each of them we shall develop operational ($\mathcal{O}$) and denotational ($\mathcal{D}$) semantics, and discuss the relationships between $\mathcal{O}$ and $\mathcal{D}$. The first two languages (L_0, L_1) are what may be called schematic or *uniform*: the elementary actions are uninterpreted symbols from some alphabet, and the meanings assigned to the language constructs concerned will have the flavour of formal (tree) languages. Next, we shall discuss three *non-uniform* languages (L_2, L_3, L_4), where the elementary actions are (primarily) assignments. These have state transformations as meanings, and the domains needed to handle them involve state-transforming functions in a variety of ways.

The domains employed to define the *operational semantics* for L_0 to L_4 are comparatively easy. For L_0, L_1 we introduce the domain of *streams*, that is, of finite or infinite sequences over the relevant alphabets. Finite sequences end in ϵ (δ) signalling proper (improper or deadlock) termination. Meanings of statements in L_0, L_1 will be (non-empty compact) sets of such streams, and the corresponding domains will be denoted by R_0, R_1. In order to bring out the (dis)similarities between the operational and denotational models, the stream domains R_0, R_1 are defined here as well, through domain equations. (At this stage, the reader may want to refer to the table in Section 5.3, surveying all domain equations.) For L_2 to L_4, the operational semantics domains (R_2 to R_4) are functions from states to sets of streams of states. Altogether, all operational models have streams as their basic constituents, and they may be collectively called *linear time* (LT) models.

The situation is rather different for the various denotational models. For L_0, L_1 we use (purely) *branching time* (BT) models, that is, we use the domain of 'trees' over some alphabets. 'Trees' are not just ordinary trees: they are *commutative* (no order on the successors of any node), what may be called *absorptive* (nodes have sets rather than multisets as successors), and *compact* (for this we omit a precise definition, since we use the technical framework of Section 5.2 anyhow). These properties taken together ensure that the domain of 'trees' does indeed fit into the general domain theory of Section 5.2. From now on, we use the term 'processes' (elements of a domain P solving $P \cong F(P)$) rather than 'trees'. (For a discussion concerning the relationship between the process domains and the class of process graphs modulo bisimulation we refer to (Bergstra and Klop 1989), where, under some mild conditions, isomorphism of the two structures is established.) The processes in P_0 and P_1, serving as models for L_0 and L_1, have as special elements the nil process $\{\epsilon\}$ and the empty process $\emptyset$. Again, these model proper and improper termination. For the languages L_2 to L_4, we introduce domains of processes (P_2 to P_4) which in some manner involve function spaces. Domain P_2 is the simplest of these: it consists of all non-empty compact subsets of a domain Q_2, where Q_2 is built recursively from itself and constant domains using the operators $\to$, $\times$, and $\cup$, but without the use of the power domain operator. Though slightly different from P_2, P_3 shares with P_2 the property that the power domain operator does not appear in a recursive way. Only when we define P_4 do we have that the power domain operator occurs combined with recursion. Since this kind of combination constitutes the essence of a domain being branching time, we are justified in calling P_4 a non-uniform BT model, whereas P_2, P_3 are, though non-uniform, more of the LT variety.

(In previous papers such as (de Bakker and Zucker 1982, de Bakker et al. 1988, de Bakker and Meyer 1988, de Bakker 1989) we have always considered, for the non-uniform case, only domains which are fully BT (such as P_4). The present models P_2, P_3 are new for us. A major motive for their introduction is our desire to understand full abstractness issues better. Domains which are fully branching time are likely to provide too much information to qualify as fully abstract. We shall return to these matters below.)

We use five languages to illustrate the use of domains as outlined above. For our present purposes, the languages themselves are not our primary concern. Our first aim is to present a representative sample of the variety of domains one may employ in semantic design. Secondly, we want to emphasize the resemblance between the definitional tools. Throughout, (unique) fixed points of (contracting) higher-order mappings play a central role. For f a contracting mapping on a complete metric space, let fix f denote its unique fixed point (which exists by Banach's theorem, cf. Section 5.2). For the operational semantics definitions we shall, for $i = 0, \ldots, 4$, define $\mathcal{O}_i = \text{fix}\,\Psi_i$, for suitable operators Ψ_i. In the definitions of the Ψ_i, we shall make fruitful use of transition systems in the sense of Plotkin's structured operational semantics (SOS), from (Hennessy and Plotkin 1979, Plotkin 1981, 1983). In the denotational case, we put $\mathcal{D}_i = \text{fix}\,\Phi_i$, $i = 0, \ldots, 4$. Here Φ_i is defined (on appropriate domains) using semantic operators such as sequential ($\circ$) and parallel ($\|$) composition. In the definition of those operators as well, use is made of the definitional technique in terms of higher-order mappings. In four out of the five cases considered, $\mathcal{O}_i$ is not *compositional*. That is, in these cases we do not have that, for each syntactic operator $\mathbf{op}_{\text{syn}}$, there exists a corresponding semantic operator $\mathbf{op}_{\text{sem}}$ such that, for all s_1, s_2, $\mathcal{O}[\![s_1\,\mathbf{op}_{\text{syn}}\,s_2]\!] = \mathcal{O}[\![s_1]\!]\,\mathbf{op}_{\text{sem}}\,\mathcal{O}[\![s_2]\!]$. (For example, for L_2 and L_4, $\|$ violates this condition.) In order to obtain compositionality, we have to add information to the codomains concerned: in going from $\mathcal{O}_i$ to $\mathcal{D}_i$, we replace R_i by P_i, and P_i is more complex than R_i. In this way we manage to define $\mathcal{D}_i$ in a compositional way, but we have lost the equivalence $\mathcal{O}_i = \mathcal{D}_i$, $i = 0, \ldots, 4$. Rather, we shall apply *abstraction* mappings $abs_i : P_i \to R_i$, $i = 0, \ldots, 4$. These mappings delete information from the P_i, and they enable us to establish that

$$\mathcal{O}_i = abs_i \circ \mathcal{D}_i \qquad i = 0, \ldots, 4 \qquad (10)$$

The question concerning full abstractness asks whether these $\langle \mathcal{D}_i, abs_i \rangle$ are the best possible (in a sense to be defined precisely below). Not much is known on this question. Apart from a few negative results ($\mathcal{D}_i$ is not fully abstract on the basis of known facts), essentially all we have to report here is a few open problems.

We conclude this introduction with a listing of the programming notions appearing in languages L_0 to L_4.

L_0, L_1 (**the uniform case**). Both have elementary actions, sequential composition, non-deterministic choice, and guarded recursion. Guardedness is a syntactic restriction reminiscent of Greibach normal form for context-free grammars. It is imposed to ensure contractivity (of an operator corresponding to (the declarations of) the program). Moreover:

- L_0 has parallel composition;
- L_1 has process creation and (CCS-like) synchronization.

L_2, L_3, L_4 (**the non-uniform case**). Each language has assignment, sequential composition, the conditional statement, and (arbitrary) recursion. In addition:

- L_2 has parallel composition;
- L_3 has process creation and (a form of) local variables;
- L_4 has parallel composition and (CSP-like) communication.

In each of L_0 to L_4, a program consists of a (main) statement s and a set D of declarations. This set 'declares' procedure variables x with corresponding bodies g (the guarded case) or s (the general case). These declarations are (therefore) *simultaneous* and they may involve mutually recursive constructs. Note that we do not utilize some form of μ-notation (in the form of $\mu x[s]$, say) to introduce recursion syntactically. The simultaneous format has technical advantages here (the interested reader may want to compare the technicalitites of (Kok and Rutten 1988) with those of (de Bakker and Meyer 1988)).

L_0: a uniform language with parallel composition

Our first language, L_0, is quite simple. It is introduced for the purpose of illustrating the definitional techniques on an elementary case. We shall design LT operational and BT denotational models for L_0. The motivation for using a BT model for L_0 is solely didactic: we want to explain the somewhat complicated machinery of BT models first for a very simple language (for which even the operational semantics $\mathcal{O}_0$ is already compositional, thus obviating the need for a more complex domain $\mathcal{D}_0$).

(From now on we employ the terminology 'let $(x \in) M$ be ...' to introduce a set M with a variable x ranging over M.) Let $(a \in) A$ be an alphabet of *elementary actions*, and let $(x \in) Pvar$ be an alphabet of *procedure variables*. We introduce the language L_0 and its guarded version L_0^g in the following.

Definition 14. $(s \in) L_0$, $(g \in) L_0^g$ and $(D \in) Decl_0$ are given by

1. $s ::= a \mid x \mid s_1 ; s_2 \mid s_1 + s_2 \mid s_1 \parallel s_2$

2. $g ::= a \mid g ; s \mid g_1 + g_2 \mid g_1 \parallel g_2$

3. A declaration D consists of a set of pairs (x, g) and a program consists of a pair (D, s).

Remarks.

1. We find it convenient not to worry about the ambiguity in the syntax for L_0 (L_0^g) — and the other languages we shall define in the sequel. If required, the reader may add parentheses around the composite constructs, or assign priorities to the operators.

2. In a guarded g, each occurrence of a procedure variable x is 'guarded' by a sequentially preceding occurrence of some $a \in A$.

We proceed with the definitions leading up to the operational semantics $\mathcal{O}_0$ for L_0. Let E be a new symbol (not in A or $Pvar$) with as connotation 'the terminated statement', and let $(r \in) L_0^+ = L_0 \cup \{E\}$. Transitions are four-tuples of the form $\langle s, a, D, r \rangle$, with $s \in L_0$, $a \in A$, $D \in Decl_0$, $r \in L_0^+$. A transition relation $\to$ is any subset of $L_0 \times A \times Decl_0 \times L_0^+$. Instead of $\langle s, a, D, r \rangle \in \to$ we write $s \xrightarrow{a}_D r$. From now on, we shall suppress explicit mention of D in our notation. For example, we shall use $s \xrightarrow{a} r$ rather than $s \xrightarrow{a}_D r$, and, at later stages, we use $\mathcal{O}[\![s]\!]$ rather than $\mathcal{O}[\![(D, s)]\!]$, etc. We feel free to do so since D is in no way manipulated in our considerations. Each time, where relevant, some fixed D may be assumed.

As the next step, we introduce a specific transition relation $\to_0$ in terms of what may be called a formal *transition system* T_0 (consisting of some *axioms* and some *rules*).

Definition 15. $\to_0$ is the least relation satisfying the following system T_0:

1. $a \xrightarrow{a}_0 E$

2. If $s \xrightarrow{a}_0 r$ then

$$s \,;\, \bar{s} \xrightarrow{a}_0 r \,;\, \bar{s}$$
$$s \parallel \bar{s} \xrightarrow{a}_0 r \parallel \bar{s}$$
$$\bar{s} \parallel s \xrightarrow{a}_0 \bar{s} \parallel r$$
$$s + \bar{s} \xrightarrow{a}_0 r$$
$$\bar{s} + s \xrightarrow{a}_0 r$$

3. If $g \xrightarrow{a}_0 r$ then $x \xrightarrow{a}_0 r$, where $(x, g) \in D$.

Remark. In Clause 2 we use the convention that (in the case $r = E$) $E \,;\, \bar{s} = E \parallel \bar{s} = \bar{s} \parallel E = \bar{s}$.

We now introduce the operational domains $(r \in) R_0$, $(u \in) S_0$, and show how to define $\mathcal{O}_0 : L_0 \to R_0$.

Definition 16.

1. $R_0 = \mathbb{P}_{nc}(S_0)$, $S_0 = (A \times S_0) \cup \{\delta, \epsilon\}$

2. Let $(F \in) M_0 = L_0^+ \to R_0$, and let $\Psi_0 : M_0 \to M_0$ be defined as follows:

$$\Psi_0(F)(E) = \{\epsilon\}$$
$$\Psi_0(F)(s) = \begin{cases} \{\langle a, u\rangle \mid s \xrightarrow{a}_0 r \wedge u \in F(r)\} & \text{if this set is non-empty} \\ \{\delta\} & \text{otherwise} \end{cases}$$

3. $\mathcal{O}_0 = \text{fix } \Psi_0$

Remarks.

1. In Clause 1, ϵ and δ are new symbols which denote proper and improper termination respectively.

2. By the definition of $\to_0$, $\{\delta\}$ will never be delivered in Clause 2. We have included this case for consistency with later definitions, where the set $\{\langle a, u\rangle \mid \cdots\}$ may well be empty.

3. For each F and s, $\Psi_0(F)(s)$ is a non-empty compact set (this follows from the definition of T_0). Moreover, Ψ_0 is a contracting operator (on the complete metric space M_0). This depends essentially on our convention (see the remark following Theorem 13) that in a domain equation such as that for S_0, recursive occurrences are implicitly proceeded by the $id_{1/2}$ operator.

Examples 17.

1. $\mathcal{O}[\![(a_1\,;\,a_2)+a_3]\!] = \{\langle a_1,\langle a_2,\epsilon\rangle\rangle, \langle a_3,\epsilon\rangle\}$

2. $\mathcal{O}[\![((x,(a\,;\,x)+b),x)]\!]$
   $= \{\underbrace{\langle a,\langle a,\ldots\rangle\rangle}_{\omega \text{ times } a}\} \cup \{\underbrace{\langle a,\langle a,\ldots,\langle b,\epsilon\rangle\ldots\rangle\rangle}_{i \text{ times } a} \mid i = 0,1,\ldots\}$

   (In a less cumbersome notation, we would write $\{a^\omega\} \cup a^*b$.)

We continue with the denotational definition for L_0. We shall, here and subsequently, follow a fixed pattern, in that we first introduce the denotational domains, then define the necessary semantic operators, and finally define a higher-order mapping Φ_i which has the desired $\mathcal{D}_i$ as fixed point.

Definition 18.

1. $P_0 = \mathbb{P}_{co}(Q_0) \cup \{\{\epsilon\}\}$, $Q_0 = A \times P_0$

2. Let $(\phi\in)\, \mathbb{P}_0 = P_0 \times P_0 \to P_0$. The operator $+ \in \mathbb{P}_0$ is defined by $p + \{\epsilon\} = \{\epsilon\} + p = p$, and, for $p_1, p_2 \neq \{\epsilon\}$, $p_1 + p_2$ is the set-theoretic union of p_1 and p_2. Also, the operators $\circ$ and $\|$ are defined by $\circ = \text{fix}\,\Omega_\circ$, $\| = \text{fix}\,\Omega_\|$, where $\Omega_\circ, \Omega_\| : \mathbb{P}_0 \to \mathbb{P}_0$ are given by

$$\Omega_\circ(\phi)(p_1,p_2) = \begin{cases} p_2 & \text{if } p_1 = \{\epsilon\} \\ \{\langle a, \phi(p')(p_2)\rangle \mid \langle a,p'\rangle \in p_1\} & \text{if } p_1 \neq \{\epsilon\} \end{cases}$$

$$\Omega_\|(\phi)(p_1,p_2) = \Omega_\circ(\phi)(p_1,p_2) + \Omega_\circ(\phi)(p_2,p_1)$$

3. Let $(F\in)\, N_0 = L_0 \to P_0$, and let $\Phi_0 : N_0 \to N_0$ be given by

(for $g \in L_0^g$)

$$\Phi_0(F)(a) = \{\langle a,\{\epsilon\}\rangle\}$$
$$\Phi_0(F)(g\,;\,s) = \Phi_0(F)(g) \circ F(s)$$
$$\Phi_0(F)(g_1 + g_2) = \Phi_0(F)(g_1) + \Phi_0(F)(g_2)$$
$$\Phi_0(F)(g_1 \| g_2) = \Phi_0(F)(g_1) \| \Phi_0(F)(g_2)$$

(for $s \in L_0$)

$$\Phi_0(F)(a) = \{\langle a,\{\epsilon\}\rangle\}$$
$$\Phi_0(F)(x) = \Phi_0(F)(g) \quad \text{with } (x,g) \in D$$
$$\Phi_0(F)(s_1\,;\,s_2) = \Phi_0(F)(s_1) \circ \Phi_0(F)(s_2)$$

and similarly for $s_1 + s_2$, $s_1 \| s_2$.

4. Let $\mathcal{D}_0 = \text{fix } \Phi_0$.

Examples 19.

1. We use an abbreviated notation for processes in P_0: we write $a \cdot p$ for $\langle a, p \rangle$, we omit final $\cdot \{\epsilon\}$, and we write $q_1 + q_2 + \cdots$ for process $p (\neq \{\epsilon\})$ with elements $q_1, q_2, \ldots$. Examples of elements in P_0 are $\emptyset$, $\{\epsilon\}$, $(a_1 \cdot a_2) + (a_1 \cdot a_3)$, $a_1 \cdot (a_2 + a_3)$, $a_1 \cdot (a_2 \cdot a_3 + a_3 \cdot a_2) + a_3 \cdot a_1 \cdot a_2$, and the processes p', p'', p''' defined by

$$\begin{aligned} p' &= \lim_i p'_i & p'_0 &= \{\epsilon\} & p'_{i+1} &= a \cdot p'_i \\ p'' &= \lim_i p''_i & p''_0 &= \{\epsilon\} & p''_{i+1} &= a \cdot p''_i + b \\ p''' &= \lim_i p'''_i & p'''_0 &= \{\epsilon\} & p'''_{i+1} &= a \cdot p'''_i \end{aligned}$$

2. Putting $\mathsf{L} = \Omega_0(\|)$, we have $p_1 \| p_2 = (p_1 \mathbin{\mathsf{L}} p_2) + (p_2 \mathbin{\mathsf{L}} p_1)$. Also, $(a_1 \cdot a_2) \| a_3 = a_1 \cdot (a_2 \cdot a_3 + a_3 \cdot a_2) + a_3 \cdot a_1 \cdot a_2$. Moreover, $\emptyset + p = p + \emptyset = p$, $\emptyset \circ p = \emptyset$ (but $p \circ \emptyset = \emptyset$ only if $p = \{\epsilon\}$ or $p = \emptyset$). Also, for p', p'', p''' as in 19(1), we have, for any p, $p' \circ p = p'$, $p'' \circ p = a \cdot p'' + b \cdot p$, and $p''' \circ p = p'''$.

3.

$$\begin{aligned} \mathcal{D}_0[\![a_1 \, ; (a_2 + a_3)]\!] &= a_1 \cdot (a_2 + a_3) \\ \mathcal{D}_0[\![(a_1 \, ; a_2) + (a_1 \, ; a_3)]\!] &= (a_1 \cdot a_2) + (a_1 \cdot a_3) \\ \mathcal{D}_0[\![(a_1 \, ; a_2) \| a_3]\!] &= a_1 \cdot ((a_2 \cdot a_3) + (a_3 \cdot a_2)) + a_3 \cdot a_1 \cdot a_2 \\ \mathcal{D}_0[\![((x, a \, ; x), x)]\!] &= p' \quad \text{as in 19(1)} \\ \mathcal{D}_0[\![((x, a \, ; x + b), x)]\!] &= p'' \quad \text{as in 19(1)} \\ \mathcal{D}_0[\![((x, a \, ; x + b \, ; x), x)]\!] &= p''' \quad \text{as in 19(1)} \end{aligned}$$

Remark. Well-definedness of Φ_0 follows by induction on the complexity of first g and then any s. Contractivity follows, essentially, from the way we have defined $\Phi_0(F)(g; s)$, together with the fact that, for d the metric as determined by the definitions in Section 5.2, we have that $d(p \circ p_1, p \circ p_2) \leq d(p_1, p_2)/2$, for $p \neq \{\epsilon\}$.

We now discuss how to relate $\mathcal{O}_0$ and $\mathcal{D}_0$, using the *abstraction* mapping $abs_0 : P_0 \to R_0$. We shall define abs_0 in such a way that each process p is mapped onto the set of all its 'paths'. For compact p, we have that $abs_0(p)$ is indeed a non-empty compact set; hence $abs_0(p)$ is a well-defined element of R_0. (We refer to (de Bakker et al. 1984) for a discussion including full proofs of these issues.)

Definition 20.

1. Let $(\pi \in)\, PR_0 = P_0 \to R_0$, and let $\Delta_0 : PR_0 \to PR_0$ be given by

$$\Delta_0(\pi)(\emptyset) = \{\delta\}$$
$$\Delta_0(\pi)(\{\epsilon\}) = \{\epsilon\}$$
$$\Delta_0(\pi)(p) = \{\langle a, u\rangle \mid \langle a, p'\rangle \in p \,\wedge\, u \in \pi(p')\} \quad \text{for } p \neq \emptyset, \{\epsilon\}$$

2. Let $abs_0 = \mathrm{fix}\,\Delta_0$.

Example 21.

$$abs_0((a_1 \cdot a_2) + (a_1 \cdot a_3)) = abs_0(a_1 \cdot (a_2 + a_3))$$
$$= \{\langle a_1, \langle a_2, \epsilon\rangle\rangle, \langle a_1, \langle a_3, \epsilon\rangle\rangle\}$$

Also, $abs_0(\emptyset) = \{\delta\}$.

We need one slight extension to $\mathcal{D}_0$ before we can relate $\mathcal{D}_0$ and $\mathcal{O}_0$. Let $\hat{\mathcal{D}}_0 : L_0^+ \to P_0$ be given by: $\hat{\mathcal{D}}_0[\![E]\!] = \{\epsilon\}$, $\hat{\mathcal{D}}_0[\![s]\!] = \mathcal{D}_0[\![s]\!]$. We have the following theorem.

Theorem 22. $\mathcal{O}_0 = abs_0 \circ \hat{\mathcal{D}}_0$.

Proof (outline). First we introduce an intermediate operational semantics $\mathcal{I} : L_0^+ \to P_0$, defined as follows. Let $(F \in)\, N_0^+ = L_0^+ \to P_0$, and let $\Psi_{\mathcal{I}} : N_0^+ \to N_0^+$ be given by

$$\Psi_{\mathcal{I}}(F)(E) = \{\epsilon\}$$
$$\Psi_{\mathcal{I}}(F)(s) = \{\langle a, F(r)\rangle \mid s \xrightarrow{a}_0 r\}$$

Let $\mathcal{I} \stackrel{\text{def}}{=} \mathrm{fix}\,\Psi_{\mathcal{I}}$. Following (de Bakker and Meyer 1988, Kok and Rutten 1988) we may show that $\mathcal{I} = \hat{\mathcal{D}}_0$ by establishing that $\Psi_{\mathcal{I}}(\hat{\mathcal{D}}_0) = \hat{\mathcal{D}}_0$ (followed by an appeal to Banach's theorem). Next, we have, by the various definitions,

$$\Psi_0(abs_0 \circ F)(r) = abs_0(\Psi_{\mathcal{I}}(F)(r))$$

Hence $\Psi_0(abs_0 \circ \mathcal{I})(r) = abs_0(\Psi_{\mathcal{I}}(\mathcal{I})(r)) = (abs_0 \circ \mathcal{I})(r)$. Thus, $abs_0 \circ \mathcal{I} = abs_0 \circ \hat{\mathcal{D}}_0$ is a fixed point of Ψ_0, and $abs_0 \circ \hat{\mathcal{D}}_0 = \mathcal{O}_0$ follows. □

L_1: a uniform language with process creation and synchronization

We next consider the language L_1 embodying two important variations on L_0. Firstly, the construct of parallel composition is replaced by that of process creation (here 'process' refers to a programming concept, and not to a mathematical process p in some domain P). Secondly, we add a notion of (CCS-like) synchronization. We now take the set of elementary actions A to consist of two disjoint subsets $(b \in) B$ and $(c \in) C$, where the actions in B may be taken as *independent*. Moreover, for each c in C we assume a counterpart $\bar{c}$ in C (where $\bar{\bar{c}} = c$), with the understanding that execution of c in some component has to *synchronize* with execution of $\bar{c}$ in a parallel component (and then delivers a special action τ in B as a result). Process creation is expressed through the construct **new**(s): its execution amounts to the creation of a new process which has the task of executing s in parallel with the execution of the already existing processes (each with its already associated task). In addition, we stipulate that termination of a number of parallel processes requires termination of all its components. This brief description of the meaning of **new**(s) (many details are given in (America and de Bakker 1988)) is elaborated in the formal definitions to follow.

Definition 23. $(s \in) L_1$, $(g \in) L_1^g$ and the auxiliary $(h \in) L_1^h$ are defined by:

1. $s ::= a \mid x \mid s_1 \,;\, s_2 \mid s_1 + s_2 \mid \mathbf{new}(s)$

2. $g ::= h \mid g_1 \,;\, g_2 \mid g_1 + g_2 \mid \mathbf{new}(g)$

3. $h ::= a \mid h \,;\, s \mid h_1 + h_2$

4. A program is a pair (D, s), where D consists of pairs (x, g)

Remark. Using only $g \in L_1^g$ (and no $h \in L_1^h$) would lead us to the definition $g ::= a \mid g \,;\, s \mid g_1 + g_2 \mid \mathbf{new}(g)$. Then $\mathbf{new}(a) \,;\, x$ would qualify as guarded, which is undesirable since this will obtain the same effect as the L_0-statement $a \parallel x$ (which is unguarded since it may start with execution of x).

We proceed with the definitions for the operational semantics $\mathcal{O}_1$.

Definition 24.

1. $(r \in) L_1^+$ is given by $r ::= E \mid s \,;\, r$ (r may be seen as a syntactic continuation). $(\rho \in) Par_1$ is given by $\rho ::= \langle r_1, r_2, \ldots, r_n \rangle$, $n \geqslant 1$. We shall identify $\langle r \rangle$ and r. Concatenation of tuples ρ_1, ρ_2 will be denoted by $\rho_1 : \rho_2$.

Concurrency semantics

2. Transitions are written as $\rho_1 \xrightarrow{a}_1 \rho_2$, where $\rightarrow_1$ is the smallest relation satisfying the formal system T_1 given by

3. $a\,;r \xrightarrow{a}_1 r$

If $s\,;r \xrightarrow{a}_1 \rho$ then $(s+\bar{s})\,;r \xrightarrow{a}_1 \rho$ and $(\bar{s}+s)\,;r \xrightarrow{a}_1 \rho$

If $g\,;r \xrightarrow{a}_1 \rho$ then $x\,;r \xrightarrow{a}_1 \rho$, where $(x,g) \in D$

If $s_1\,;(s_2\,;r) \xrightarrow{a}_1 \rho$ then $(s_1\,;s_2)\,;r \xrightarrow{a}_1 \rho$

If $\langle r,s;E\rangle \xrightarrow{a}_1 \rho$ then $\mathbf{new}(s)\,;r \xrightarrow{a}_1 \rho$

If $\rho_1 \xrightarrow{a}_1 \rho_2$ then $\rho:\rho_1 \xrightarrow{a}_1 \rho:\rho_2$ and $\rho_1:\rho \xrightarrow{a}_1 \rho_2:\rho$

If $\rho_1 \xrightarrow{c}_1 \rho'$ and $\rho_2 \xrightarrow{\bar{c}}_1 \rho''$ then $\rho_1:\rho_2 \xrightarrow{\tau}_1 \rho':\rho''$

We present the next definition of (the domain for) $\mathcal{O}_1$.

Definition 25.

1. $R_1 = \mathbb{P}_{nc}(S_1)$, $S_1 = (B \times S_1) \cup \{\delta, \epsilon\}$

2. Let $(F\in) M_1 = Par_1 \rightarrow R_1$, and let $\Psi_1 : M_1 \rightarrow M_1$ be given by

$$\Psi_1(F)(\rho) = \{\epsilon\} \quad \text{for } \rho = \langle E, \ldots, E\rangle$$

Otherwise

$$\Psi_1(F)(\rho) = \begin{cases} \{\langle a,u\rangle \mid \rho \xrightarrow{a}_1 \rho' \wedge u \in F(\rho') \wedge a \in B\}, \\ \qquad \text{if this set is non-empty} \\ \{\delta\} \quad \text{otherwise} \end{cases}$$

3. $\mathcal{O}_1 = \text{fix } \Psi_1$

Examples 26.

1.
$$\mathcal{O}_1[\![b\,;E]\!] = \mathcal{O}_1[\![\mathbf{new}(b)\,;E]\!] = \{\langle b,\epsilon\rangle\}$$
$$\mathcal{O}_1[\![b_1\,;b_2\,;E]\!] = \{\langle b_1,\langle b_2,\epsilon\rangle\rangle\}$$
$$\mathcal{O}_1[\![\mathbf{new}(b_1)\,;b_2\,;E]\!] = \{\langle b_1,\langle b_2,\epsilon\rangle\rangle, \langle b_2,\langle b_1,\epsilon\rangle\rangle\}$$

2.
$$\mathcal{O}_1[\![c\,;E]\!] = \mathcal{O}_1[\![\bar{c}\,;E]\!] = \{\delta\}$$
$$\mathcal{O}_1[\![\langle c\,;E,\bar{c}\,;E\rangle]\!] = \{\tau\}$$

3.
$$\mathcal{O}_1[\![\mathbf{new}(c)\,;\,b_1\,;\,\mathbf{new}(\bar{c})\,;\,b_2\,;\,E]\!] \;=\; \{\langle b_1,\langle \tau,\langle b_2,\epsilon\rangle\rangle\rangle, \langle b_1,\langle b_2,\langle \tau,\epsilon\rangle\rangle\rangle\}$$

4.
$$\mathcal{O}_1[\![(b_1\,;\,b_2) + (b_1\,;\,c)]\!] \;=\; \{\langle b_1,\langle b_2,\epsilon\rangle\rangle, \langle b_1,\delta\rangle\}$$

From the examples we see that $\mathcal{O}_1$ is not compositional (Examples 26(1) and 26(2) show this with respect to ; and :). We remedy this as follows: in order to handle : we introduce the BT domain P_1 (refining R_1). P_1 is the same as P_0 from the previous section, but now its branching structure is indeed exploited. Process creation (and the ensuing problems with ;) is dealt with in a different way, namely by using the technique of so-called semantic continuations. We shall define $\mathcal{D}_1 : L_1 \to (P_1 \to P_1)$, rather than just $\mathcal{D}_1 : L_1 \to P_1$. Details follow in the next definition.

Definition 27.

1. $P_1 = P_0$, $Q_1 = Q_0$

2. Let $(\phi \in)\, \mathbb{P}_1 = P_1 \times P_1 \to P_1$. We define $+ \in \mathbb{P}_1$, and $\Omega_\circ : \mathbb{P}_1 \to \mathbb{P}_1$ as in Definition 18. Also, $\Omega_{\|} : \mathbb{P}_1 \to \mathbb{P}_1$ is given by $\Omega_{\|}(\phi)(p_1,p_2) = \Omega_\circ(\phi)(p_1,p_2) + \Omega_\circ(\phi)(p_2,p_1) + \Omega_|(\phi)(p_1,p_2)$, where

$$\Omega_|(\phi)(p_1,p_2) \;=\; \{\langle \tau, \phi(p',p'')\rangle \mid \langle c,p'\rangle \in p_1 \,\wedge\, \langle \bar{c},p''\rangle \in p_2\}$$

Let $\| = \mathrm{fix}\,\Omega_{\|}$.

3. In the definition of Φ_1 we use an extra argument (from P_1), namely the semantic continuation. Let $(F \in)\, N_1 = L_1 \to (P_1 \to P_1)$, and let $\Phi_1 : N_1 \to N_1$ be given by

(for $h \in L_1^h$)

$$\begin{aligned}
\Phi_1(F)(a)(p) &= \{\langle a,p\rangle\} \\
\Phi_1(F)(h\,;\,s)(p) &= \Phi_1(F)(h)(F(s)(p)) \\
\Phi_1(F)(h_1 + h_2)(p) &= \Phi_1(F)(h_1)(p) + \Phi_1(F)(h_2)(p)
\end{aligned}$$

(for $g \in L_1^g$)

$$\begin{aligned}
\Phi_1(F)(h)(p) &= \text{as above} \\
\Phi_1(F)(g_1\,;\,g_2)(p) &= \Phi_1(F)(g_1)(\Phi_1(F)(g_2)(p)) \\
\Phi_1(F)(g_1 + g_2)(p) &= \Phi_1(F)(g_1)(p) + \Phi_1(F)(g_2)(p) \\
\Phi_1(F)(\mathbf{new}(g))(p) &= \Phi_1(F)(g)(\{\epsilon\}) \,\|\, p
\end{aligned}$$

(for $s \in L_1$)

$$\Phi_1(F)(x)(p) = \Phi_1(F)(g)(p) \quad \text{where } (x,g) \in D$$
$$\Phi_1(F)(\mathbf{new}(s))(p) = \Phi_1(F)(s)(\{\epsilon\}) \parallel p$$

The cases $s \equiv a$, $s_1 ; s_2$, $s_1 + s_2$ are similar to the above.

4. $\mathcal{D}_1 = \text{fix } \Phi_1$.

Examples 28.

1. $\mathcal{D}_1[\![c]\!](p) = \{\langle c, p \rangle\}$, and, using the abbreviated notation for processes in P_0 ($= P_1$) from the previous section, $\mathcal{D}_1[\![\mathbf{new}(c) ; \bar{c}]\!](\{\epsilon\}) = c \cdot \bar{c} + \bar{c} \cdot c + \tau$.

2. $\mathcal{D}_1[\![\mathbf{new}(b_1) ; b_2]\!](p) = \{\langle b_1, \{\langle b_2, p \rangle\}\rangle, \langle b_2, \{\langle b_1, \{\epsilon\}\rangle\} \parallel p\rangle\}$.

We see that $\mathcal{D}_1$ makes more distinctions than does $\mathcal{O}_1$: $\mathcal{O}_1[\![c_1 ; E]\!] = \{\delta\} = \mathcal{O}_1[\![c_2 ; E]\!]$, whereas $\mathcal{D}_1[\![c_1]\!] = \lambda p \cdot \{\langle c_1, p \rangle\} \neq \lambda p \cdot \{\langle c_2, p \rangle\} = \mathcal{D}_1[\![c_2]\!]$. Also, $\mathcal{O}_1[\![b ; E]\!] = \{\langle b, \{\epsilon\}\rangle\} = \mathcal{O}_1[\![\mathbf{new}(b) ; E]\!]$, whereas $\mathcal{D}_1[\![b]\!] = \lambda p \cdot \{\langle b, p \rangle\} \neq \lambda p \cdot (\{\langle b, \{\epsilon\}\rangle\} \parallel p) = \mathcal{D}_1[\![\mathbf{new}(b)]\!]$.

We next introduce the abstraction mapping $abs_1 : P_1 \to R_1$, which will be used to relate $\mathcal{O}_1$ and $\mathcal{D}_1$.

Definition 29. Let $(\pi \in) PR_1 = P_1 \to R_1$, and let $\Delta_1 : PR_1 \to PR_1$ be given by

$$\Delta_1(\pi)(\{\epsilon\}) = \{\epsilon\}$$

and, for $p \neq \{\epsilon\}$,

$$\Delta_1(\pi)(p) = \begin{cases} \{\langle a, u \rangle \mid \langle a, p' \rangle \in p \land u \in \pi(p') \land a \in B\}, \\ \quad \text{if this set is non-empty} \\ \{\delta\} \quad \text{otherwise} \end{cases}$$

Let $abs_1 = \text{fix } \Delta_1$.

Remark. $abs_1(p)$ yields the set of all paths from p which involve no c-steps.

Since not only the codomains, but also the domains of $\mathcal{O}_1$ and $\mathcal{D}_1$ differ, we first introduce an auxiliary semantic mapping $\mathcal{E}_1$, and then relate $\mathcal{O}_1$ and $\mathcal{E}_1$. We define $\mathcal{E}_1 : Par_1 \to P_1$ by putting $\mathcal{E}_1[\![E]\!] = \{\epsilon\}$, $\mathcal{E}_1[\![s ; r]\!] = \mathcal{D}_1[\![s]\!](\mathcal{E}_1[\![r]\!])$, and $\mathcal{E}_1[\![\langle r_1, \ldots, r_n \rangle]\!] = \mathcal{E}_1[\![r_1]\!] \parallel \cdots \parallel \mathcal{E}_1[\![r_n]\!]$. We have the following theorem.

Theorem 30. $\mathcal{O}_1 = abs_1 \circ \mathcal{E}_1$

Proof (Sketch). First introduce an intermediate operational semantics $\mathcal{I}_1$ (in the style of the $\mathcal{I}$ of the previous section), and show that $\mathcal{I}_1 = \mathcal{E}_1$ (the reader may consult (de Bakker and Meyer 1988) for this). Then prove that $\mathcal{D}_1 = abs_1 \circ \mathcal{I}_1$ by an argument as in the proof of Theorem 22. □

We conclude this section with a few remarks concerning the question whether $\mathcal{D}_1$ is the 'best possible' with respect to $\mathcal{O}_1$. In technical terms, we ask whether $\mathcal{D}_1$ is *fully abstract* with respect to $\mathcal{O}_1$. Recall that we added information in the denotational domain P_1 (as compared with R_1) in order to make $\mathcal{D}_1$ compositional. In principal, it may be envisaged that more information has been added than is necessary to achieve this purpose. For a language with parallel composition (rather than process creation) and synchronization this is indeed the case. A so-called failure set model (which preserves less information than the full BT model) suffices. See (Brookes et al. 1984) for the notion of failure set model; (Rutten 1989) gives a theorem from (Bergstra et al. 1988) stating that this model is fully abstract is translated into a metric setting. This result makes it likely that, for L_1 as well, we do not have that $\mathcal{D}_1$ is fully abstract with respect to $\mathcal{O}_1$. A rigorous formulation of this fact (see (Rutten 1989) for alternative formulations and further discussion) is the following. We expect that it is *not* true that, for each $s_1, s_2 \in L_1$, the following two facts are equivalent:

1. $\mathcal{D}_1[\![s_1]\!] = \mathcal{D}_1[\![s_2]\!]$;

2. for each 'context' $C[\bullet]$ we have that $\mathcal{O}_1[\![C[s_1]]\!] = \mathcal{O}_1[\![C[s_2]]\!]$.

Here a context $C[\bullet]$ is a text with a 'hole' such that $C[s]$, the result of filling the hole with s, is a well-formed element of Par_1.

Clearly, it would already be of some interest to investigate these questions for a language L'_1 with only process creation (and no synchronization).

L_2: a non-uniform language with parallel composition

We now engage upon the discussion of a number of languages of the non-uniform variety. In the first (L_2) elementary actions are replaced by assignments $v := e$, where ($v \in$) *Ivar* is the set of *individual variables*, and ($e \in$) *Exp* is the set of *expressions*. We also introduce the set ($b \in$) *Test*, which is the set of logical expressions. We assume a simple syntax (not specified here) for e, b. 'Simple' ensures at least that no side effects or non-termination occurs in their evaluation. Furthermore, we introduce a set of states ($\sigma \in$) $\Sigma = Ivar \to V$, where ($\alpha \in$) V is some set of *values*. It is convenient (for later purposes) to postulate that $V \subseteq Exp$. The notation $\sigma[\alpha/v]$

Concurrency semantics

denotes a state such that $\sigma[\alpha/v](v') =$ **if** $v = v'$ **then** α **else** $\sigma(v')$ **fi**. Finally, note that for non-uniform languages we shall not distinguish guarded recursion from the general case. (Contractivity of the operator corresponding to the program will be ensured by (semantically) proceeding each call of a procedure by the equivalent of a skip statement.)

The syntax for L_2 is given in the following definition.

Definition 31.

1. $(s\in) L_2$ is given by

$$ s ::= v := e \mid x \mid s_1 \,;\, s_2 \mid \textbf{if } b \textbf{ then } s_1 \textbf{ else } s_2 \textbf{ fi} \mid s_1 \parallel s_2 $$

2. Declarations D are sets of pairs (x, s), and a program is a pair (D, s).

The operational semantics $\mathcal{O}_2$ is given in terms of a relation $\to_2$: transitions are now of the form $\langle s, \sigma \rangle \to_2 \langle r, \sigma' \rangle$, with $\sigma, \sigma' \in \Sigma$, $s \in L_2$, $r \in L_2^+ = L_2 \cup \{E\}$, and $\to_2$ the smallest relation satisfying the transition system T_2 given in the following definition.

Definition 32. $\langle v := e, \sigma \rangle \to_2 \langle E, \sigma[\alpha/v] \rangle$, where $\alpha = \llbracket e \rrbracket(\sigma)$
$\langle x, \sigma \rangle \to_2 \langle s, \sigma \rangle$, where $(x, s) \in D$
If $\langle s, \sigma \rangle \to_2 \langle r, \sigma' \rangle$ then

$$\langle s\,;\,\bar{s}, \sigma \rangle \to_2 \langle r\,;\,\bar{s}, \sigma' \rangle$$
$$\langle s \parallel \bar{s}, \sigma \rangle \to_2 \langle r \parallel \bar{s}, \sigma' \rangle$$
$$\langle \bar{s} \parallel s, \sigma \rangle \to_2 \langle \bar{s} \parallel r, \sigma' \rangle$$

with the convention that $E\,;\,\bar{s} = E \parallel \bar{s} = \bar{s} \parallel E = \bar{s}$.
If $\langle s, \sigma \rangle \to_2 \langle r, \sigma' \rangle$, then

if $\llbracket b \rrbracket(\sigma) = tt$ then $\langle \textbf{if } b \textbf{ then } s \textbf{ else } s_2 \textbf{ fi}, \sigma \rangle \to_2 \langle r, \sigma' \rangle$
if $\llbracket b \rrbracket(\sigma) = \textit{ff}$ then $\langle \textbf{if } b \textbf{ then } s_1 \textbf{ else } s \textbf{ fi}, \sigma \rangle \to_2 \langle r, \sigma' \rangle$

The operational domains and semantics are given in the following definition.

Definition 33.

1. $R_2 = \Sigma \to \mathbb{P}_{nc}(S_2)$, $S_2 = (\Sigma \times S_2) \cup \{\delta, \epsilon\}$
2. Let $(F\in) M_2 = L_2^+ \to R_2$, and let $\Psi_2 : M_2 \to M_2$ be given by

$$\Psi_2(F)(E) = \lambda \sigma \cdot \{\epsilon\}$$
$$\Psi_2(F)(s) = \lambda \sigma \cdot \begin{cases} \{\langle \sigma', u \rangle \mid \langle s, \sigma \rangle \to_2 \langle r, \sigma' \rangle \wedge u \in F(r)(\sigma')\}, \\ \qquad \text{if this set is non-empty} \\ \{\delta\} \quad \text{otherwise} \end{cases}$$

3. $\mathcal{O}_2 = \text{fix } \Psi_2$

Example 34.

$$\mathcal{O}_2[\![v := 0\,;\,v := v+1]\!] = \mathcal{O}_2[\![v := 0\,;\,v := 1]\!]$$
$$= \lambda\sigma \cdot \{\langle\sigma[0/v], \langle\sigma[1/v], \epsilon\rangle\rangle\}$$

but

$$\mathcal{O}_2[\![(v := 0\,;\,v := v+1) \parallel (v := 2)]\!] \neq \mathcal{O}_2[\![(v := 0\,;\,v := 1) \parallel (v := 2)]\!]$$

From this example we see that $\mathcal{O}_2$ is not compositional. We therefore add information to the domains R_2, S_2 obtaining P_2, Q_2 in such a way that $\mathcal{D}_2$ is indeed compositional. The definitions are collected below.

Definition 35.

1. $P_2 = \mathbb{P}_{nc}(Q_2)$, $Q_2 = (\Sigma \to (\Sigma \times Q_2)) \cup \{\epsilon\}$

2. Let $(\phi \in) \mathbb{P}_2 = P_2 \times P_2 \to P_2$. The operator $+ \in \mathbb{P}_2$ is defined by: $\{\epsilon\} + p = p + \{\epsilon\} = p$, and, for $p_1, p_2 \neq \{\epsilon\}$, $p_1 + p_2$ is the set-theoretic union of p_1 and p_2. The mappings $\Omega_\circ, \Omega_\parallel : \mathbb{P}_2 \to \mathbb{P}_2$ are given by

$$\Omega_\circ(\phi)(p_1,p_2) = \bigcup\{\tilde\phi(q_1)(q_2) \mid q_1 \in p_1 \wedge q_2 \in p_2\}$$
$$\tilde\phi(\epsilon)(q) = \{q\}$$

and, for $q_1 \neq \epsilon$,

$$\tilde\phi(q_1)(q_2) = \{q \mid \forall\sigma \cdot q(\sigma) \in \hat\phi(q_1(\sigma))(q_2)\}$$
$$\hat\phi(\langle\sigma,q'\rangle)(q) = \{\langle\sigma,\bar q\rangle \mid \bar q \in \phi(\{q'\})(\{q\})\}$$

Also, $\Omega_\parallel(\phi)(p_1,p_2) = \Omega_\circ(\phi)(p_1,p_2) + \Omega_\circ(\phi)(p_2,p_1)$, $\circ = \text{fix } \Omega_\circ$, and $\parallel = \text{fix } \Omega_\parallel$.

3. Let $(F\in) N_2 = L_2 \to P_2$, and let $\Phi_2 : N_2 \to N_2$ be given by

$$\begin{array}{ll}\Phi_2(F)(v := e) = \{\lambda\sigma \cdot \langle\sigma[\alpha/v], \epsilon\rangle\} & \alpha = [\![e]\!](\sigma) \\ \Phi_2(F)(x) = \{\lambda\sigma \cdot \langle\sigma, \epsilon\rangle\} \circ F(s) & (x,s) \in D \\ \Phi_2(F)(s_1\,;\,s_2) = \Phi_2(F)(s_1) \circ \Phi_2(F)(s_2) & \end{array}$$

and similarly for $\parallel$

$$\Phi_2(F)(\text{if } b \text{ then } s_1 \text{ else } s_2 \text{ fi})$$
$$= \left\{ \begin{array}{l} \lambda\sigma \cdot \text{if } [\![b]\!](\sigma) \text{ then } q_1(\sigma) \text{ else } q_2(\sigma) \text{ fi} \\ \mid q_1 \in \Phi_2(F)(s_1) \wedge q_2 \in \Phi_2(F)(s_2) \end{array} \right\}$$

4. $\mathcal{D}_2 = \text{fix } \Phi_2$

We conclude this section with the introduction of the abstraction operator $abs_2 : P_2 \to R_2$.

Definition 36. *(The structure of this definition slightly deviates from the previous abstraction definitions.)*

1. Let $(\pi \in) Q\Sigma S_2 = Q_2 \to (\Sigma \to S_2)$. We define $\Delta_2' : Q\Sigma S_2 \to Q\Sigma S_2$ by putting

$$\Delta_2'(\pi)(\epsilon) = \lambda \sigma \cdot \epsilon$$

and, for $q \neq \epsilon$,

$$\Delta_2'(\pi)(q) = \lambda \sigma \cdot \hat{\pi}(q(\sigma))$$
$$\hat{\pi}(\langle \sigma, q \rangle) = \langle \sigma, \pi(q)(\sigma) \rangle$$

2. Let $abs_2' = \text{fix } \Delta_2'$. Let $abs_2 : P_2 \to P_2$ be given by

$$abs_2(p) = \lambda \sigma \cdot \begin{cases} \{ abs_2'(q)(\sigma) \mid q \in p \} & \text{if this set is non-empty} \\ \{\delta\} & \text{otherwise} \end{cases}$$

We have (putting $\hat{\mathcal{D}}_2[\![E]\!] = \{\lambda \sigma \cdot \epsilon\}$, $\hat{\mathcal{D}}_2[\![s]\!] = \mathcal{D}_2[\![s]\!]$) the following theorem.

Theorem 37. $\mathcal{O}_2 = abs_2 \circ \hat{\mathcal{D}}_2$

The proof is a non-essential variation on previously given proofs (in turn relying on (Kok and Rutten 1988) and (de Bakker and Meyer 1988)). For the intermediate semantics definition we use the clauses

$$\Psi_\mathcal{I}(F)(E) = \{\lambda \sigma \cdot \epsilon\}$$
$$\Psi_\mathcal{I}(F)(s) = \{q \mid \forall \sigma \cdot q(\sigma) \in \{\langle \sigma', \bar{q} \rangle \mid \langle s, \sigma \rangle \to_2 \langle r, \sigma' \rangle \wedge \bar{q} \in F(r)\}\}$$

As before, we have the issue of full abstractness. Is it true that, for all s_1, s_2, $\mathcal{D}_2[\![s_1]\!] = \mathcal{D}_2[\![s_2]\!]$ iff, for all contexts $C[\bullet]$, $\mathcal{O}_2[\![C[s_1]]\!] = \mathcal{O}_2[\![C[s_2]]\!]$? It has been shown by E. Horita that the answer to this question is negative.

L_3: a non-uniform language with process creation and locality

We continue with the treatment of the language L_3 which has process creation (as for L_1, but this time without some form of synchronization) and the notion of local declaration of an individual variable. We find it convenient to discuss only *initialized declarations* (cf. (de Bakker 1980, Chapter 6)). Our first aim in this section is to motivate a type of domain of the form $P_3 = \Sigma \to \mathbb{P}_{nc}(Q_3)$, rather than the previous case $P_2 = \mathbb{P}_{nc}(Q_2)$: the elements of P_3 are (apart from special cases) of the form $\lambda\sigma \cdot \{\cdots, \langle\sigma', q'\rangle, \cdots\}$, where the 'resumptions' q' depend, in general, on the argument σ. With L_3 we intend to illustrate the need for this type of construction.

The syntax of L_3 is given in the next definition.

Definition 38.

1. $(s \in) L_3$ is given by

    $s ::= v := e \mid x \mid s_1 ; s_2 \mid \text{if } b \text{ then } s_1 \text{ else } s_2 \text{ fi} \mid \text{new}(s)$
    $ \mid \text{begin int } v := e ; s \text{ end}$ where v does not occur in e

2. Declarations and programs are as usual.

The operational semantics domains for L_3 are the same as those for L_2. We again (cf. the section on L_1) introduce $(\rho \in) Par_3$, where $\rho = \langle r_1, \ldots, r_n \rangle$, $n \geq 1$ (and where we identify $\langle r \rangle$ and r). Also, $r (\in L_3^+)$ is given by $r ::= E \mid s ; r$.

The transition system T_3 employs transitions of the form $\langle \rho, \sigma \rangle \to_3 \langle \rho', \sigma' \rangle$, where $\to_3$ is the least relation satisfying the following.

Definition 39.

1. $\langle v := e ; r, \sigma \rangle \to_3 \langle r, \sigma[\alpha/v] \rangle$, where $\alpha = [\![e]\!](\sigma)$

2. $\langle x ; r, \sigma \rangle \to_3 \langle s ; r, \sigma \rangle$, where $(x, s) \in D$

3. If $\langle s_1 ; (s_2 ; r), \sigma \rangle \to_3 \langle \rho, \sigma' \rangle$ then $\langle (s_1 ; s_2) ; r, \sigma \rangle \to_3 \langle \rho, \sigma' \rangle$

4. If $\langle \langle s ; E, r \rangle, \sigma \rangle \to_3 \langle \rho, \sigma' \rangle$ then $\langle \text{new}(s) ; r, \sigma \rangle \to_3 \langle \rho, \sigma' \rangle$

5. If $\langle v := e ; s ; v := \sigma(v) ; r, \sigma \rangle \to_3 \langle \rho, \sigma' \rangle$
    then $\langle \text{begin int } v := e ; s \text{ end} ; r, \sigma \rangle \to_3 \langle \rho, \sigma' \rangle$

6. if ... fi: omitted

7. If $\langle \rho_1, \sigma_1 \rangle \to_3 \langle \rho_2, \sigma_2 \rangle$ then

$$\langle \rho_1 : \rho, \sigma_1 \rangle \to_3 \langle \rho_2 : \rho, \sigma_2 \rangle$$
$$\langle \rho : \rho_1, \sigma_1 \rangle \to_3 \langle \rho : \rho_2, \sigma_2 \rangle$$

$\mathcal{O}_3$ is obtained from T_3 in the usual manner.

Definition 40.

1. Let $(F \in) \, Par R_3 = Par_3 \to R_3$, and let $\Psi_3 : Par R_3 \to Par R_3$ be given by

$$\Psi_3(F)(\langle E, \ldots, E \rangle) \;=\; \lambda \sigma \cdot \{\epsilon\}$$

and, for $\rho \neq \langle E, \ldots, E \rangle$,

$$\Psi_3(F)(\rho) \;=\; \lambda \sigma \cdot \begin{cases} \{\langle \sigma', u \rangle \mid \langle \rho, \sigma \rangle \,\wedge\, u \in F(\rho')(\sigma')\}, \\ \qquad \text{if this set is non-empty} \\ \{\delta\} \quad \text{otherwise} \end{cases}$$

2. $\mathcal{O}_3 = \text{fix } \Psi_3$

Example 41.

$\mathcal{O}_3 \llbracket \text{begin int } v := 0 \,;\, \text{begin int } v := 1 \,;\, v' := v \text{ end} \,;\, v' := v \text{ end} \,;\, E \rrbracket$
$= \lambda \sigma \cdot \{[\sigma[0/v], [\sigma[1/v], [\sigma[1/v][1/v'], [\sigma[0/v][1/v'], [\sigma[0/v][0/v'],$
$\qquad [\sigma[0/v][0/v'][\sigma(v)/v], \epsilon]]]]]\}$

$\mathcal{O}_3$ is not compositional (cf. the discussion for $\mathcal{O}_1$), and we resort to a more complex domain for the denotational semantics. In the remainder of this section we shall employ the following notation.

Notation 42. Let $f : A \to \mathbb{P}(B)$ be a function from A to subsets of B. We then put

$$f^\dagger \;=\; \{g : A \to B \mid \forall a \cdot g(a) \in f(a)\}$$

The denotational definitions are collected in the next definition.

Definition 43.

1. $P_3 = \Sigma \to \mathbb{P}_{nc}(Q_3)$, $Q_3 = (\Sigma \times (\Sigma \to Q_3)) \cup \{\epsilon\}$. We shall use X to range over $\mathbb{P}_{nc}(Q_3)$, and ξ to range over $\Sigma \to Q_3$.

2. Let $(\phi \in) \mathbb{P}_3 = P_3 \times P_3 \to P_3$, and let $\Omega_\circ, \Omega_\| : \mathbb{P}_3 \to \mathbb{P}_3$ be given as follows

$$\Omega_\circ(\phi)(p_1, p_2) = \lambda\sigma \cdot \tilde{\phi}(p_1(\sigma))(p_2)$$
$$\tilde{\phi}(X)(p) = \bigcup \{\hat{\phi}(q)(p) \mid q \in X\}$$
$$\hat{\phi}(\epsilon)(p) = \{\langle \sigma, \xi \rangle \mid \sigma \in \Sigma \wedge \xi \in p^\dagger\}$$
$$\hat{\phi}(\langle \sigma, \xi \rangle)(p) = \{\langle \sigma, \bar{\xi} \rangle \mid \bar{\xi} \in \phi(\lambda\bar{\sigma} \cdot \{\xi(\bar{\sigma})\})(p)^\dagger\}$$
$$\Omega_\|(\phi)(p_1, p_2)(\sigma) = \Omega_\circ(\phi)(p_1, p_2)(\sigma) \cup \Omega_\circ(\phi)(p_2, p_1)(\sigma)$$

Let $\circ = \text{fix } \Omega_\circ$, $\| = \text{fix } \Omega_\|$.

3. Let $(F \in) N_3 = L_3 \to (P_3 \to P_3)$, and let $\Phi_3 : N_3 \to N_3$ be given by

$$\Phi_3(F)(v := e)(p) = \lambda\sigma \cdot \{\langle \sigma[\alpha/v], \xi \rangle \mid \xi \in p^\dagger\}$$
$$\text{where } \alpha = \llbracket e \rrbracket(\sigma)$$
$$\Phi_3(F)(x)(p) = \lambda\sigma \cdot \{\langle \sigma, \xi \rangle \mid \xi \in F(s)(p)^\dagger\}$$
$$\text{where } (x, s) \in D$$
$$\Phi_3(F)(s_1 \,;\, s_2)(p) = \Phi_3(F)(s_1)(\Phi_3(F)(s_2)(p))$$
$$\Phi_3(F)(\textbf{if} \ldots \textbf{fi})(p) = \lambda\sigma \cdot \textbf{if } \llbracket b \rrbracket(\sigma)$$
$$\textbf{then } \Phi_3(F)(s_1)(p)(\sigma)$$
$$\textbf{else } \Phi_3(F)(s_2)(p)(\sigma)$$
$$\textbf{fi}$$
$$\Phi_3(F)(\textbf{new}(s))(p) = \Phi_3(F)(s)(\lambda\sigma \cdot \{\epsilon\}) \| p$$

and

$$\Phi_3(F)(\textbf{begin int } v := e \,;\, s \textbf{ end})(p)$$
$$= \lambda\sigma \cdot \Phi_3(F)(v := e \,;\, s)(\lambda\bar{\sigma} \cdot \{\langle \bar{\sigma}[\sigma(v)/v], \xi \rangle \mid \xi \in p^\dagger\})(\sigma)$$

4. Let $\mathcal{D}_3 = \text{fix } \Phi_3$, and let $\mathcal{E}_3 : Par_3 \to P_3$ be obtained from $\mathcal{D}_3$ similar to the definitions of $\mathcal{E}_1$ for L_1 (where $\mathcal{E}_3 \llbracket E \rrbracket = \lambda\sigma \cdot \{\epsilon\}$).

We finally relate $\mathcal{O}_3$ and $\mathcal{E}_3$ in the usual manner through the abstraction function abs_3.

Definition 44.

1. Let $(\pi \in) QS_3 = Q_3 \to S_3$, and let $\Delta'_3 : QS_3 \to QS_3$ be given by

$$\Delta'_3(\pi)(\epsilon) = \epsilon$$
$$\Delta'_3(\pi)(\langle \sigma, \xi \rangle) = \langle \sigma, \pi(\xi(\sigma)) \rangle$$

Concurrency semantics

2. Let $abs_3' = \text{fix } \Delta_3'$, and let $abs_3 : P_3 \to P_3$ be given as

$$abs_3(p) = \lambda\sigma \cdot \begin{cases} \{abs_3'(q) \mid q \in p(\sigma)\} & \text{if this set is non-empty} \\ \{\delta\} & \text{otherwise} \end{cases}$$

We have the, now familiar, result.

Theorem 45. $\mathcal{O}_3 = abs_3 \circ \mathcal{E}_3$

We do not know whether $\mathcal{E}_3$ is fully abstract with respect to $\mathcal{O}_3$.

L_4: a non-uniform language with parallel composition and communication

The language L_4 is an extension of L_2 in that now (CSP-like) communication over channels $c (\in \textit{Chan})$ is added. A send statement has the form $c!e$, a receive statement has the form $c?v$, and synchronized execution of these (in two parallel components) amounts to the execution of the assignment $v := e$.

The syntax for L_4 is given in the next definition.

Definition 46.

1. $(s \in) L_4$ has as syntax

$$s ::= v := e \mid x \mid s_1 \, ; \, s_2 \mid \textbf{if } b \textbf{ then } s_1 \textbf{ else } s_2 \textbf{ fi} \mid s_1 \parallel s_2 \mid c \, ? \, v \mid c \, ! \, e$$

2. Declarations and programs are as usual.

The operational semantics for L_4 employs the sets

$$(\gamma \in) \Gamma = \{c \, ? \, v \mid c \in \textit{Chan} \wedge v \in \textit{Ivar}\} \cup \{c \, ! \, \alpha \mid c \in \textit{Chan} \wedge \alpha \in V\}$$
$$(\eta \in) H = \Sigma \cup \Gamma$$

Transitions are of the form $\langle s, \sigma \rangle \to_4 \langle r, \eta \rangle$, with $r \in L_4^+ = L_4 \cup \{E\}$. The transition system T_4 is given in the following definition.

Definition 47.

1.

$$\langle v := e, \sigma \rangle \to_4 \langle E, \sigma[\alpha/v] \rangle \quad \alpha \text{ as usual}$$
$$\langle c \, ? \, v, \sigma \rangle \to_4 \langle E, c \, ? \, v \rangle$$
$$\langle c \, ! \, e, \sigma \rangle \to_4 \langle E, c \, ! \, \alpha \rangle \quad \alpha \text{ as usual}$$

144   Metric concurrency semantics

2. The rules for x, $;$, **if**...**fi**, $\|$ are as those in T_2 (with $\to_4$ replacing $\to_2$). For $\|$ we have in addition the following rule.

3. If $\langle s_1, \sigma \rangle \to_4 \langle r', c\,?\,v \rangle$ and $\langle s_2, \sigma \rangle \to_4 \langle r'', c\,!\,\alpha \rangle$ then $\langle s_1 \| s_2, \alpha \rangle \to_4 \langle r' \| r'', \sigma[\alpha/v] \rangle$. (We assume the usual convention that $E\|r = r\|E = r$.)

The operational domains and semantics are given in the next definition.

Definition 48.

1. $R_4 = \Sigma \to \mathbb{P}_{nc}(S_4)$, $S_4 = (\Sigma \times S_4) \cup \{\delta, \epsilon\}$

2. Let $(F\in) M_4 = L_4^+ \to R_4$, and let $\Psi_4 : M_4 \to M_4$ be given by

$$\Psi_4(F)(E) = \lambda\sigma \cdot \{\epsilon\}$$

$$\Psi_4(F)(s) = \lambda\sigma \cdot \begin{cases} \{\langle s', u \rangle \mid \langle s, \sigma \rangle \to_4 \langle r, \sigma' \rangle \wedge u \in F(r')(s')\}, \\ \qquad \text{if this set is non-empty} \\ \{\delta\} \quad \text{otherwise} \end{cases}$$

3. $\mathcal{O}_4 = \text{fix}\,\Psi_4$.

Remark. Note that, in the definition of $\Psi_4(F)(s)(\sigma)$, no contributions are made by steps $\langle s, \sigma \rangle \to_4 \langle r, \gamma \rangle$.

Once more $\mathcal{O}_4$ is not compositional. The denotational definitions assume a domain P_4 which combines the BT structure of P_1 with the non-uniform structure of P_3.

Definition 49.

1. $P_4 = (\Sigma \to \mathbb{P}_{co}(Q_4)) \cup \{\{\epsilon\}\}$, $Q_4 = (\Sigma \cup \Gamma) \times P_4$

2. Let X range over $\mathbb{P}_{co}(Q_4)$. Let $(\phi\in)\mathbb{P}_4 = P_4 \times P_4 \to P_4$, and let $\Omega_\circ$, $\Omega_\| : \mathbb{P}_4 \to \mathbb{P}_4$ be given by

$$\Omega_\circ(\phi)(p_1, p_2) = p_2 \quad \text{if } p_1 = \{\epsilon\}$$
$$= \lambda\sigma \cdot \hat{\phi}(p_1(\sigma))(p_2) \quad \text{if } p_1 \neq \{\epsilon\}$$
$$\hat{\phi}(X)(p) = \{\tilde{\phi}(q)(p) \mid q \in X\}$$
$$\tilde{\phi}(\langle \eta, p' \rangle)(p) = \langle \eta, \phi(p')(p) \rangle$$
$$\Omega_\|(\phi)(p_1, p_2) = \lambda\sigma \cdot \begin{pmatrix} \Omega_\circ(\phi)(p_1, p_2)(\sigma) \\ \cup\, \Omega_\circ(\phi)(p_2, p_1)(\sigma) \\ \cup\, \Omega_|(\phi)(p_1, p_2)(\sigma) \end{pmatrix}$$

where

$$\Omega_|(\phi)(p_1,p_2)(\sigma)
= \lambda\sigma \cdot \left\{ \begin{array}{l} \langle \sigma[\alpha/v], \phi(p')(p'') \rangle \\ \mid \langle c\,?\,v, p' \rangle \in p_1 \,\wedge\, \langle c\,!\,\alpha, p'' \rangle \in p_2 \text{ or vice versa} \end{array} \right\}$$

$\circ = \mathrm{fix}\,\Omega_\circ$, $\| = \mathrm{fix}\,\Omega_\|$.

3. Let $(F\in)\,N_4 = L_4 \to P_4$, and let $\Phi_4 : N_4 \to N_4$ be given by

$$\begin{aligned}
\Phi_4(F)(v := e) &= \lambda\sigma \cdot \{\langle \sigma[\alpha/v], \{\epsilon\}\rangle\} \quad \alpha \text{ as usual} \\
\Phi_4(F)(c\,?\,v) &= \lambda\sigma \cdot \{\langle c\,?\,v, \{\epsilon\}\rangle\} \\
\Phi_4(F)(c\,!\,e) &= \lambda\sigma \cdot \{\langle c\,!\,\alpha, \{\epsilon\}\rangle\} \quad \alpha \text{ as usual}
\end{aligned}$$

$s \equiv s_1\,;\,s_2$, $s_1 \parallel s_2$, **if** ... **fi**: omitted

$$\Phi_4(F)(x) = \lambda\sigma \cdot \{\langle \sigma, F(s)\rangle\} \quad (x,s) \in D$$

4. Let $\mathcal{D}_4 = \mathrm{fix}\,\Phi_4$.

We conclude with the abstraction mapping between $\mathcal{O}_4$ and $\mathcal{D}_4$.

Definition 50. Let $(\pi\in)\,PR_4 = P_4 \to R_4$, and let $\Delta_4 : PR_4 \to PR_4$ be defined as follows:

$$\Delta_4(\pi)(\{\epsilon\}) = \lambda\sigma \cdot \{\epsilon\}$$

and, for $p \neq \{\epsilon\}$,

$$\begin{aligned}
\Delta_4(\pi)(p) &= \lambda\sigma \cdot \left\{ \begin{array}{ll} \bigcup\{\tilde{\pi}(q) \mid q \in p(\sigma)\} & \text{if this set is non-empty} \\ \{\delta\} & \text{otherwise} \end{array} \right. \\
\tilde{\pi}(\langle \sigma, p \rangle) &= \{\langle \sigma, q \rangle \mid q \in \pi(p)(\sigma)\} \\
\tilde{\pi}(\langle \gamma, p \rangle) &= \emptyset
\end{aligned}$$

Let $abs_4 = \mathrm{fix}\,\Delta_4$.

We have (for $\hat{\mathcal{D}}_4$ similar to $\hat{\mathcal{D}}_2$) the following theorem.

Theorem 51. $\mathcal{O}_4 = abs_4 \circ \hat{\mathcal{D}}_4$

As to the question of full abstractness, since $\mathcal{D}_1$ is (probably) not fully abstract with respect to $\mathcal{O}_1$ (cf. the discussion for L_1), there is no reason to expect $\mathcal{D}_4$ to be fully abstract with respect to $\mathcal{O}_4$. (In (Horita et al. 1990) full abstractness with respect to a non-uniform version of the failure set model is shown.)

Conclusion

We conclude with a table which surveys the domain equations encountered in Section 5.3.

	Operational	Denotational
Uniform		
L_0	$R_0 = \mathbb{P}_{nc}(S_0)$	$P_0 = \mathbb{P}_{co}(Q_0) \cup \{\{\epsilon\}\}$
	$S_0 = (A \times S_0) \cup \{\delta, \epsilon\}$	$Q_0 = A \times P_0$
L_1	$R_1 = \mathbb{P}_{nc}(S_1)$	$P_1 = \mathbb{P}_{co}(Q_1) \cup \{\{\epsilon\}\}$
	$S_1 = (B \times S_1) \cup \{\delta, \epsilon\}$	$Q_1 = (B \cup C) \times P_1$
Non-uniform		
L_2	$R_2 = \Sigma \to \mathbb{P}_{nc}(S_2)$	$P_2 = \mathbb{P}_{nc}(Q_2)$
	$S_2 = (\Sigma \times S_2) \cup \{\delta, \epsilon\}$	$Q_2 = (\Sigma \to (\Sigma \times Q_2)) \cup \{\epsilon\}$
L_3	$R_3 = R_2$	$P_3 = \Sigma \to \mathbb{P}_{nc}(Q_3)$
	$S_3 = S_2$	$Q_3 = (\Sigma \times (\Sigma \to Q_3)) \cup \{\epsilon\}$
L_4	$R_4 = R_2$	$P_4 = (\Sigma \to \mathbb{P}_{co}(Q_4)) \cup \{\{\epsilon\}\}$
	$S_4 = S_2$	$Q_4 = (\Sigma \cup \Gamma) \times P_4$

5.4 Labelled transition systems and bisimulation

In this section we shall use the domain P_0 of the previous section to give a general model for *bisimulation* equivalence (Park 1981), a well-known notion in the theory of concurrency. (The same result holds for P_1. For the domains used for the non-uniform languages some further study is still needed.) It is based on the basic notion of a *labelled transition system* (LTS).

Definition 52. (LTS) *A labelled transition system is a triple* $\mathcal{A} = (S, L, \to)$ *consisting of a set of states S, a set of labels L, and a transition relation* $\to \; \subseteq S \times L \times S$. *We shall write* $s \xrightarrow{a} s'$ *for* $(s, a, s') \in \to$. *Following the approach of the previous section, we assume the presence of a special element* $E \in S$ *that syntactically denotes successful termination. An LTS is called finitely branching if for all* $s \in S$, $\{(a, s') \mid s \xrightarrow{a} s'\}$ *is finite.*

Every LTS induces a *bisimulation* equivalence.

Definition 53. *Let* $\mathcal{A} = (S, L, \to)$ *be an LTS. A relation* $R \subseteq S \times S$ *is called a (strong) bisimulation if it satisfies for all* $s, t \in S$ *and* $a \in A$:

$$(s \, R \, t \wedge s \xrightarrow{a} s') \Rightarrow \exists t' \in S \cdot t \xrightarrow{a} t' \wedge s' \, R \, t'$$

and

$$(s \, R \, t \wedge t \xrightarrow{a} t') \Rightarrow \exists s' \in S \cdot s \xrightarrow{a} s' \wedge s' \, R \, t'$$

Labelled transition systems and bisimulation   147

We require that $E \, R \, s$ or $s \, R \, E$ implies $s = E$. Two states are bisimilar in $\mathcal{A}$, notation $s \leftrightarrows t$, if there exists a bisimulation relation R with $s \, R \, t$. (Note that bisimilarity is an equivalence relation on states.)

Next we define, for every LTS $\mathcal{A}$, a model assigning to every state a process in P_0.

Definition 54. Let $\mathcal{A} = (S, L, \to)$ be a finitely branching LTS. Here we have taken for the set of labels the alphabet A of elementary actions used in the definition of P_0. We define a model $\mathcal{M}_\mathcal{A} : S \to P_0$ by

$$\mathcal{M}_\mathcal{A}[\![s]\!] = \{\langle a, \mathcal{M}_\mathcal{A}[\![s']\!]\rangle \mid s \xrightarrow{a} s'\}$$

if $s \neq E$, and by $\mathcal{M}_\mathcal{A}[\![E]\!] = \{\epsilon\}$.

We can justify this recursive definition by taking $\mathcal{M}_\mathcal{A}$ as the unique fixed point (Banach's theorem) of a contraction $\Phi : (S \to^1 P) \to (S \to^1 P)$ defined by

$$\Phi(F)(s) = \{\langle a, F(s')\rangle \mid s \xrightarrow{a} s'\}$$

if $s \neq E$, and by $\Phi(F)(E) = \{\epsilon\}$. The fact that Φ is a contraction can be easily proved. The compactness of the set $\Phi(F)(s)$ is an immediate consequence of the fact that $\mathcal{A}$ is finitely branching.

As an example we can take in the above definition the LTS of Definition 15. We then obtain the function $\mathcal{I}$ given in the proof of Theorem 22.

This model is of interest because it assigns the same meaning to bisimilar states. This we prove next.

Theorem 55. Let $\leftrightarrows \, \subseteq S \times S$ denote the bisimilarity relation induced by the labelled transition system $\mathcal{A} = (S, A, \to)$. Then

$$\forall s, t \in S \cdot s \leftrightarrows t \iff \mathcal{M}_\mathcal{A}[\![s]\!] = \mathcal{M}_\mathcal{A}[\![t]\!]$$

Proof. Let $s, t \in S$.

$\Leftarrow$ Suppose $\mathcal{M}_\mathcal{A}[\![s]\!] = \mathcal{M}_\mathcal{A}[\![t]\!]$. We define a relation $\equiv \, : S \times S$ by

$$s' \equiv t' \iff \mathcal{M}_\mathcal{A}[\![s']\!] = \mathcal{M}_\mathcal{A}[\![t']\!]$$

From the definition of $\mathcal{M}_\mathcal{A}$ it is straightforward that $\equiv$ is a bisimulation relation on S. Suppose $s' \equiv t'$ and $s' \xrightarrow{a} s''$. Then $\langle a, \mathcal{M}_\mathcal{A}[\![s'']\!]\rangle \in \mathcal{M}_\mathcal{A}[\![s']\!] = \mathcal{M}_\mathcal{A}[\![t']\!]$; thus there exists $t'' \in S$ with $t' \xrightarrow{a} t''$ and $\mathcal{M}_\mathcal{A}[\![s'']\!] = \mathcal{M}_\mathcal{A}[\![t'']\!]$, that is, $s'' \equiv t''$. Symmetrically, the second property of a bisimulation relation holds. From the hypothesis we have $s \equiv t$. Thus we have $s \leftrightarrows t$.

⇒ Let $R \subseteq S \times S$ be a bisimulation relation with $s\,R\,t$. We define

$$\epsilon = \sup_{s',t' \in S} \{d(\mathcal{M}_\mathcal{A}[\![s']\!], \mathcal{M}_\mathcal{A}[\![t']\!]) \mid s'\,R\,t'\}$$

We prove that $\epsilon = 0$, from which $\mathcal{M}_\mathcal{A}[\![s]\!] = \mathcal{M}_\mathcal{A}[\![t]\!]$ follows, by showing that $\epsilon \leqslant \epsilon/2$. We prove for all s', t' with $s'\,R\,t'$ that $d(\mathcal{M}_\mathcal{A}[\![s']\!], \mathcal{M}_\mathcal{A}[\![t']\!]) \leqslant \epsilon/2$. Consider $s', t' \in S$ with $s'\,R\,t'$. From the definition of the Hausdorff metric on P it follows that it suffices to show

$$d(x, \mathcal{M}_\mathcal{A}[\![t']\!]) \leqslant \epsilon/2 \ \text{and} \ d(y, \mathcal{M}_\mathcal{A}[\![s']\!]) \leqslant \epsilon/2$$

for all $x \in \mathcal{M}_\mathcal{A}[\![s']\!]$ and $y \in \mathcal{M}_\mathcal{A}[\![t']\!]$. We shall only show the first inequality; the second is similar. Consider $\langle a, \mathcal{M}_\mathcal{A}[\![s'']\!]\rangle$ in $\mathcal{M}_\mathcal{A}[\![s']\!]$ with $s' \xrightarrow{a} s''$. (The case that $\mathcal{M}_\mathcal{A}[\![s']\!] = \{\epsilon\}$ is trivial.) Because $s'\,R\,t'$ and $s' \xrightarrow{a} s''$ there exists $t'' \in S$ with $t' \xrightarrow{a} t''$ and $s''\,R\,t''$. Therefore

$$d(\langle a, \mathcal{M}_\mathcal{A}[\![s'']\!]\rangle, \mathcal{M}_\mathcal{A}[\![t']\!])$$
$$= d(\langle a, \mathcal{M}_\mathcal{A}[\![s'']\!]\rangle, \{\langle \bar{a}, \mathcal{M}_\mathcal{A}[\![\bar{t}]\!]\rangle \mid t' \xrightarrow{\bar{a}} \bar{t}\})$$
$$\leqslant [\text{we have: } d(x, Y) = \inf\{d(x, y) \mid y \in Y\}]$$
$$d(\langle a, \mathcal{M}_\mathcal{A}[\![s'']\!]\rangle, \langle a, \mathcal{M}_\mathcal{A}[\![t'']\!]\rangle)$$
$$= d(\mathcal{M}_\mathcal{A}[\![s'']\!], \mathcal{M}_\mathcal{A}[\![t'']\!])/2$$
$$\leqslant [\text{because } s''\,R\,t'']$$
$$\epsilon/2$$

□

The proof above makes convenient use of the Hausdorff metric on P. It was first given in (Rutten 1989). An alternative proof, using so-called non-well-founded sets, can be found in (van Glabbeek and Rutten 1989, Rutten 1990b).

References

America, P., and de Bakker, J. W. (1988). Designing equivalent semantic models for process creation, *Theoretical Computer Science*, Vol. 60, pp. 109–176.

America, P., and Rutten, J. J. M. M. (1989a). Solving reflexive domain equations in a category of complete metric spaces, *Journal of Computer and System Sciences*, Vol. 35, No. 3, pp. 343–375.

America, P., and Rutten, J. J. M. M. (1989b). A parallel object-oriented language: Design and semantic foundations, in: J. W. de Bakker (Ed.), *Languages for Parallel Architectures: Design, Semantics, Implementation Models*, Wiley Series in Parallel Computing, pp. 1–49.

References

America, P., de Bakker, J. W., Kok, J. N., and Rutten, J. J. M. M. (1989). Denotational semantics of a parallel object-oriented language, *Information and Computation*, Vol. 83, No. 2, pp. 152–205.

de Bakker, J. W. (1980). *Mathematical Theory of Program Correctness*, Prentice Hall International.

de Bakker, J. W. (1988). *Comparative Semantics for Flow of Control in Logic Programming without Logic*, CWI Report CS-R8840, to appear in *Information and Computation*.

de Bakker, J. W. (1989). Designing concurrency semantics, in: G. X. Rytter (Ed.), *Proc. 11th World Computer Congress*, North-Holland, pp. 591–598.

de Bakker, J. W., and Kok, J. N. (1988). Uniform abstraction, atomicity and contractions in the comparative semantics of Concurrent Prolog, in: *Proc. Int. Conference on Fifth Generation Computer Systems*, Institute for New Generation Computer Technology, pp. 347–355.

de Bakker, J. W. and Kok, J. N. (1990). Comparative semantics for Concurrent Prolog, *Theoretical Computer Science*, to appear.

de Bakker, J. W., and Meyer, J.-J. Ch. (1988). Metric semantics for concurrency, *BIT 28*, pp. 504–529.

de Bakker, J. W., and Zucker, J. I. (1982). Processes and the denotational semantics of concurrency, *Information and Control*, Vol. 54, pp. 70–120.

de Bakker, J. W., Bergstra, J. A., Klop, J. W., and Meyer, J.-J. Ch. (1984). Linear time and branching time semantics for recursion with merge, *Theoretical Computer Science*, Vol. 34, pp. 135–156.

de Bakker, J. W., Meyer, J.-J. Ch., and Olderog, E.-R. (1987). Infinite streams and finite observations in the semantics of uniform concurrency, *Theoretical Computer Science*, Vol. 49, pp. 87–112.

de Bakker, J. W., Meyer, J.-J. Ch., Olderog, E.-R., and Zucker, J. I. (1988). Transition systems, metric spaces and ready sets in the semantics of uniform concurrency, *Journal of Computer and System Sciences*, Vol. 36, pp. 158–224.

de Bakker, J. W., Kok, J. N., Meyer, J.-J. Ch., Olderog, E.-R., and Zucker, J. I. (1986). Contrasting themes in the semantics of imperative concurrency, in: J. W. de Bakker *et al.* (Eds.), *Current Trends in Concurrency: Overviews and Tutorials*, Lecture Notes in Computer Science, Vol. 224, Springer-Verlag, pp. 51–121.

Bergstra, J. A., and Klop, J. W. (1989). Bisimulation semantics, in: J. W. de Bakker *et al.* (Eds.), *Linear Time, Branching Time and Partial Order in Logics and Models for Concurrency*, Lecture Notes in Computer Science, Vol. 354, Springer-Verlag, pp. 50–122.

Bergstra, J. A., Klop, J. W., and Olderog, E.-R. (1988). Readies and failures in the algebra of communicating processes, *SIAM J. of Computing*, Vol. 17, No. 6, pp. 1134–1177.

Brookes, S. D., Hoare, C. A. R., and Roscoe, A. W. (1984). A theory of communicating sequential processes, *J. ACM*, Vol 31, pp. 499–560.

Dugundji, J. (1966). *Topology*, Allen and Bacon, Rockleigh, N. J.

Enkelking, R. (1977). *General Topology*, Polish Scientific Publishers.

van Glabbeek, R. J., and Rutten, J. J. M. M. (1989). The processes of de Bakker and Zucker represent bisimulation equivalence classes, in: *J. W. de Bakker, 25 jaar semantiek*, Centre for Mathematics and Computer Science, Amsterdam.

Hennessy, M., and Plotkin, G. D. (1979). Full abstraction for a simple parallel programming language, in: J. Bečvář (Ed.), *Proceedings 8th MFCS*, Lecture Notes in Computer Science, Vol. 74, Springer-Verlag, pp. 108–120.

Horita, E., de Bakker, J. W., and Rutten, J. J. M. M. (1990). *Fully abstract denotational models for non-uniform concurrent languages*, Report CS-R90?, Centre for Mathematics and Computer Science, Amsterdam.

Kok, J. N., and Rutten, J. J. M. M. (1988). Contractions in comparing concurrency semantics, in: T. Lepisto and A. Salomaa (Eds.), *Proc. 15th ICALP*, Lecture Notes in Computer Science, Vol. 317, Springer-Verlag, pp. 317–332. (To appear in *Theoretical Computer Science*.)

Mac Lane, S. (1971). *Categories for the Working Mathematician*, Springer-Verlag.

Majster-Cederbaum, M. E. (1988). The contraction property is sufficient to guarantee the uniqueness of fixed points of endofunctors in a category of complete metric spaces, *Information Processing Letters*, Vol. 29, pp. 277–281.

Majster-Cederbaum, M. E. (1989). On the uniqueness of fixed points of endofunctors in a category of complete metric spaces, *Information Processing Letters*, Vol. 33, pp. 15–20.

Majster-Cederbaum, M. E., and Zetzsche, F. (199?). Towards a foundation for semantics in complete metric spaces, *Information and Computation*, to appear.

Michael, E. (1951). Topologies on spaces of subsets, *Trans. AMS*, Vol. 71, pp. 152–182.

Nivat, M. (1979). Infinite words, infinite trees, infinite computations, in: J. W. de Bakker and J. van Leeuwen (Eds.), *Foundations of Computer Science III.2*, Mathematical Centre Tracts 109, pp. 3–52.

Park, D. M. R. (1981). Concurrency and automata on infinite sequences, in: *Proc. 5th GI Conference*, Lecture Notes in Computer Science, Vol. 104, Springer-Verlag, pp. 15–32.

Plotkin, G. D. (1981). *A structural approach to operational semantics*, Report DAIMI FN-19, Computer Science Department, Aarhus University.

Plotkin, G. D. (1983). An operational semantics for CSP, in: D. Björner

(Ed.), *Formal Description of Programming Concepts II*, North-Holland, pp. 199-223.

Rutten, J. J. M. M. (1989). Correctness and full abstraction of metric semantics for concurrency, in: J. W. de Bakker *et al.* (Eds.), *Linear Time, Branching Time and Partial Order in Logics and Models for Concurrency*, Lecture Notes in Computer Science, Vol. 354, Springer-Verlag, pp. 628–659.

Rutten, J. J. M. M. (1989). *Deriving Denotational Models for Bisimulation from Transition System Specifications*, CWI Report CS-R8955, Amsterdam. (To appear in the proceedings of IFIP TC2 Working Conference, Israel, 1990.)

Rutten, J. J. M. M. (1990a). Semantic correctness for a parallel object-oriented language, *SIAM J. of Computing*, Vol. 19, No. 2, pp. 341–383.

Rutten, J. J. M. M. (1990b). *Nonwellfounded sets and programming language semantics*, CWI Report CS-R90?, Amsterdam.

6
On topological characterization of behavioural properties

MARTA Z. KWIATKOWSKA

Abstract

A topological characterization of behavioural properties in a non-interleaving model for concurrency is presented. The model used is an extension of Mazurkiewicz's trace theory to allow for infinite behaviours. Safety and liveness properties are semantically characterized with respect to the Scott topology of the domain of traces. Fairness properties are defined as a subclass of liveness properties. The relationship between safety, liveness, and fairness is investigated.

6.1 Introduction

Behavioural properties of programs are usually expressed as formulae of temporal logic interpreted over a suitable structure that models the program's behaviour (for example, a set, a tree, or a partial order). Each formula thus gives rise to a subset of behaviours that have that property. An interesting question arises as to how the sets of behaviours corresponding to different classes of behavioural properties may be characterized *semantically*, as opposed to a syntactic characterization in terms of the language of a given logic.

As observed by Smyth (1983), topology can provide elegant means for a semantic characterization of *properties* of data objects and *specifications* of programs that operate on these data. A topological space X should be viewed as a 'data type', with the open sets representing the *computable* properties defined on that type. A property P is computable if there exists a uniform procedure for determining in finite time, for any element x, that $P(x)$ holds whenever this is true. This identification of (computable) properties with open sets is a consequence of the existence of an interpretation of predicates as continuous maps over the space X, under which every open set may be viewed as a representation of some predicate. Given a topological space X, define the predicate $p : X \to B$ as a map into the Boolean cpo $B = \{\bot \sqsubseteq tt, \bot \sqsubseteq f\!f\}$ (with the Scott topology) by taking, for any subset S of X, $p(x) = tt$ if $x \in S$, and $\bot$ otherwise. This map is continuous iff S is open.

A specification is a finite or countably infinite list of (computable) properties that the program is to satisfy. A program satisfies the specification iff it satisfies each property. The topological notion that corresponds to specifications is a G_δ-set, that is, a countable (finite or infinite) intersection of a family of open sets.

The aim of the chapter is to apply the above ideas to obtain a characterization of *behavioural properties* of concurrent systems. The classification of behavioural properties (also called 'temporal') is attributed to Lamport (1977). Two classes of behavioural properties are distinguished: *safety* and *liveness* properties. Safety properties are informally described as stating 'nothing bad will happen'; examples of safety properties are *partial correctness* and *mutual exclusion*. Liveness properties are complementary in the sense that they require that 'something good will happen', and include properties such as *termination*, *guaranteed response* and *starvation freedom* (fairness). This informal classification is motivated by different proof techniques applied in each case; while the verification of safety is based on the invariance argument, the verification of liveness requires the method of well-founded sets. In the more recent reformulation (Lamport 1989), safety was described as requiring that 'some action *must not* occur', while liveness is to guarantee that 'some action *must* occur'.

Although it seems intuitive to accept the distinction between safety and liveness on the grounds of the fundamental difference in the verification method, the informal definitions above have been the cause of some confusion and differences over what exactly distinguishes each class of properties (see, for example, (Pnueli 1986)). In addition, the actual behavioural structure chosen may have an effect on liveness (Mazurkiewicz et al. 1989) and fairness (Kwiatkowska 1989a).

The chapter presents a topological characterization of the classes of safety and liveness properties in a non-interleaving model for concurrency. Fairness properties are included as a subclass of liveness properties, and the relationship of fairness with the remaining classes of properties is investigated. The model used is a natural extension of (labelled) transition systems with the notion of independency of actions originally introduced in (Mazurkiewicz 1977). Two actions are independent if they can happen concurrently. The independency relation gives rise to trace congruence for strings which identifies strings up to commutations of consecutive independent actions; equivalence classes of strings are called *traces*. The quotient poset of finite traces can be extended with infinite traces to form a domain in the sense of Scott (Kwiatkowska 1989b, 1989c), thus providing a representation of behaviours of concurrent systems. The approach is a *non-interleaving* one, as it does not reduce concurrency to nondeterministic choice.

Safety, liveness, and fairness have attracted a lot of attention recently. A topological characterization of safety and liveness was formulated by Alpern and Schneider (1985); it relies on an interleaving representation of behaviours in terms of sets of infinite sequences of states. A syntactic and semantic characterization of properties in linear temporal logic can be found in (Manna and Pnueli 1989). The work presented here investigates behavioural properties in the context of partially ordered spaces.

The chapter is largely self-contained, and is organized as follows. Section 6.2 recalls standard definitions. Sections 6.3–6.5 give a comprehensive overview of the abstract model used, illustrated by suitable examples. Sections 6.6–6.11 present the characterization of behavioural properties.

6.2 Notation and basic notions

This section introduces the notation and recalls standard definitions, which may be found in (Kuratowski 1966, Gierz et al. 1980, Lawson 1988) for example.

Let A denote a (finite) set of symbols. A^* denotes the set of all finite *strings* (sequences) of symbols in A, A^ω denotes the set of all infinite strings, and A^∞ is the union of A^* and A^ω. We shall refer to the monoid $(A^*, ., \epsilon)$ of finite strings; ϵ is the empty string, and concatenation will be denoted by juxtaposition. The notation of regular expressions extended with ω-iteration will be used to describe infinitary string languages. For example, the expression $\{(a^*ba^*b)^*a^*\}$ denotes the set of all finite strings over the alphabet $\{a, b\}$ that contain an even number of occurrences of the symbol b, while $\{(a^*bb^*)^\omega\}$ is the set of all infinite strings over $\{a, b\}$ which contain an infinite number of occurrences of the symbol b.

A *topology* on a set S is a collection of subsets of S that contains $\emptyset$ and S, and is closed under finite intersection and arbitrary union. A set S together with a topology $\mathcal{T}$ on S is called a *topological space*; the elements of $\mathcal{T}$ are the *open sets* of the space. Equivalently, a topological space is a set S and a function called *closure* assigning to each set $X \subseteq S$ a set $\text{Cl}(X)$ satisfying the following four axioms:

1. $\text{Cl}(X \cup Y) = \text{Cl}(X) \cup \text{Cl}(Y)$
2. $X \subseteq \text{Cl}(X)$
3. $\text{Cl}(\emptyset) = \emptyset$
4. $\text{Cl}(\text{Cl}(X)) = \text{Cl}(X)$

A *base* of the topology is a subset $\mathcal{B} \subseteq \mathcal{T}$ such that every open set is the union of elements of $\mathcal{B}$. A *subbase* of the topology $\mathcal{T}$ is a family of open sets such that the family of all finite intersections of elements of the subbase is a base of $\mathcal{T}$. A set X is *closed* iff its complement $S \setminus X$ is open. A set is *dense* iff its closure is the set S. Equivalently, a set is dense iff its complement contains no non-empty open sets. A G_δ-*set* is a countable (finite or infinite) intersection of a family of open sets. For a subset T of S, we say $X \subseteq T$ is open (closed) *relative to* T iff X is an intersection of T and some open (closed) set. $X \subseteq T$ is dense *relative to* T iff $\mathrm{Cl}(X) \cap T = T$.

Let $(P, \sqsubseteq)$ be a partially ordered set (*poset*). If it exists, the *least upper bound* (also called *supremum* or *sup*) w.r.t. $\sqsubseteq$ of a subset $X \subseteq P$ will be denoted by $\bigsqcup X$; similarly, the *greatest lower bound* (the *infimum* or *inf*) is denoted by $\bigsqcap X$. For $x, y \in P$ it is customary to write $x \sqcup y$ for $\bigsqcup\{x, y\}$ and $x \sqcap y$ for $\bigsqcap\{x, y\}$.

Let $(P, \sqsubseteq)$ be a poset. Elements $x, y \in P$ are *comparable* iff either $x \sqsubseteq y$ or $y \sqsubseteq x$, and *incomparable* otherwise. $X \subseteq P$ is *totally ordered*, or a *chain*, iff every $x, y \in X$ are comparable. $X \subseteq P$ is a *directed set* iff it is non-empty and every pair $x, y \in X$ has a bound $z \in X$.

Let $(P, \sqsubseteq)$ be a poset. For $X \subseteq P$ and $x \in P$ write

$$\downarrow X = \{y \in P \mid y \sqsubseteq x \text{ for some } x \in X\}$$
$$\uparrow X = \{y \in P \mid x \sqsubseteq y \text{ for some } x \in X\}$$
$$\downarrow x = \downarrow\{x\}$$
$$\uparrow x = \uparrow\{x\}$$

$X \subseteq P$ is a *lower set* (also *prefix-* or *downward-closed*) iff $X = \downarrow X$. $X \subseteq P$ is an *upper set* (or *upward-closed*) iff $X = \uparrow X$. $X \subseteq P$ is an *ideal* iff it is a directed lower set.

Let $(P, \sqsubseteq)$ be a poset. P is a *lattice* iff every pair of elements $x, y \in P$ has a least upper bound $x \sqcup y$ and a greatest lower bound $x \sqcap y$. P is a *complete lattice* iff every $X \subseteq P$ has a least upper bound and a greatest lower bound. A lattice P is *distributive* iff for all elements $x, y, z \in P$ we have $x \sqcap (y \sqcup z) = (x \sqcap y) \sqcup (x \sqcap z)$. A lattice is *completely distributive* if it is a complete lattice and distributivity holds in its strongest form.

$(P, \sqsubseteq)$ is a *complete partial order* (cpo) iff P has a least element and every directed subset $X \subseteq P$ has a least upper bound $\bigsqcup X$ in P. Let $(P, \sqsubseteq)$ be a poset, and let $\mathrm{Id}(P)$ denote the set of ideals of P ordered by inclusion. Then $(\mathrm{Id}(P), \subseteq)$ is a cpo, often called the *ideal completion* (or the *order completion*) of P. For example, the set A^* of all finite strings over the alphabet A forms a poset together with the usual string prefix ordering $\leqslant$. The ideal completion of $(A^*, \leqslant)$ can be identified with the cpo $(A^\infty, \leqslant)$ of all finite and infinite strings; the infinite strings correspond to infinite ideals in A^*. For example, the string $(abb)^\omega$ can be identified with the ideal $\{\epsilon, a, ab, abb, abba, \ldots\}$.

Let $(D, \sqsubseteq)$ be a cpo and $x, y \in D$. We say x is *essentially below* (or *way below*) y, denoted $x \sqsubseteq_{\mathrm{fin}} y$, iff given a directed set $M \subseteq D$ such that

$y \sqsubseteq \bigsqcup M$ there exists $z \in M$ such that $x \sqsubseteq z$. Intuitively, x is essentially below y if x is some finite approximation of y. For $X \subseteq D$ and $x \in D$ write

$$\downarrow_{\text{fin}} X = \{y \in D \mid y \sqsubseteq_{\text{fin}} x \text{ for some } x \in X\}$$
$$\uparrow_{\text{fin}} X = \{y \in D \mid x \sqsubseteq_{\text{fin}} y \text{ for some } x \in X\}$$
$$\downarrow_{\text{fin}} x = \downarrow_{\text{fin}} \{x\}$$
$$\uparrow_{\text{fin}} x = \uparrow_{\text{fin}} \{x\}$$

Let $(D, \sqsubseteq)$ be a cpo. Say $x \in D$ is a *finite element* (also called *compact*) iff $x \sqsubseteq_{\text{fin}} x$. The set of all finite elements of D is denoted B_D. D is *algebraic* iff, for every $x \in D$, the set $M = \{y \in B_D \mid y \sqsubseteq x\} = \downarrow_{\text{fin}} x$ is directed and $\bigsqcup M = x$. D is *consistently complete* iff every $X \subseteq D$ with an upper bound has a least upper bound. D is a *(Scott) domain* iff D is algebraic, consistently complete, and B_D is countable. The set of all finite elements B_D of an algebraic cpo D is often called the *finitary basis* in the case of B_D countable. The cpo $(A^\infty, \leqslant)$ of finite and infinite strings forms a Scott domain; the set A^* is the finitary basis. Thus, the domain-theoretic notion of a finite approximation naturally corresponds to the intuitive notion of a finite approximation of a string.

Let $(D, \sqsubseteq)$ be a Scott domain. The *Scott* topology is the topology consisting of all sets U such that U is upward-closed and, for every directed set $M \subseteq D$, if $\bigsqcup M \in U$ then some element of M is in U. The basis of the Scott topology is the family of the sets $\uparrow x$, where $x \in B_D$. A set L is *Scott-closed* iff it is a lower set closed under suprema of directed subsets. The *Lawson* topology of a domain is defined as the topology generated by a sub-basis containing all Scott-open sets, together with the sets $D \setminus \uparrow x$, for $x \in B_D$.

For example, in the domain $(A^\infty, \leqslant)$ of strings, the family of sets $\uparrow x$, where x is a finite string, forms the basis of the Scott topology. The family of Scott-open sets in this domain, together with the family of complements of the basic Scott-open sets, forms a subbasis of the Lawson topology.

6.3 The domain of traces

Let A be a finite alphabet of action symbols, and let $\iota \subseteq A \times A$ be a symmetric and irreflexive relation (called the *independency*). Intuitively, two actions are *independent* if they can happen concurrently without affecting the result, and *dependent* otherwise. For example, two actions local to distinct distributed processes are independent, while two actions appearing on either side of a nondeterministic choice operator are dependent. Thus, $\iota = \emptyset$ corresponds to behaviours that are sequential, but possibly nondeterministic, whereas $\iota \neq \emptyset$ permits non-sequential behaviours. Define *trace equivalence* $\equiv$ (Mazurkiewicz 1977) as the least congruence in the

Fig. 6.1 The poset of all prefixes of [abbca] with a ι b

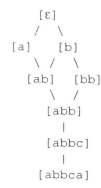

monoid $(A^*, ., \epsilon)$ of finite strings such that:

$$a \; \iota \; b \;\Rightarrow\; ab \equiv ba$$

The trace equivalence should be viewed as equality of strings up to commutations of consecutive independent action symbols. For example, if $\iota = \{(a,b),(b,a)\}$ then $abc \equiv bac$, but $abc \not\equiv acb$. An equivalence class σ of strings is called a *trace*. Each trace consists of all the possible sequentializations of some non-sequential behaviour.

The quotient algebra $(A^*, ., \epsilon)_\equiv$ of traces is a monoid (Mazurkiewicz 1989). Denote $A^*_\equiv$ by Θ^*. Define *trace prefix ordering* $\sqsubseteq$ in the monoid of finite traces by

$$\sigma \sqsubseteq \tau \iff \exists \gamma \in \Theta^* \cdot \sigma\gamma = \tau$$

for all $\sigma, \tau \in \Theta^*$ ($\sigma\gamma$ denotes the concatenation of traces σ and γ). $(\Theta^*, \sqsubseteq)$ is a poset (Mazurkiewicz 1989, Kwiatkowska 1989b). Each finite trace may be viewed as an *event*, and the trace prefix ordering as a (reflexive) *causality* ordering on events, in the sense that $\sigma \sqsubseteq \tau$ means τ may not occur unless σ occurs first. Figure 6.1 shows an example poset of trace prefixes (we use square brackets to denote equivalence classes). Incomparable prefixes of a trace (for example, $[a]$ and $[b]$) correspond to concurrent events.

Let $\text{Id}(\Theta^*)$ denote the set of ideals of the poset $(\Theta^*, \sqsubseteq)$. Then $\text{Id}(\Theta^*)$ ordered by inclusion can be identified with all finite and infinite traces (Kwiatkowska 1989c); in particular, infinite traces correspond to infinite ideals. It can be shown that $(\text{Id}(\Theta^*), \subseteq)$ is a domain in the sense of Scott, Θ^* constitutes the (countable) finitary basis, and the inherited order agrees with the new order on Θ^* (Kwiatkowska 1989c). Denote the ideal completion of $(\Theta^*, \sqsubseteq)$ by $(\Theta^\infty, \sqsubseteq)$ the set of infinite traces by Θ^ω.

The ideal completion of the poset of finite traces allows for infinitary behaviours, which are particularly useful in the study of fairness. Figure 6.2

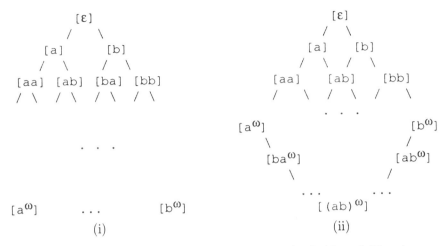

Fig. 6.2 Two domains of traces over the alphabet $\{a,b\}$: (i) $\iota = \emptyset$ (ii) $a \iota b$

depicts two orderings of traces over the set of action symbols $\{a,b\}$: observe that (i) is a tree ordering, whereas (ii) is a complete lattice. Note that if $a \iota b$ then $[ba^i] = \{ba^i, aba^{i-1}, \ldots, a^i b\}$ and $[ba^\omega] = \{ba^\omega, aba^\omega, \ldots, a^i ba^\omega, \ldots\}$, that is, a commutes with b.

A subset $T \subseteq \Theta^\infty$ is called an *infinitary trace language*.

It can be shown (Kwiatkowska 1989b) that if $\iota = \emptyset$ then $(\Theta^\infty, \sqsubseteq)$ is isomorphic to the domain $(A^\infty, \leqslant)$ of strings with the usual string prefix ordering of (Boasson and Nivat 1980). The concatenation of finite traces extends to a (non-associative) concatenation of possibly infinite traces (Kwiatkowska 1989c).

6.4 Asynchronous transition systems

Recall that a (*rooted*) *labelled transition system* is a tuple $(Q, A, \rightarrow, q_0)$ where:

1. Q is a non-empty countable set of (global) *states* (or *configurations*);
2. A is a non-empty finite set of *action labels*;
3. $\rightarrow \subseteq Q \times A \times Q$ is a *transition relation*;
4. $q_0 \in Q$ is the *initial state*.

$q \rightarrow^a q'$ means that a transition labelled with the symbol a is possible in the state q and the resulting state is q'. A system is said to be *unambiguous* iff, for every $a \in A$, $\rightarrow^a$ is a (partial) function on states.

Definition 1. A (rooted) asynchronous transition system *(ATS)* is a quintuple $\Sigma = (Q, A, \rightarrow, \iota, q_0)$ such that $(Q, A, \rightarrow, q_0)$ is an unambiguous rooted labelled transition system, $\iota \subseteq A \times A$ is an independency relation, and

$$\forall q, q', q'' \in Q \cdot \forall a, b \in A \cdot$$
$$(q \rightarrow^a q' \rightarrow^b q'' \wedge a \iota b) \Rightarrow \exists q''' \cdot q \rightarrow^b q''' \rightarrow^a q'' \qquad (1)$$

An asynchronous transition system is called forward stable (Bednarczyk 1987) iff, in addition, the following holds:

$$\forall q, q', q'' \in Q \cdot \forall a, b \in A \cdot$$
$$(q \rightarrow^a q' \wedge q \rightarrow^b q'' \wedge a \iota b) \Rightarrow \exists q''' \cdot (q' \rightarrow^b q''' \wedge q'' \rightarrow^a q''') \qquad (2)$$

We shall only consider forward stable systems. Conditions (1) and (2) are independent and reduce concurrency to *commutativity*. Condition (1) states that any two consecutive transitions labelled with independent action symbols may be permuted. Condition (2), a variant of the Church–Rosser property, states that if two transitions labelled with *independent* actions are simultaneously enabled, then there must exist a state reachable through the two actions invoked in either order (uniqueness of this state follows from the assumption of unambiguity).

Asynchronous transition systems (Shields 1985, Bednarczyk 1987, Kwiatkowska 1989b) were originally introduced to serve as abstract representations of concurrent systems, in the same manner as labelled transition systems were intended to serve as an abstraction of nondeterminism. We use a notation based on CCS (Milner 1989) to demonstrate examples of asynchronous systems. Let A denote the set of action symbols. We assume a, b, c range over A, B, C range over $\mathbb{P} A$ and X, Y are process variables. The syntax of the CCS expressions is

$$p ::= X \mid \text{NIL} \mid ap \mid p + p \mid \text{fix } X \cdot p \mid p \,{}_B\|_C\, p$$

We shall only consider unambiguous (that is, deterministic) expressions. The usual CCS synchronization is replaced by synchronization on shared events in the style of TCSP (Brookes et al. 1984). The (modified) transition rules are:

1. $ap \rightarrow^a p$
2. $p \rightarrow^a p'$ implies $(p + r) \rightarrow^a p'$, $(r + p) \rightarrow^a p'$
3. $p \rightarrow^a p'$ and $a \in B \setminus C$ implies $(p \,{}_B\|_C\, q) \rightarrow^a (p' \,{}_B\|_C\, q)$
4. $q \rightarrow^a q'$ and $a \in C \setminus B$ implies $(p \,{}_B\|_C\, q) \rightarrow^a (p \,{}_B\|_C\, q')$
5. $p \rightarrow^a p'$, $q \rightarrow^a q'$ and $a \in B \cap C$ implies $(p \,{}_B\|_C\, q) \rightarrow^a (p' \,{}_B\|_C\, q')$

6. $p[\text{fix } X \cdot p/X] \to^a p'$ implies $\text{fix } X \cdot p \to^a p'$ where $[./.]$ denotes substitution.

Note that interprocess communication is enforced by means of sharing an action symbol, rather than the restriction operator. A process expression in this notation can be viewed as an asynchronous transition system (strictly speaking, this is true for unambiguous expressions that are closed in the sense of not containing free variables). The corresponding ATS is as follows: the *states* are identified with (closed) expressions defining processes, the *initial state* is the original expression, and the transition rules constitute the *transition relation*. Roughly speaking, two (distinct) actions are independent if they appear on the opposite sides of the ${}_B\|_C$ operator.

Formally, let p be a closed deterministic expression. Define the set $Act(p)$ of *actions* of the process p inductively by:

1. $Act(\text{NIL}) = \emptyset$
2. $Act(ap) = \{a\} \cup Act(p)$
3. $Act(p+q) = Act(p) \cup Act(q)$
4. $Act(p\ {}_B\|_C\ q) = (Act(p) \cap B) \cup (Act(q) \cap C)$
5. $Act(\text{fix } X \cdot p) = Act(p)$

For the purpose of this chapter $p \| q$ will be used to denote $p\ {}_{Act(p)}\|_{Act(q)}\ q$. We shall also distinguish the set $Inits(p)$ of *initial* actions of the process p defined by

$$Inits(p) = \{a \mid \exists p' \cdot p \to^a p'\}$$

Define the *dependency* relation $\delta(p) \subseteq Act(p) \times Act(p)$ inductively by:

1. $\delta(\text{NIL}) = \emptyset$
2. $\delta(ap) = SymR(\{a\}, Inits(p)) \cup \delta(p)$
3. $\delta(p+q) = SymR(Inits(p), Inits(q)) \cup \delta(p) \cup \delta(q)$
4. $\delta(p\ {}_B\|_C\ q) = (\delta(p) \cap B \times B) \cup (\delta(q) \cap C \times C)$
5. $\delta(\text{fix } X \cdot p) = \delta(p)$

where

$$SymR(A, B) = (A \times B) \cup (B \times A) \cup (A \times A) \cup (B \times B)$$

Fig. 6.3 The ATS arising from the process
$p = \text{fix } X . (aX + b\text{NIL})$

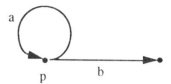

Fig. 6.4 The ATS arising from the process
$p \parallel q$, where $p = \text{fix } X . aX$, and $q = b\text{NIL}$

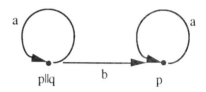

It is easy to see that $\delta(p)$ is symmetric and reflexive. Now we can define the *independency* $\iota(p)$ by

$$\iota(p) = Act(p) \times Act(p) \setminus \delta(p)$$

Finally, the rooted unambiguous ATS $\Sigma(p)$ determined by the closed deterministic expression p is

$$\Sigma(p) = (Q(p), Act(p), \rightarrow, \iota(p), p)$$

where $Q(p)$ is the set of process expressions that may be obtained from p by any finite sequence of transitions, and $\rightarrow$ is the usual transition relation restricted to $Act(p)$.

Figure 6.3 shows the ATS determined by the process $p = \text{fix } X . (aX + b\text{NIL})$. At every step, p can choose between (local) actions a and b; it terminates only if b chosen. Here, a and b are dependent. To simplify, we have identified $p \parallel \text{NIL}$ with p (likewise, $\text{NIL} \parallel p$ with p).

Figure 6.4 depicts the ATS determined by the process $p \parallel q$, where $p = \text{fix } X . aX$, and $q = b\text{NIL}$. In this example, a and b are independent.

Figure 6.5 includes the ATS determined by the process $p \parallel q \parallel r$, where $p = \text{fix } X . (aX + c\text{NIL})$, $q = \text{fix } X . (bX + c\text{NIL})$, $r = c\text{NIL}$. At every step, p and q may choose between performing a local action (a or b respectively) and a communication c; the system terminates only if c is chosen. Here, a and b are independent, but a and c (respectivly b and c) are dependent.

We have defined the independency relation for an arbitrary deterministic expression p. It is perhaps worth stressing that certain undesirable identifications may arise owing to the *dynamic*, or *context-dependent*, character of concurrency in CCS-like process algebras, as opposed to the *static*

Fig. 6.5 The ATS arising from the process $p \parallel q \parallel r$, where $p = \text{fix } X . (aX + c\text{NIL})$, $q = \text{fix } X . (bX + c\text{NIL})$, $r = c\text{NIL}$

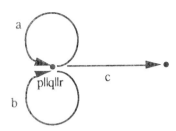

character of the independency relation in asynchronous transition systems. Consider, for example, the process $p = ab(a\text{NIL} \parallel b\text{NIL})$; here, a and b are concurrent *after* the sequence ab of transitions, but not before. Observe that a and b are dependent in p because, by definition of $\delta(p)$, if two actions are dependent in some subterm, but independent in another, the dependency overrides the independency. As a result, the rooted ATS determined by the process p above will have the independency relation identical to the independency of the process $q = ab(ab\text{NIL} + ba\text{NIL})$. Note that the feature of independency being overridden by the dependency is intuitively the right one when dealing with synchronization; for example, a and b are dependent in $(a\text{NIL} \parallel b\text{NIL}) \parallel ab\text{NIL}$, although they are independent in the left subterm. Here, the a on the left and the a on the right are viewed as giving rise to the *same* action instance; thus the subterm on the right simply introduces an external constraint on the subterm on the left.

Formally, the subset of closed deterministic CCS expressions whose dependency relation $\delta(p)$ is static can be characterized inductively as follows:

1. $\delta(\text{NIL})$ is static;

2. if $\delta(p)$ is static and

$$a \in Act(p) \Rightarrow SymR(\{a\} \times Inits(p)) \subseteq \delta(p)$$

   then $\delta(Ap)$ is static;

3. if $\delta(p), \delta(q)$ are static and the conditions

$$a, b \in Act(p) \cap Act(q) \Rightarrow a, b \text{ consistent in } p, q$$
$$SymR(Inits(p) \times Inits(q)) \subseteq \delta(p) \cup \delta(q)$$

   hold, then $\delta(p + q)$ is static;

4. if $\delta(p), \delta(q)$ are static then $\delta(p \,{}_B\|_C\, q)$ is static;

5. if $\delta(p)$ is static then $\delta(\text{fix } X \, . \, p)$ is static.

In the above, a, b *consistent in* p, q iff

$$(a, b) \in \delta(p) \iff (a, b) \in \delta(q)$$

An investigation of the precise relationship of nondeterministic CCS expressions and asynchronous transition systems is currently in progress and will be presented separately. A relabelling in the style of (Boudol and Castellani 1989) will be required. A context-dependent independency relation has been dealt with in (Katz and Peled 1990).

6.5 Non-interleaving semantics

Let $\Sigma = (Q, A, \to, \iota, q_0)$ be a rooted ATS. We say that $u \in A^\infty$ is a *derivation* of Σ iff $u = \epsilon$ or there exists a sequence of states (q_i), $i \in \{0, 1, \ldots, \text{len}(u)\}$ such that

$$q_0 \to^{u(1)} q_1 \to^{u(2)} \cdots \to^{u(i-1)} q_{i-1} \to^{u(i)} \cdots$$

where $u(i)$ denotes the ith symbol of the string u. Thus, an infinite string is a derivation iff all its finite (string) prefixes are derivations. It is easy to see that every derivation u of an (unambiguous) rooted ATS uniquely determines a sequence of states (q_i) such that, for all i, $q_{i-1} \to^{u(i)} q_i$.

The set of all finite derivations (often called *runs*) of Σ is denoted by $D^*(\Sigma)$ and infinite derivations by $D^\omega(\Sigma)$, and $D^\infty(\Sigma) = D^*(\Sigma) \cup D^\omega(\Sigma)$. We refer to $D^\infty(\Sigma)$ as the *interleaving* semantics. The set $T^\infty(\Sigma)$ of *traces* of Σ (the *non-interleaving semantics*) is defined as the set of equivalence classes determined by the derivations of Σ, namely $T^\infty(\Sigma) = [D^\infty(\Sigma)]$. $T^*(\Sigma)$ and $T^\omega(\Sigma)$ denote the sets of all finite and infinite traces respectively.

It is clear from the definition of trace semantics that each derivation of Σ is contained in some trace of Σ. In fact, a stronger result can be shown (Kwiatkowska 1989c), namely that *each string contained in a trace of Σ is a derivation of Σ, and all equivalent finite derivations lead to the same state.* Observe that we can write $q \to^\sigma q'$ whenever $q \to^x q'$ for some $x \in D^*(\Sigma)$.

Asynchronous transition systems accept infinitary trace languages (Kwiatkowska 1989c), as shown below. An infinite trace σ is a trace of a rooted ATS Σ iff finite $\sqsubseteq$-prefixes of σ are also traces of Σ.

Proposition 2. *Let $\Sigma = (Q, A, \to, \iota, q_0)$ be a rooted ATS.*

1. *$T^\infty(\Sigma)$ is a Scott-closed infinitary trace language.*

2. *If $T \subseteq \Theta^\infty$ is Scott-closed, then there exists a rooted ATS Σ such that $T^\infty(\Sigma) = T$.*

Properties and systems 165

Corollary 3. $T^\infty(\Sigma)$ is Lawson-closed.

Proof. Since $T^\infty(\Sigma)$ is Scott-closed, it is a lower set closed under directed suprema. Thus, $T^\infty(\Sigma)$ is Lawson-closed because a lower set is Lawson-closed iff it is closed under suprema of directed sets (Gierz et al. 1980). □

If the independency is empty, then the domain of traces is isomorphic to the domain of strings. Thus, the set $D^\infty(\Sigma)$ of derivations of Σ is both Scott- and Lawson-closed in the domain $(A^\infty, \leqslant)$ of strings.

An advantage of using the non-interleaving, as opposed to the interleaving, approach is that infinite behaviours that are intuitively 'incomplete', or 'extendable', can be excluded; for example, the infinite trace $[a^\omega]$ of the system shown in Figure 6.4 is incomplete because it can be extended with the symbol b giving the trace $[ba^\omega]$ (note that $[ba^\omega]$ may be viewed as a $\sqsubseteq$-extension of $[a^\omega]$ because $[a^\omega] \sqsubseteq [ba^\omega]$ (see, for example, Figure 6.2(ii)). Only *maximal* traces are 'non-extendable' in this sense.

We introduce the notion of admissibility for traces. The set of traces maximal w.r.t. $\sqsubseteq$ in some infinitary trace language $T \subseteq \Theta^\infty$ is denoted Max(T). A trace $\tau \in T^\infty(\Sigma)$ is *maximal* in a rooted ATS Σ, denoted Max(Σ), if it is maximal in $T^\infty(\Sigma)$. A trace $\tau \in T^\infty(\Sigma)$ is *admissible* in Σ iff $\tau \in T^*(\Sigma)$ or τ is maximal in $T^\infty(\Sigma)$. Maximality not only allows the exclusion of traces which are 'non-extendable' in the above sense, but also corresponds to a certain notion of fairness (Kwiatkowska 1989a). For example, the trace $[a^\omega]$ of the system shown in Figure 6.4 is *not* admissible because $[a^\omega]$ is not maximal (see, for example, Figure 6.2(ii)); neither is it *(weakly) process fair* w.r.t. q. The set of maximal traces of the system in Figure 6.5 is $\{[(ab)^\omega]\} \cup \{[a^*b^*c]\}$. Observe that maximality is not sufficient as a fairness notion as the (maximal) trace $[(ab)^\omega]$ is *not* (weakly) process fair w.r.t. r.

It should be stressed that the set of all admissible traces of Σ is not, in general, Scott-closed, as it is not closed under suprema of directed subsets (consider, for example, the set $\{[a^*]\} \cup \{[b]\} \cup \{[ba^\omega]\}$ of admissible traces of the system shown in Figure 6.4 and the directed set $\{[a^*]\}$). Consequently, the set of admissible traces of a rooted ATS is not closed w.r.t. the Lawson topology. (This lack of closedness also applies to fairness notions.)

6.6 Properties and systems

We now introduce a formal definition of a property, and what it means for a system to satisfy a given property.

Let $(Q^\infty, \sqsubseteq)$ denote the domain of traces over the alphabet A and the independency $\iota \subseteq A \times A$. A *property* is defined as a subset $\Psi \subseteq \Theta^\infty$. A property $\Psi \subseteq \Theta^\infty$ is a *finitary* property iff

$$\sigma \in \Psi \cap \Theta^\omega \iff (\exists \tau \in \Psi \cdot \tau \sqsubseteq_{\text{fin}} \sigma)$$

and *finitary* otherwise. Thus, a finitary property may contain an infinite trace iff some finite prefix of that trace is in the property.

A trace $\sigma \in \Theta^\infty$ *has the property* Ψ iff σ is contained in Ψ.

Let $\Sigma = (Q, A, \rightarrow, \iota, q_0)$ be an unambiguous rooted ATS. Σ is said to *satisfy* the property Ψ iff $\text{Max}(\Sigma) \subseteq \Psi$. The maximal traces of Σ correspond to 'complete' behaviour.

Examples of properties in the domain shown in Figure 6.2(ii) are (note that a commutes with b):

$$F(b) = \uparrow\{[b]\}$$
$$WPF = \{[ba^\omega]\}$$

$F(b)$ is a finitary property, whereas WPF is an infinitary property. The system shown in Figure 6.4 satisfies $F(b)$ (that is, b eventually occurs). It also satisfies WPF (that is, it is weakly process fair w.r.t. both p and q).

6.7 The 'good' and 'bad' things

According to the informal definition of safety and liveness (Lamport 1977), a system satisfies a safety property if it prevents some 'bad' thing from occurring; on the other hand, it satisfies a liveness property if it ensures that some 'good' thing occurs. What is the nature of these 'things'? Are 'good' things fundamentally different from 'bad' things?

As a state transition system progresses, the 'things' that occur during execution correspond to state transitions, which are caused by actions (events) performed by the system. A trace of an asynchronous transition system corresponds to its computation, finite $\sqsubseteq$-prefixes of this computation may be identified with events, and the relation $\sqsubseteq$ constitutes the (reflexive) causality relation. We are looking for a characterization of the sets of computations (traces) which correspond to some 'thing' occurring (or not occurring) during system execution.

The definitions in this section have been influenced by the work of Plotkin (1990) and Smyth (1983). Let σ denote a trace of some rooted ATS. It would seem reasonable to assume that if some 'thing' occurred in a trace σ, then it must have been observed at finite time; in other words, there must exist a finite $\sqsubseteq$-prefix of σ in which the 'thing' has been observed. Moreover, if some 'thing' has occurred in σ, then it must also have occurred in any $\sqsubseteq$-*extension* of σ. For these reasons, 'things' are identified with $\uparrow_{\text{fin}}$-closed subsets $U \subseteq \Theta^*$; let us call such subsets U *observations*. Formally, an observation $U \subseteq \Theta^*$ is said to be true of a trace σ, denoted $\sigma \vDash U$, iff there exists $\beta \in U$ such that $\beta \sqsubseteq_{\text{fin}} \sigma$. It then follows that

$$(\sigma \sqsubseteq \tau \wedge \sigma \vDash U) \Rightarrow \tau \vDash U$$

Note that observations are closed under (arbitrary) union and intersection, but are not closed under complement. Examples of observations in the domain shown in Figure 6.2(i) are

$$U(a,b) = \uparrow_{\text{fin}}\{[a^i b] \mid i \in \mathbb{N}\} \quad a \text{ until } b$$
$$F(a) = \uparrow_{\text{fin}}\{[xa] \mid x \in A^*\} \quad \text{eventually } a$$

The set $G(a) = [a^*] \cup [a^\omega]$ (always a) is not an observation, as it is not $\uparrow_{\text{fin}}$-closed. It denotes the *absence* (complement in Θ^∞) of the observation $\uparrow_{\text{fin}}\{[a^i b] \mid i \in \mathbb{N} \wedge b \neq a\}$.

It is generally recognized that (finite or countably infinite) *collections* of observations should be allowed. Given a family of observations $R = \{U_i \mid i \in \mathbb{N}\}$, where each $U_i \subseteq \Theta^*$ is $\uparrow_{\text{fin}}$-closed, we say R is true of a trace σ, denoted $\sigma \models R$, iff for all $i \in \mathbb{N}$, $\sigma \models U_i$. Note that if the family R is finite then there exists an observation U (intersection of observations in R) such that $\sigma \models R$ iff $\sigma \models U$.

Examples of countably infinite families of observations are

$$GF(a) = \{\uparrow_{\text{fin}}\{[x] \in X_i\} \mid i \in \mathbb{N}\} \quad \text{infinitely often } a$$

where $X_i \subseteq A^*$ denotes the set of all finite strings in which the symbol a occurs i times, and

$$GF(\{a,b\}) = \{\uparrow_{\text{fin}}\{[y] \in Y_i\} \mid i \in \mathbb{N}\} \quad \text{infinitely often } \{a,b\}$$

where $Y_i \subseteq A^*$ denotes the set of all finite strings in which each symbol in the set $\{a,b\}$ occurs i times.

It should be stressed that observations are irrevocable, in the sense that if an observation is true of some trace, it must also be true of all its extensions. As a consequence, there are subsets of Θ^* which cannot be represented as families of observations; consider for example, the set $\text{Even}(a) = [(aa)^* b^*]$ in the domain shown in Figure 6.2(ii) (a occurred an even number of times). Finer properties may be introduced on the basis of the Lawson topology, and will be discussed separately.

6.8 Uniform behavioural properties

Assuming 'things' are identified with observations, which subsets of Θ^∞ correspond to safety and liveness properties? The 'good' things, for example, entry to a critical section, should be guaranteed to happen, or *must* happen, during system execution. Given an observation (the 'good' thing) $U \subseteq \Theta^*$, the set $L_U = \{\tau \in \Theta^\infty \mid \tau \models U\}$ corresponds to the *presence* of U, in the sense that it contains all behaviours that exhibit U. It is easy to see that L_U is Scott-open. The 'bad' things, for example, two processes visiting a critical section at the same time, should be prevented during system execution; in other words, the 'bad' things *must not* happen. Given an observation (the 'bad' thing) $U \subseteq \Theta^*$, the set $S_U = \{\tau \in \Theta^\infty \mid \tau \not\vDash U\}$ corresponds to the *absence* of U, in the sense that it contains all behaviours that do not exhibit U. Clearly, S_U is the complement of a Scott-open set, and hence it is Scott-closed. (Note that the absence of U does *not* necessarily correspond to the logical negation of U.)

While the 'bad' things are discrete, it is generally accepted that the 'good' things should be a finite or countably infinite collection of observations to allow for properties such as *starvation freedom* (the 'good' thing is a process making progress infinitely often) to be liveness properties.

Thus, define $S \subseteq \Theta^\infty$ to be a *safety property* iff S is Scott-closed. Define $L \subseteq \Theta^\infty$ to be a *liveness property* iff L is a Scott G_δ-set. Note that the class of liveness properties includes *finitary* liveness properties which correspond precisely to the Scott-open sets. The following motivates our definitions.

Proposition 4.

1. If S is a safety property, then there exists observation U_S such that

$$\forall \tau \in \Theta^\infty \cdot \tau \in S \iff \tau \not\vDash U_S$$

2. If L is a finitary liveness property, then there exists observation U_L such that

$$\forall \tau \in \Theta^\infty \cdot \tau \in L \iff t \vDash U_L$$

3. If L is a liveness property, then there exists a countable family of observations R_L such that

$$\forall \tau \in \Theta^\infty \cdot \tau \in L \iff t \vDash R_L$$

Proof.

1. $U_S = (\Theta^\infty \setminus S) \cap \Theta^*$.
2. $U_L = L \cap \Theta^*$.
3. Observe that $L = \bigcap \{P_i \mid i \in \mathbb{N} \wedge P_i \text{ Scott-open}\}$ and define $R_L = \{U_i \mid i \in \mathbb{N}\}$ where $U_i - P_i \cap \Theta^*$.

□

Thus, a safety property contains precisely all those traces in which some 'bad' thing does not occur, whereas a liveness property contains precisely all those traces in which some family of 'good' things does occur. Safety is sometimes described as *once lost, never regained*, and liveness as *once gained, never lost*.

The following properties of the classes of safety and liveness can be shown.

Proposition 5.

1. Every safety property S is the complement of some finitary liveness property L, and vice versa.

2. The class of all finitary liveness properties is a completely distributive lattice (w.r.t. inclusion) closed under finite intersections and arbitrary unions.

3. The class of all safety properties is a completely distributive lattice (w.r.t. inclusion) closed under arbitrary intersections and finite unions.

4. The class of all liveness properties is closed under countable intersections and finite unions.

Proof.

1. Immediate from definitions.

2. See, for example, (Gierz et al. 1980).

3. As 2.

4. See, for example, (Kuratowski 1966).

□

The above classes of safety and liveness properties are *uniform*, in the sense that they identify *all* the possible behaviours that correspond to the absence of some observation (respectively, the presence of some family of observations). In practice, we would need to establish whether a particular action *will* or *will not* occur during execution of a given system. We now characterize those properties which are safety and liveness properties *relative* to a given ATS.

6.9 Safety properties in Σ

Let $\Sigma = (Q, A, \rightarrow, \iota, q_0)$ be a rooted ATS. Define a safety property S to be a *safety property in* Σ iff $S \cap T^\infty(\Sigma)$ is Scott-dense relative to $T^\infty(\Sigma)$.

The following can be shown.

Proposition 6. *Let S denote a safety property, $\Sigma = (Q, A, \rightarrow, \iota, q_0)$ a rooted ATS. Then Σ satisfies S iff S is a safety property in Σ.*

Proof.

$\Rightarrow$ Σ satisfies S, that is, $\mathrm{Max}(\Sigma) \subseteq S$ (by definition); so $T^\infty(\Sigma) \subseteq S$ (because $X \subseteq Y \Rightarrow \mathrm{Cl}(X) \subseteq \mathrm{Cl}(Y)$, S is Scott-closed, and $T^\infty(\Sigma)$ is the Scott-closure of $\mathrm{Max}(\Sigma)$). Therefore, $S \cap T^\infty(\Sigma) = T^\infty(\Sigma)$ and so $S \cap T^\infty(\Sigma)$ is Scott-dense relative to $T^\infty(\Sigma)$, and hence by definition S is a safety property in Σ.

$\Leftarrow$ $S \cap T^\infty(\Sigma)$ is Scott-dense relative to $T^\infty(\Sigma)$, so (by definition) $\mathrm{Cl}(S) \cap T^\infty(\Sigma) = T^\infty(\Sigma)$; hence $S \cap T^\infty(\Sigma) = T^\infty(\Sigma)$ (because S is Scott-closed). Thus, $\mathrm{Max}(\Sigma) \subseteq S$ (because $\mathrm{Max}(\Sigma) \subseteq T^\infty(\Sigma)$) and so by definition Σ satisfies S.

□

Thus, the safety property does indeed state that some action *must not* occur during execution of the system Σ. Observe that all safety properties are finitary. An example of a safety property in the domain shown in Figure 6.2(i) is

$$G(a) = \{[a^*]\} \cup \{[a^\omega]\} \text{ always } a$$

where the 'bad' thing is $\uparrow_{\mathrm{fin}}\{[a^i b] \mid i \in \mathbb{N}\}$ (that is, action b occurring at some stage of computation). It is easy to see that $G(a)$ is not a safety property in the system p shown in Figure 6.3 because there exist maximal traces of p, namely $\{[a^*b]\}$, which are not contained in $G(a)$.

The following proposition shows that the verification of safety properties may be based on examining finite behaviours only.

Proposition 7. *Let S denote a safety property, $\Sigma = (Q, A, \rightarrow, \iota, q_0)$ a rooted ATS. Then Σ satisfies S iff $T^*(\Sigma) \subseteq S$.*

Proof.

$\Rightarrow$ Σ satisfies S, that is, $\mathrm{Max}(\Sigma) \subseteq S$ (by definition); so $\downarrow \mathrm{Max}(\Sigma) \subseteq \downarrow S$ (because $X \subseteq Y \Rightarrow \downarrow X \subseteq \downarrow Y$). Since $\downarrow \mathrm{Max}(\Sigma) = T^\infty(\Sigma)$ and $T^*(\Sigma) \subseteq T^\infty(\Sigma)$, it follows that $T^*(\Sigma) \subseteq \downarrow S = S$ (because S is Scott-closed).

$\Leftarrow$ Suppose $T^*(\Sigma) \subseteq S$. Observe that $T^\infty(\Sigma)$ is the least Scott-closed set containing $T^*(\Sigma)$, and S is by definition Scott-closed. Thus, $T^*(\Sigma) \subseteq S$ implies $T^\infty(\Sigma) \subseteq S$ (because $X \subseteq Y \Rightarrow \mathrm{Cl}(X) \subseteq \mathrm{Cl}(Y)$). Finally, since $\mathrm{Max}(\Sigma) \subseteq T^\infty(\Sigma)$, it follows that $\mathrm{Max}(\Sigma) \subseteq S$.

□

Thus, if it is established that all finite trace prefixes of some *infinite* trace are 'safe', then it can be deduced that the infinite trace is 'safe' as well. This is the essence of the invariance argument, which is based on verifying that some property holds initially, and that it is preserved by every transition of the system.

Proposition 8. *The class of all safety properties in Σ is closed under arbitrary intersection and finite union.*

Proof. Direct from definition and Proposition 5.3. □

6.10 Liveness properties in Σ

We now investigate the class of liveness properties that a given system satisfies.

Let $\Sigma = (Q, A, \rightarrow, \iota, q_0)$ be a rooted ATS. Define a liveness property L to be a *liveness property in Σ* iff $L \cap T^\infty(\Sigma)$ is Scott-dense relative to $T^\infty(\Sigma)$.

Observe that for a liveness property L to be a liveness property in a given system Σ, L must not only be a Scott G_δ-set, but also a dense set. This requirement ensures that L does indeed express that some 'good' thing *must* happen. Without density, L would only state that the 'good' thing *may* happen, which would mean that some (maximal) traces of Σ do not have the property L. Thus, L would be an *avoidable* property.

The reason we have not adopted density as the sole characteristic feature of liveness (see, for example, (Alpern and Schneider 1985)) is that this would result in liveness not being closed under intersection. Scott-dense sets are not closed under intersection; consider, for example, the sets $FG(a) = \{(a^*b^*)^*a^\omega\}$ and $FG(b) = \{(a^*b^*)^*b^\omega\}$, whose intersection is empty.

The following may be shown.

Proposition 9.

1. Let L denote a liveness property, $\Sigma = (Q, A, \rightarrow, \iota, q_0)$ a rooted ATS. If Σ satisfies L then L is a liveness property in Σ.

2. Let L denote a finitary liveness property, $\Sigma = (Q, A, \rightarrow, \iota, q_0)$ a rooted ATS. Then Σ satisfies L iff L is a liveness property in Σ.

Proof.

1. Since L is by definition a Scott G_δ-set, we have $L = \bigcap \{P_i \mid i \in \mathbb{N}\}$ where, for each i, P_i is Scott-open. So, $L \cap T^\infty(\Sigma) = \bigcap \{P_i \mid i \in \mathbb{N}\} \cap T^\infty(\Sigma) = \bigcap \{P_i \cap T^\infty(\Sigma) \mid i \in \mathbb{N}\}$, that is, $L \cap T^\infty(\Sigma)$ is an intersection of Scott-open sets relative to $T^\infty(\Sigma)$, and hence it is a Scott G_δ-set relative to $T^\infty(\Sigma)$. It remains to show that $L \cap T^\infty(\Sigma)$ is dense. Suppose Σ satisfies L, that is, $\text{Max}(\Sigma) \subseteq L$; then $\downarrow \text{Max}(\Sigma) = T^\infty(\Sigma) \subseteq \downarrow L$ (because $X \subseteq Y \Rightarrow \downarrow X \subseteq \downarrow Y$). Hence, $L \cap T^\infty(\Sigma)$ must be dense relative to $T^\infty(\Sigma)$.

2. (By contraposition) Suppose L is a finitary liveness property in Σ such that $\text{Max}(\Sigma) \not\subseteq L$. Thus there exists $\sigma \in \text{Max}(\Sigma)$ such that $\sigma \notin L$. Then $\sigma \notin L \cap T^\infty(\Sigma)$ and so $\sigma \notin \text{Cl}(L \cap T^\infty(\Sigma)) \subseteq \text{Cl}(L) \cap \text{Cl}(T^\infty(\Sigma)) = \downarrow L \cap T^\infty(\Sigma)$. Thus, $L \cap T^\infty(\Sigma)$ is not dense relative to $T^\infty(\Sigma)$, and hence L is not a liveness property in Σ. □

Note that the converse of implication 1 does not hold in general. Consider the domain shown in Figure 6.2(i) and the set $GF(a) = \{[(b^*aa^*)^\omega]\}$. Then $GF(a)$ is a Scott-dense G_δ-set in Θ^∞, but $\text{Max}(\Theta^\infty) \not\subseteq \Theta^\infty$.

The following are examples of liveness properties in the domain shown in Figure 6.2(i):

$$F(b) = \uparrow\{[a^*b]\} \quad \text{eventually } b$$
$$GF(a) = \{[(b^*aa^*)^\omega]\} \quad \text{infinitely often } a$$

$F(b)$ is not a liveness property in the system shown in Figure 6.3. On the other hand, the equivalent of the property $F(b)$ in the domain shown in Figure 6.2(ii) is a liveness property in the system shown in Figure 6.4. $GF(a)$ is an infinitary liveness property, and it is satisfied by both above-mentioned systems.

Proposition 10. *The class of liveness properties in Σ is closed under countable intersection and finite union.*

Proof. The class of G_δ-sets is closed under countable intersection and finite union (Kuratowski 1966). The union of two dense sets is obviously dense. Suppose Ψ is the intersection of a countable family of liveness properties in S; then Ψ is the intersection of a countable family $\{P_i\}$ of Scott-open sets relative to Σ, each of which must contain Ψ. Since Ψ is dense, each P_i is dense. Finally, Ψ must be dense on the basis of Baire's theorem (Gierz et al. 1980). □

6.11 Fairness properties

Fairness properties are often considered a subclass of liveness properties. The collection of 'good' things for a (process) fairness property would be some process making progress infinitely often.

Let $\Sigma = (Q, A, \rightarrow, \iota, q_0)$ be a rooted ATS. Define $F \subseteq \text{Max}(\Sigma)$ to be a *fairness property in* Σ iff F is a liveness property in Σ. Observe that if Σ satisfies F then F is a liveness property in Σ (Proposition 6).

Note that by this definition $\text{Max}(\Sigma)$ is a fairness property, which is the weakest (in the sense of inclusion) fairness property for a given ATS Σ. The fact that maximality corresponds to a notion of fairness is one of the advantages of using a non-interleaving, as opposed to an interleaving, approach.

Fairness is often viewed as an external *constraint* placed on the infinite computations of the system, rather than a *property* the system must satisfy. We therefore introduce the notion of satisfaction under a fairness constraint. Σ is said to *satisfy a property* Ψ *under the fairness constraint* F iff $\text{Max}(\Sigma) \cap F \subseteq \Psi$.

The above definition of fairness allows a variety of fairness properties to be expressed including process fairness and fairness of choice. In this chapter, we shall only present a few examples; for more details on fairness refer to (Kwiatkowska 1990b).

The following is a fairness property in the ATS determined by the process $p \parallel q \parallel r$ shown in Figure 6.5 (note that a commutes with b):

$$WPF = \uparrow\{[a^*b^*c]\}$$

WPF is a weak process fairness property, which ensures that *no (sub)process that is enabled continuously from some point onwards is delayed indefinitely*. Observe that traces such as $[a^\omega]$, $[b^\omega]$ are non-maximal. The maximal trace $[(ab)^\omega]$ was excluded because it indefinitely delays process r.

Consider the ATS shown in Figure 6.5. The process $p \parallel q$ determines the same ATS as the process $p \parallel q \parallel r$. The following is a fairness property in the ATS determined by $p \parallel q$:

$$WPF = \uparrow\{[a^*b^*c]\} \cup \{[(ab)^\omega]\}$$

The trace $[(ab)^\omega]$ is weakly process fair w.r.t. both p and q because neither of them is indefinitely delayed.

Fairness properties are said to affect liveness properties, but not safety properties. This statement can be motivated in the following way.

Proposition 11. *Let* $\Sigma = (Q, A, \rightarrow, \iota, q_0)$ *be a rooted ATS, L be a liveness property in* Σ, *S be a safety property in* Σ, *and F be a fairness property in* Σ.

1. If Σ satisfies L then Σ satisfies L under the fairness constraint F.

2. Σ satisfies S iff Σ satisfies S under the fairness constraint F.

Proof.

1. Suppose Σ satisfies L, then $\text{Max}(\Sigma) \subseteq L$ (by definition) and $F \subseteq \text{Max}(\Sigma)$ (because F is a fairness property in Σ); hence $\text{Max}(\Sigma) \cap F \subseteq L$.

2. Suppose Σ satisfies S, then $\text{Max}(\Sigma) \subseteq S$ (by definition) and $F \subseteq \text{Max}(\Sigma)$ (because F is a fairness property in Σ); hence $\text{Max}(\Sigma) \cap F \subseteq S$. Suppose Σ satisfies S under the constraint F; then $\text{Max}(\Sigma) \cap F \subseteq S$, and $F \subseteq \text{Max}(\Sigma)$ is dense relative to $T^\infty(\Sigma)$. Thus, $\text{Max}(\Sigma) \cap F = F \subseteq S$ and so $T^\infty(\Sigma) \subseteq S$ (because $X \subseteq Y \Rightarrow \text{Cl}(X) \subseteq \text{Cl}(Y)$, S is closed) and Σ clearly satisfies S.

□

Note that the converse of implication 1 does not hold in general. Consider the property $Trm = \uparrow\{[a^*b^*c]\}$ for the system shown in Figure 6.5 and the fairness property $WPF = \uparrow\{[a^*b^*c]\}$. Then the system satisfies Trm under the fairness constraint WPF (in other words, it terminates), but it does not satisfy Trm.

6.12 Conclusion

A topological characterization of safety, liveness, and fairness properties in a non-interleaving model for concurrency has been presented. The approach concentrates on the order-theoretic properties of the domain of traces used as an abstract representation of behaviours of concurrent systems, and the associated Scott topology. Safety properties are shown to be precisely those sets of traces which correspond to the *absence* of some action, while liveness corresponds to the sets of traces which exhibit the *presence* of some (finite or infinitely countable) family of actions. Thus, safety states that some 'bad' thing *must not* occur, and liveness that some 'good' thing *must* occur. Fairness properties form a subclass of liveness; in general, fairness states that some countably infinite family of 'good' things *must* occur. Note that a safety property is the complement of a finitary liveness property in the set-theoretic sense, but set-theoretic complement does not necessarily coincide with logical negation.

It should be noted that we have been rather discriminating about the choice of what constitutes a 'good' thing by restricting the 'good' things to a countable (finite or infinite) collection of observations. Consider the domain of strings over the alphabet $\{a, b\}$, which is isomorphic to the domain of traces over the same alphabet if the independency is empty, and the property $FG(a) = \{(a^*b^*)^*a^\omega\}$ (*eventually always a*). $FG(a)$ is not a liveness property in our sense, as it is not a G_δ-set. As a matter of fact, $FG(a)$ is an F_σ-set, that is, it is a complement of a G_δ-set, and thus it denotes the *absence* of a countable collection of 'bad' things, rather than the 'good' thing itself. In contrast to our approach, Alpern and Schneider (1985) would view $FG(a)$ as a liveness property.

It should be stressed that there are properties which are neither safety nor liveness; again, we differ here from (Alpern and Schneider 1985), where every property is shown to be the intersection of a safety and a liveness property. Although adequate to deal with safety and liveness, this coarseness of properties may be considered too restrictive. It is not inherent to the model itself, but rather to the Scott topology. Since the domain of traces also has an associated metric (Kwiatkowska 1990a), finer properties may be introduced on the basis of the Lawson topology that arises from this metric. Since the Lawson topology is a refinement of the Scott topology, safety and liveness properties should not be affected.

The work presented here applies more generally to the models for concurrency based on partial order, for example, event structures. An attempt to relate the topological characterization to a temporal logic over a partial order would be an interesting development.

Acknowledgements

The author would like to thank Mike Smyth for many discussions on the subject, and Colin Stirling for suggesting the use of topology as a basis for the characterization. The help of Mike Shields, Mike Mislove, Fred Schneider, and Paul Warren in the preparation of this chapter is also gratefully acknowledged.

References

Alpern, B., and Schneider, F. B. (1985). Defining liveness, *Information Processing Letters 21*, pp. 181–185.

Bednarczyk, M. (1987). *Categories of Asynchronous Systems*, Ph.D. thesis, University of Sussex.

Boasson, L., and Nivat, M. (1980). Adherences of languages, *Journal of Computer and System Sciences 20*, pp. 285–309.

Boudol, G., and Castellani, I. (1989). Permutation of transitions: An event structure semantics for CCS and SCCS, in: J. W. de Bakker *et al.*, (Eds.), *Linear Time, Branching time and Partial Order in Logics and Models for Concurrency*, Lecture Notes in Computer Science, Vol. 354, Springer-Verlag.

Brookes, S. D., Hoare, C. A. R., and Roscoe, A. W. (1984). A theory of communicating sequential processes, *Journal of the ACM 31*, pp. 560–599.

Gierz, G., Hofmann, K. H., Keimel, K., Lawson, J. D., Mislove, M., and Scott, D. (1980). *A Compendium of Continuous Lattices*, Springer-Verlag.

Katz, S., and Peled, D. (1990). Defining conditional independence using collapses, in: M. Z. Kwiatkowska et al., (Eds.), *Semantics for Concurrency*, Leicester 1990, Springer-Verlag.

Kuratowski, K. (1966). *Topology*, Academic Press.

Kwiatkowska, M. Z. (1989a). Event fairness and non-interleaving concurrency, *Formal Aspects of Computing 1*, pp. 213–228. Also available as Technical Report No. 26, University of Leicester, Department of Computing Studies.

Kwiatkowska, M. Z. (1989b). *Fairness for Non-Interleaving Concurrency*, Ph.D. Thesis, University of Leicester. Also available as Technical Report No. 22, University of Leicester, Department of Computing Studies.

Kwiatkowska, M. Z. (1989c). On infinitary trace languages, submitted for publication. Also available as Technical Report No. 31, University of Leicester, Department of Computing Studies.

Kwiatkowska, M. Z. (1990a). A metric for traces, *Information Processing Letters*, Vol. 35, No. 3. Also available as Technical Report No. 28, University of Leicester, Department of Computing Studies.

Kwiatkowska, M. Z. (1990b). Defining fairness for non-interleaving concurrency, to appear in K. V. Nori (Ed.) *Proceedings of the 10th Conference on Foundations of Software Technology and Theoretical Computer Science*. Lecture Notes in Computer Science, Springer-Verlag.

Lamport, L. (1977). Proving the correctness of multiprocess programs, *IEEE Transactions on Software Engineering SE-3*, pp. 125–143.

Lamport, L. (1989). A simple approach to specifying concurrent systems, *Communications of the ACM, Vol. 32 No. 1*, pp. 32–45.

Lawson, J. D. (1988). The versatile continuous order, in: M. Main et al. (Eds.), *Mathematical Foundations of Programming Language Semantics, 3rd Workshop*, Lecture Notes in Computer Science, Vol. 298, Springer-Verlag.

Manna, Z., and Pnueli, A. (1989). The anchored version of the temporal framework, in: J. W. de Bakker et al., (Eds.), *Linear Time, Branching time and Partial Order in Logics and Models for Concurrency*, Lecture Notes in Computer Science, Vol. 354, Springer-Verlag.

Mazurkiewicz, A. (1977). *Concurrent Program Schemes and Their Interpretations*, DAIMI Report PB-78, Aarhus University.

Mazurkiewicz, A. (1989). Basic notions of trace theory, in: J. W. de Bakker et al., (Eds.), *Linear Time, Branching time and Partial Order in Log-*

ics and Models for Concurrency, Lecture Notes in Computer Science, Vol. 354, Springer-Verlag.

Mazurkiewicz, A., Ochmanski, E., and Penczek, W. (1989). Concurrent systems and inevitability, *Theoretical Computer Science 64*, pp. 281–304.

Milner, R. (1989). *Communication and Concurrency*, Prentice-Hall.

Plotkin, G. (1990). Private communication.

Pnueli, A. (1986). Applications of temporal logic to the specification and verification of reactive systems: a survey of current trends, in: J. W. de Bakker et al., (Eds.), *Current Trends in Concurrency*, Lecture Notes in Computer Science, Vol. 224, Springer-Verlag.

Shields, M. W. (1985). Deterministic asynchronous automata, in: *Formal Methods in Programming*, North-Holland.

Smyth, M. B. (1983). Powerdomains and predicate transformers: a topological view, in: J. Diaz, (ed.), *Automata, Languages and Programming, 10th Coll.*, Lecture Notes in Computer Science, Vol. 154, Springer-Verlag.

7
Order and strongly sober compactifications

J. D. LAWSON

Abstract

A useful class of spaces for 'compactifications' of T_0-spaces are those spaces which are locally compact and strongly sober; such spaces have an alternate characterization as compactly ordered spaces. Hence compactifications with respect to this class can be viewed as ordered compactifications or strongly sober compactifications. We explore the theory of such compactifications. By analogy with the Hausdorff case, notions appropriate to the T_0-context of complete regularity and uniformities play a central role, and results reminiscent of the Hausdorff setting are obtained. Applications to semicontinuous functions and an analogue of the one-point compactification, the Fell compactification, are also presented.

7.1 Introduction

The notion of compactness in Hausdorff spaces has been a fundamental property in general topology and various contexts where it has been applied. In the context of T_0-spaces the property of compactness is, in general, much weaker, and consequently of less use. For example, one can consider an infinite set with the cofinite topology or a space with a point whose only neighbourhood is the whole space. In recent years a class of spaces has emerged—the locally compact strongly sober spaces—which appears to be a suitable analogue in the T_0-category to the class of compact spaces in the Hausdorff category.

In this chapter we collect together some of the significant aspects of forming 'compactifications' with respect to this class of spaces. Among the class of dcpo's (partially ordered sets in which every directed set has a supremum), those which are continuous and are strongly sober with respect to the Scott topology have some very agreeable features. Indeed, these are precisely those for which the associated Lawson topology is compact Hausdorff (and this is what causes these spaces to behave much like compact spaces in the Hausdorff category). Such spaces arise, for example, in the work of Jung (1988) in characterizing large cartesian closed categories. They also lend themselves to powerdomain constructions (Lawson 1988). This richer topological structure is also proving of some interest in certain aspects of denotational semantics such as safety tests. Thus in certain circumstances where the property of being strongly sober may fail, one may want to look for nice embeddings into larger objects which are strongly sober.

We should mention here the recent work of Smyth (1990). He also considers compactifications in much the same spirit as appears here. His approach is via quasi-proximities, while the one here emphasizes the equivalent idea of totally bounded quasi-uniformities. Via this equivalence, there is some overlapping of results. The author is also indebted to H. Künzi for several pertinent observations about material in the chapter and especially for pointing out some of the pertinent literature related to topology and ordered spaces.

Since there is an equivalent characterization of strongly sober spaces in terms of compactly ordered spaces, we begin with a consideration of ordered spaces.

7.2 Compactly embeddable ordered spaces

In this section we consider sets X endowed with both a topology π and a partial order $\leq$. We customarily denote the triple $(X, \pi, \leq)$ simply by X and call it a *partially ordered topological space*. For $A \subseteq X$, we set $\uparrow A := \{y : x \leq y \text{ for some } x \in A\}$ and $\uparrow x := \uparrow\{x\}$. The sets $\downarrow A$ and $\downarrow x$ are defined analogously. A set A is an *upper* (*lower*) set if $A = \uparrow A$ ($A = \downarrow A$).

The order on X is *semiclosed* if $\downarrow x$ and $\uparrow x$ are closed for each $x \in X$; the order is *closed* if its graph $\mathrm{Gr}(\leq) := \{(x,y) : x \leq y\}$ is closed in $X \times X$. If X is compact T_2 and the order is closed, X is called a *compactly ordered space*. A basic example of a compactly ordered space is the unit interval equipped with its usual order and topology. There exists a fairly substantial literature on compactly ordered spaces dating back to the work of L. Nachbin in the late 1940s (Nachbin 1965, Gierz et al. 1980, Chapter VI.1) Among other things Nachbin showed that a compactly ordered space is *strongly order convex*, that is, that it has a basis of open sets each of which consists of the intersection of an open upper set and an open lower set. Note that each such set is *order convex* in the sense that if x, z are in the set and $x \leq y \leq z$, then y is in the set.

Definition 1. *Let X and Y be partially ordered topological spaces. A function $\phi: X \to Y$ is monotone if $x_1 \leq x_2$ implies $\phi(x_1) \leq \phi(x_2)$ and is an order embedding if ϕ is a homeomorphism from X into Y and $\phi(x_1) \leq \phi(x_2) \Leftrightarrow x_1 \leq x_2$ for all $x_1, x_2 \in X$. If for X there exists a compactly* ordered space Y and an order embedding from X into Y, we say that X is compactly embeddable.

Theorem 2. *A necessary condition for X to be compactly embeddable is that it be strongly order convex and that the order be closed.*

Proof. As mentioned previously, a compactly ordered space is strongly order convex, and by definition the order is closed. Since these properties are hereditary, the proposition follows. □

We want to consider necessary and sufficient conditions for an ordered space to be compactly embeddable. As originally worked out by Nachbin (1965), the theory turns out to have close parallels with finding compactifications of a space in the Hausdorff setting. We refer the reader to (Nachbin 1965), especially the Appendix, and Chapter 4 of (Fletcher and Lindgren 1982) for details of the material we sketch in the remainder of this section.

We first recall the notion of an order compactification.

Definition 3. *If $\phi\colon X \to Y$ is a continuous monotone function from a partially ordered topological space X into a compactly ordered space Y and $\phi(X)$ is dense in Y, then the pair (Y, ϕ) is called an* order compactification *of X. The order compactification (Y, ϕ)* dominates *(Z, θ) if there exists a continuous monotone mapping $g\colon Y \to Z$ such that $g \circ \phi = \theta$. The two compactifications are called* equivalent *if there exists an order embedding h from Y onto Z such that $h \circ \phi = \theta$. It is easy to show that two compactifications are equivalent if and only if they dominate each other.*

Definition 4. *We say that a partially ordered Hausdorff space X is a* completely regular ordered space *if given $a \in U$, an open set, there exist continuous functions $f, g\colon X \to [0, 1]$ such that f is monotone, g is antimonotone ($x \leqslant y$ implies $g(x) \geqslant g(y)$), $f(a) = 1 = g(a)$, and either $f(x) = 0$ or $g(x) = 0$ for $x \notin U$. An alternate equivalent definition is to require that (i) the set of inverse images of open upper sets and open lower sets on the line $\mathbb{R}$ (that is, open rays) under all continuous monotone real-valued mappings on X forms a subbasis for the topology on X and (ii) that there exist enough continuous monotone real-valued mappings ϕ so that if $x \not\leqslant y$, then $\phi(x) > \phi(y)$ for some ϕ.*

A partially ordered topological space is said to be *normally ordered* if, given a closed upper set A and a closed lower set B such that $A \cap B = \emptyset$, there exist an open upper set U and an open lower set V such that $A \subseteq U$, $B \subseteq V$, and $U \cap V = \emptyset$. By an analogue of Urysohn's lemma, one shows in a normally ordered space X that given disjoint closed sets A and B such that A is an upper set and B is a lower set, then there exists a continuous

Fig. 7.1

monotone function $f: X \to [0,1]$ such that $f(A) = 1$ and $f(B) = 0$. It follows that if X is normally ordered, strongly order convex, and the order is semiclosed, then X is a completely regular ordered space. Compactly ordered spaces turn out to be both normally ordered and strongly order convex, and hence are completely regular ordered spaces.

Consider all continuous monotone functions f from a partially ordered topological space X into $[0,1]$, and let $\prod f$ from X to $\prod [0,1]_f$ be their product. Then $\prod f$ is a continuous monotone function into the compactly ordered space consisting of a product of intervals (with the product topology and coordinate-wise ordering). The corestriction j of $\prod f$ into the closure $\beta(X)$ of the image then gives a compactification $(\beta(X), j)$ analogous to the Stone–Čech compactification in topology. We call it the Nachbin–Stone–Čech compactification. By analogy with the Stone–Čech compactification, it is characterized (up to order homeomorphism) by the property that any continuous monotone function g from X into a compactly ordered space Y factors uniquely through $j : X \to \beta(X)$ (alternately, any continuous bounded monotone function from X to $\mathbb{R}$ factors through $j: X \to \beta(X)$) (Figure 7.1).

Thus among the order compactifications of X, $\beta(X)$ is maximal or universal in the sense that it dominates all the others.

The conditions that an ordered space X be a completely regular ordered space are precisely the conditions one needs to conclude that j is an order embedding. We thus obtain the following.

Theorem 5. *Let X be a partially ordered topological space. The following are equivalent*

1. *X is compactly embeddable.*

2. *$j : X \to \beta(X)$ is an order embedding, where $\beta(X)$ is the Nachbin–Stone–Čech order compactification.*

3. *X is a completely regular ordered space (understood to be Hausdorff).*

Proof.

1⇒3 The property of being a completely regular ordered space is hereditary and (as remarked earlier) holds for compactly ordered spaces.

3⇒2 This follows along the line of the comments preceding this proposition.

2⇒1 Immediate. □

7.3 Upper and lower semicontinuous functions

We recall certain basic notions about T_0-spaces (see, for example, (Gierz et al. 1980)). Let (X, τ) be a T_0-topological space. The *specialization order* is given by

$$x \leqslant_\tau y \stackrel{\text{def}}{\Leftrightarrow} x \in \overline{\{y\}}$$

We note that open sets are upper sets, closed sets are lower sets, and $\downarrow x = \overline{\{x\}}$ for each $x \in X$. The specialization order of a T_0-space is an important topological invariant, but does not determine the space. Indeed, given a partial order on a set there is a complete lattice of topologies that have that order for their specialization order. The finest of these is the *Alexandroff discrete* (or *A-discrete*) topology, which consists of all upper sets. The coarsest of these is the *weak* topology which has the sets $\downarrow x$, $x \in X$, as a subbasis for the closed sets.

Definition 6. *Two T_0-topologies τ and σ are* complementary *if $\leqslant_\tau = \geqslant_\sigma$, and (X, τ, σ) is a* complemented bitopological space. *In this case all τ-open sets are upper sets and all σ-open sets are lower sets for the order $\leqslant_\tau$.*

Again for a given topology τ, there exists a complete lattice of complementary topologies, namely all those topologies for which the specialization order is the order dual of $\leqslant_\tau$.

A complemented bitopological space gives rise to a partially ordered topological space via the next construction. The following assertions about the patch and split spaces are readily apparent.

Construction 7. *The* patch space *of a complemented bitopological space (X, τ, σ) is the partially ordered topological space $(X, \tau \vee \sigma, \leqslant_\tau)$. The topology $\tau \vee \sigma$, the join of τ and σ, is called the* patch topology.

The patch space satisfies:

1. *(semiclosedness)* $\forall x \in X$, $\uparrow x$ and $\downarrow x$ are closed;

2. *(strong order convexity)* the open upper sets and open lower sets form a subbasis for the topology.

Next we have a type of converse construction.

Construction 8. Let $(X, \pi, \leqslant)$ be a topological space with a semiclosed partial order. The *split space* of $(X, \pi, \leqslant)$ is the complemented bitopological space $(X, \pi^\sharp, \pi^\flat)$, where $\pi^\sharp$ is the collection of open upper sets and $\pi^\flat$ is the collection of lower open sets. The patch space of $(X, \pi^\sharp, \pi^\flat)$ is the original $(X, \pi, \leqslant)$ if and only if $(X, \pi, \leqslant)$ is strongly order convex.

We need a slightly modified version of the earlier ordered analogue of a completely regular space.

Definition 9. Let X be a partially ordered topological space. We say that X is a *strictly completely regular ordered space* if:

1. the order on X is semiclosed;

2. X is strongly order convex;

3. given a closed lower (upper) set A and a point $x \notin A$, there exists a continuous monotone function $f: X \to [0, 1]$ such that $f(A) = 0$ and $f(x) = 1$ ($f(A) = 1$ and $f(x) = 0$).

Remark 10. If X is a strictly completely regular ordered space, then it is a completely regular ordered space. If X is normally ordered and satisfies (1) and (2), then X is a strictly completely regular space.

Proof. Let O be a open set containing x. We find an open upper set U and an open lower set V such that $x \in U \cap V \subseteq O$. Let $f, g: X \to [0, 1]$ be continuous monotone mappings such that $f(x) = 1$, $f(X \setminus U) = 0$, $g(x) = 0$, and $g(X \setminus V) = 1$. The two functions f and $1 - g$ are then the functions required in the definition of a completely regular ordered space.

If X is normally ordered and satisfies (1), then there exists a continuous monotone function separating a closed lower set from the closed upper set $\uparrow x$ for a point x not in the closed lower set and dually. □

Problem 11. Find (if possible) an example of a completely regular ordered space which is not a strictly completely regular ordered space. Under what hypotheses do the two notions coincide? A special case of the problem is the case that P be a continuous dcpo endowed with the Lawson topology.

After the circulation of a preprint of this chapter H. Künzi found a counterexample to the preceding problem. His example is non-metrizable and one might still consider the problem in the metric case. Künzi also pointed out that a completely regular ordered space X is strictly completely regular if and only if the associated split space (Construction 8) is pairwise completely regular as a bitopological space. We recall that a bitopological space (X, τ, σ) is called *pairwise completely regular* (Lane 1967) if, for each τ-closed set A not containing $x \in X$, there is a bicontinuous function $f: (X, \tau, \sigma) \to ([0,1], \pi^{\#}, \pi^{b})$, where π is the usual topology on the ordered space $[0,1]$, such that $f(x) = 1$ and $f(A) = 0$ and a dual condition holds for σ-closed sets.

Theorem 12. *Let X be a strongly order convex space. The following are equivalent.*

1. *X is a strictly completely regular ordered space.*

2. *The mapping $j: (X, \pi^{\#}, \pi^{b}) \to (\beta(X), \mu^{\#}, \mu^{b})$ is a bitopological embedding, where $j: (X, \pi, \leqslant) \to (\beta(X), \mu, \leqslant)$ is the Nachbin–Stone–Čech compactification.*

3. *The mapping $j: (X, \pi^{\#}, \pi^{b}) \to (Y, \tau^{\#}, \tau^{b})$ is a bitopological embedding for some order compactification $j: (X, \pi, \leqslant) \to (Y, \tau, \leqslant)$.*

Proof.

1⇒2 By Remark 10, X is a completely regular ordered space and so $j: X \to \beta(X)$ is an order embedding by Theorem 5. If V is a $\mu^{\#}$ open set, then it is an upper set and open in the μ topology. Since j is continuous, its inverse image is open, and since j is monotone, its inverse image is an upper set. Hence the inverse image is open for the $\pi^{\#}$ topology. Thus $j: (X, \pi^{\#}) \to (\beta(X), \mu^{\#})$ is continuous.

Now let U be an open upper set containing x. Since X is a strictly completely regular ordered space, there exists a continuous monotone mapping $f: X \to [0,1]$ such that $f(x) = 1$ and $f(X \setminus U) = 0$. Extend f to a monotone mapping $F: \beta(X) \to [0,1]$ such that $F \circ j = f$. Then $F^{-1}((0,1])$ is an open set containing $j(x)$, and the intersection of this open set with $j(X)$ maps into U under j^{-1}. Hence $j: (X, \pi^{\#}) \to (\beta(X), \mu^{\#})$ is an embedding. The case for the topology π^{b} is similar.

2⇒3 Immediate.

3⇒1 By hypothesis j is injective. The hypothesis (3) also yields directly that j is an embedding for the corresponding patch topologies on X and Y. Since X is strongly order convex by hypothesis and since

Y is compactly ordered and hence strongly order convex, the patch topologies agree with the original topologies. By Theorem 2 the order on X is closed and hence semiclosed.

Let $x \in U$, an open upper set in X. Since $j:(X,\pi^\sharp) \to (Y,\tau^\sharp)$ is an embedding, there exists an open upper set W in Y such that $W \cap j(X) = j(U)$. Since Y is compactly ordered, there exists a monotone continuous function $F: Y \to [0,1]$ such that $F(j(x)) = 1$ and $F(Y \backslash W) = 0$. Then $f = F \circ j$ satisfies $f(x) = 1$ and $f(X \backslash U) = 0$. This and the dual argument show that X is a strictly completely regular ordered space.

□

Let τ denote the usual topology on the real line. Then $\tau^\sharp$ consists of all open right rays and $\tau^\flat$ of all open left rays. Note that a function from a topological space X into $\mathbb{R}$ is lower semicontinuous if and only if it is continuous as a function into $(\mathbb{R}, \tau^\sharp)$. If X is a partially ordered topological space and if f is a monotone continuous function, then the inverse image of any open upper set in $\mathbb{R}$ will be an open upper set in X. Thus f is continuous as a map from $(X, \pi^\sharp)$ to $(\mathbb{R}, \tau^\sharp)$.

Theorem 13. *Let X be strongly convex ordered space for which the order is semiclosed. The following are equivalent.*

1. *X is a strictly completely regular ordered space.*

2. *Every lower (upper) semicontinous monotone function from X to $\mathbb{R}^* = [-\infty, \infty]$ is the supremum (infimum) of a set of continuous monotone functions.*

Proof.

$1 \Rightarrow 2$ Let $f: X \to \mathbb{R}^*$ be a lower semicontinuous monotone function. Let $x \in X$ and $\epsilon > 0$. Then $U = f^{-1}((f(x) - \epsilon, \infty))$ is an open upper set containing x. The function $(f(x) - \epsilon) \cdot \chi_U$ which takes the value $f(x) - \epsilon$ on U and $-\infty$ on $X \setminus U$ is lower semicontinuous, and f is the supremum of such simple functions as we let x and $\epsilon > 0$ range. Thus it suffices to prove the assertion for them.

Now consider a function of the form $t \cdot \chi_U$ which takes on the value t on some open upper set U and $-\infty$ on its complement. Since X is a strictly completely regular ordered space, we can choose a continuous monotone function which takes the maximum value t at a given $x \in U$ and the value $-\infty$ on $X \setminus U$. Then $t \cdot \chi_U$ is the supremum of all such continuous functions. The case for upper semicontinuous functions is completely analogous.

2⇒1 Let A be a closed descending set and let y be a member of its complement U. Then the function χ_U which takes the value 1 on U and the value 0 on its complement is lower semicontinuous and monotone. Hence it is the supremum of continuous monotone mappings. Pick one such f such that $f(y) > 1/2$. Then $g(x) := 2(f(x) \vee 0) \wedge 1$ is the desired function into $[0,1]$. The symmetric condition follows analogously.

□

Beer (1986) has considered the circle of ideas of the preceding proposition for the case that X is a locally (order) convex topological lattice (indeed, it was his work that suggested to the author the more general context considered here). In the lattice case Beer showed that condition (2) was equivalent to X admitting an order embedding into a compact topological lattice. This latter condition is easily seen to be equivalent to X being a completely regular ordered space. Thus in this case the notion of being a strictly completely regular ordered space is equivalent to being a completely regular ordered space. Künzi, in a recent memorandum to the author, has generalized Beer's result.

Notions about semicontinuous functions generalize nicely to continuous lattice contexts (see, for example, (Keimel and Gierz 1982)). By means of standard continuous lattice theory one can significantly generalize aspects of Theorem 13. We only sketch the development since it is somewhat technical and outside our primary focus of interest in this chapter. Consider the function space $[X \to L]$ of all Scott-continuous monotone functions from X into a continuous lattice L (where Scott-continuous means continuous into L with the Scott topology). These are just the continuous functions from $(X, \pi^\sharp)$ to (L, Scott). If $(X, \pi^\sharp)$ is locally compact, then $[X \to L]$ forms a continuous lattice (Gierz et al. 1980, Chapter II.4, particularly Exercise 4.20). One can then ask for conditions that guarantee that any such function is the supremum of monotone functions into L which are continuous from (X, π) to L equipped with the Lawson topology. Suppose that L is connected in the Lawson topology; then for $y \in L$, there exists an order dense complete chain (a *chain* is a totally ordered set) from the least element $\perp$ to y (Gierz et al. 1980, Chapter VI.5). We assume further that all chains in L are separable (that is, have countable subsets that are dense in the order topology of the chain); this will be true, for example, if the Scott topology has a countable base. Then there is a monotone homeomorphism from $[0,1]$ to any complete order dense chain with its order topology, which is just its relative Lawson topology. We can again write any Scott-continuous function $\alpha \in [X \to L]$ as a supremum of characteristic-like functions of the form $y \cdot \chi_U$, which takes the value $y \in L$ on an open upper set U in X and the value $\perp$ on the complement of U. Let h be a monotone homeomorphism from $[0,1]$ to a complete order dense chain from $\bot$ to y. If X is a strictly completely regular ordered space, then we can write $h^{-1} \circ y \cdot \chi_U$ as a supremum of continuous monotone functions into $[0, 1]$ (as in the proof of the preceding proposition), and composing with h we obtain $y \cdot \chi_U$ as a supremum of Lawson continuous monotone functions into L. We thus have the following theorem.

Theorem 14. *Let X be a strictly completely regular ordered space, and let L be a continuous lattice which is connected in the Lawson topology and for which every chain is separable. Then any Scott-continuous monotone function from X to L is the supremum of Lawson-continuous monotone functions.*

Under the appropriate circumstances one can 'interpolate' continuous functions between lower semicontinuous functions and functions appropriately beneath it. The next result may be viewed as a generalization of results in Nachbin (1965, Appendix).

Corollary 15. *Let X be a compactly ordered space and let L be a continuous lattice such that with respect to the Lawson topology L is a connected topological lattice (for example, $L = [0, 1]$ or L is an order dense completely distributive lattice) and chains are separable. If $f: X \to L$ is a Scott continuous monotone function and if $g: X \to L$ is an arbitrary function with the property that, for each $x \in X$, there exists an open set U containing x and a $w \ll f(x)$ in L such that $g(U) \subseteq \downarrow w$, then there exists a continuous monotone function $\beta: X \to L$ such that $g(x) \ll \beta(x) \leq f(x)$ for all $x \in X$.*

Proof. In this setting the supremum of two Lawson continuous functions will again be continuous, and so by Theorem 14 the function f is a directed supremum of continuous monotone functions. Let $x \in X$. By hypothesis there exists an open set U, $x \in U$, and a $w \ll f(x)$ such that $g(U) \subseteq \downarrow w$. Pick a continuous monotone function $\beta = \beta_x$ such that $w \ll \beta(x)$ (this is possible since $f(x)$ is the directed supremum of all such $\beta(x)$). By continuity, there exists an open set V containing x such that $\beta(V) \subseteq \Uparrow w = \{z: w \ll z\}$, since the latter is a Scott-open set. Let $W_x = U \cap V$. Finitely many of the W_x cover X, and the finite supremum of the corresponding β_x gives the desired monotone continuous function. □

Problem 16. *To what extent do the preceding results extend to more general function spaces between continuous posets? For example, Theorem 14 easily extends to continuous semilattices with $\bot$ such that $\downarrow x$ is a complete lattice for every x.*

7.4 Quasi-uniform spaces

In recent years an asymmetric version of uniform spaces has been rather extensively investigated. Since these spaces have close connections with our considerations, we now turn to them. We refer to (Fletcher and Lindgren 1982) as a basic reference for the material of this section.

A filter $\mathcal{U}$ on a non-empty set X is a *quasi-uniformity* if (i) each member of $\mathcal{U}$ contains the diagonal and (ii) given $U \in \mathcal{U}$, there exists $V \in \mathcal{U}$ such that $V \circ V \subseteq U$. The pair $(X,\mathcal{U})$ is then a *quasi-uniform space*. For $U \in \mathcal{U}$, $x \in X$, we set $U(x) := \{y \in X : (x,y) \in U\}$. The topology $\mathcal{T}(\mathcal{U})$ induced by the quasi-uniformity $\mathcal{U}$ is given by

$$\mathcal{T}(\mathcal{U}) = \{G \subseteq X : x \in G \Rightarrow \exists U \in \mathcal{U} . U(x) \subseteq G\}$$

If $\mathcal{U}$ is a quasi-uniformity on X, then so is $\mathcal{U}^{-1} := \{U^{-1} : U \in \mathcal{U}\}$. The sets $\{U \cap U^{-1} : U \in \mathcal{U}\}$ form a basis for a uniformity $\mathcal{U}^*$, the coarsest uniformity finer than both $\mathcal{U}$ and $\mathcal{U}^*$. A filter $\mathcal{F}$ is a $\mathcal{U}^*$-Cauchy filter if and only if for each $U \in \mathcal{U}$, there exists $F \in \mathcal{F}$ with $F \times F \subseteq U$. The quasi-uniformity $\mathcal{U}$ is said to be *bicomplete* if $\mathcal{U}^*$ is complete as a uniform space.

We are particularly interested in the case that the topology $\mathcal{T}(\mathcal{U})$ is T_0.

Theorem 17. *Let $(X,\mathcal{U})$ be a quasi-uniform space. The following are equivalent.*

1. *The space $(X, \mathcal{T}(\mathcal{U}))$ is T_0.*
2. *Given $x \neq y$, there exists $U \in \mathcal{U}$ such that $(x,y) \notin U \cap U^{-1}$.*
3. *The relation $\leqslant = \bigcap\{U : U \in \mathcal{U}\}$ is a partial order.*
4. *The uniformity $\mathcal{U}^*$ is separated.*

When these conditions are satisfied, then the partial order $\leqslant$ of part (3) is the order of specialization for $\mathcal{T}(\mathcal{U})$ and the topology $\mathcal{T}(\mathcal{U}^{-1})$ is a complementary topology for $\mathcal{T}(\mathcal{U})$. The patch topology for these two topologies is the topology $\mathcal{T}(\mathcal{U}^*)$. The partially ordered topological space $(X, \mathcal{T}(\mathcal{U}^*), \leqslant)$ is strongly order convex and the order is closed.

Proof. These results can essentially be found in Chapters 1.1 and 4.3 of (Fletcher and Lindgren 1982). The assertions about the patch topology and strong order convexity follow readily from that fact that a basis can be chosen for the uniformity $\mathcal{U}$ such that $U(x)$ is open in $\mathcal{T}(\mathcal{U})$ for each $x \in X$ and for each $U \in \mathcal{U}$ which is in the basis. □

By Theorem 17 a quasi-order yields not only a T_0-space, but also a complementary topology and a partially ordered topological space. We say that two complementary topologies are *uniformly complementary* if there exists a T_0 quasi-order such that one of them is given by $\mathcal{T}(\mathcal{U})$ and the other is given by $\mathcal{T}(\mathcal{U}^{-1})$. We say that the partially ordered topological space arising in Theorem 17 is *determined* by the quasi-uniformity $\mathcal{U}$. Fletcher and Lindgren (1982) show that such spaces are compactly embeddable. We sketch the construction.

Let $(X,\mathcal{U})$ be a quasi-uniform space satisfying the conditions of Theorem 17. Then the *bicompletion* of X as a set is the uniform completion of the separated uniform space $(X,\mathcal{U}^*)$; the points of the bicompletion may be viewed either as equivalence classes of Cauchy filters or minimal Cauchy filters. The bicompletion $\tilde{X}$ admits a unique quasi-uniformity $\tilde{\mathcal{U}}$ such that (i) $(\tilde{X},\tilde{\mathcal{U}})$ is a T_0 quasi-uniform space, (ii) the restriction of $\tilde{\mathcal{U}}$ to the embedded image of X agrees with $\mathcal{U}$, and (iii) $\tilde{\mathcal{U}}$ is bicomplete. These properties characterize the bicompletion up to quasi-uniform homeomorphisms preserving the embedded image of X. Note that the natural inclusion of X into $\tilde{X}$ is a uniform homeomorphism (and hence a topological embedding) for the various quasi-uniformities $\mathcal{U}$, $\mathcal{U}^{-1}$, and $\mathcal{U}^*$.

A quasi-uniformity $\mathcal{U}$ on X is *totally bounded* if for each $U \in \mathcal{U}$, there exists a finite cover $\mathcal{A}$ of X such that $\bigcup\{A \times A : A \in \mathcal{A}\} \subseteq U$. Note that if $\mathcal{U}$ is totally bounded, then so are $\mathcal{U}^{-1}$ and $\mathcal{U}^*$.

Given any T_0 quasi-uniform space $(X,\mathcal{U})$, there exists an *associated* totally bounded quasi-uniform space $(X,\mathcal{U}_\omega)$ having as a subbase all sets of the form

$$\{X \times X \setminus A \times B : A \times B \cap U = \emptyset \text{ for some } U \in \mathcal{U}\}$$

Then $\mathcal{T}(\mathcal{U}) = \mathcal{T}(\mathcal{U}_\omega)$, and $\mathcal{U}_\omega$ is the finest totally bounded quasi-uniformity coarser than $\mathcal{U}$. Additionally, $\mathcal{T}(\mathcal{U}^{-1}) = \mathcal{T}(\mathcal{U}_\omega^{-1})$ and $\mathcal{T}(\mathcal{U}^*) = \mathcal{T}(\mathcal{U}_\omega^*)$, so that both $\mathcal{U}$ and $\mathcal{U}_\omega$ determine the same topological ordered space. If one now forms the bicompletion $(\tilde{X},\tilde{\mathcal{U}}_\omega)$ of $(X,\mathcal{U}_\omega)$, then $\tilde{X}$ is totally bounded with respect to the uniformity $\tilde{\mathcal{U}}_\omega$, and hence is compact in $\mathcal{T}(\tilde{\mathcal{U}}_\omega^*)$. One thus obtains that the partially ordered topological space determined by $(\tilde{X},\tilde{\mathcal{U}}_\omega)$ is a compactly ordered space and the inclusion from the partially ordered topological space determined by $(X,\mathcal{U})$ into it is an order compactification. This order compactification is called the one *constructed from* $\mathcal{U}$. We summarize as follows.

Theorem 18. *Let $(X,\mathcal{U})$ be a T_0 quasi-uniform space. Then the partially ordered topological space determined by $\mathcal{U}$ order embeds in the order compactification constructed from $\mathcal{U}$. This compactification arises as the compactly ordered space determined by the bicompletion $(\tilde{X},\tilde{\mathcal{U}}_\omega)$. In* particular, the partially ordered topological space determined by $\mathcal{U}$ is compactly embeddable. Furthermore, the embedding

$$(X, \mathcal{T}(\mathcal{U}), \mathcal{T}(\mathcal{U}^{-1})) \hookrightarrow (\tilde{X}, \mathcal{T}(\tilde{\mathcal{U}}_\omega), \mathcal{T}(\tilde{\mathcal{U}}_\omega^{-1}))$$

is a topological embedding of complemented bitopological spaces.

Proof. See Chapters 3.2 and 4.3 of (Fletcher and Lindgren 1982) for all but the last assertion. It follows from the characterization of $\tilde{\mathcal{U}}$ given after Theorem 17 that $(X, \mathcal{T}(\mathcal{U}_\omega)) \hookrightarrow (\tilde{X}, \mathcal{T}(\tilde{\mathcal{U}}_\omega))$ is an embedding. Since $\mathcal{T}(\mathcal{U}) = \mathcal{T}(\mathcal{U}_\omega)$, we also obtain an embedding in this case. It again follows from the characterization of $\tilde{\mathcal{U}}$ given after Theorem 17 that $\widetilde{\mathcal{U}^{-1}} = \tilde{\mathcal{U}}^{-1}$. Hence the part that we just finished can be applied to the last coordinate also to obtain a bitopological embedding. □

There is a reverse construction to the preceding. Suppose that X is a partially ordered topological space and $j\colon X \to \tilde{X}$ is an order embedding into a compactly ordered space $\tilde{X}$. Now there exists precisely one quasi-uniformity that determines a compactly ordered space, namely all neighbourhoods of the graph $\mathrm{Gr}(\leqslant) = \{(x,y)\colon x \leqslant y\}$. If this quasi-uniformity is restricted to X (embedded in $\tilde{X}$), then the restriction is shown to determine the original partially ordered topological space X. The order compactification constructed from this restriction is equivalent, as an order compactification, to the original one $j\colon X \to \tilde{X}$. Conversely, if the original partially ordered topological space was determined by a totally bounded quasi-uniformity and the compactification $\tilde{X}$ is the order compactification constructed from that quasi-uniformity, then the quasi-uniformity obtained by the reverse construction just outlined gives rise to the original quasi-uniformity.

Theorem 19. *Let X be a partially ordered topological space which is compactly embeddable. Then there exists a one-to-one correspondence between the totally bounded quasi-uniformities on X which determine it as partially ordered topological space and the order compactifications of X (up to equivalence) which are order embeddings. A totally bounded quasi-uniformity $\mathcal{U}$ which determines X gives rise to the order compactification determined by the bicompletion, and an order compactification which is an order embedding gives rise to the restriction to X of the unique quasi-uniformity on the order compactification which determines it.*

Corollary 20. *A partially ordered topological space is compactly embeddable if and only if there is a (totally bounded) quasi-uniformity on it which determines it as a partially ordered topological space.*

The preceding results appear essentially in (Fletcher and Lindgren 1982, Section 4.3), except that there the results are stated in terms of quasi-proximities instead of totally bounded quasi-uniformities. Since totally bounded quasi-uniformities and quasi-proximities uniquely determine each other in a natural way (Fletcher and Lindgren 1982, Section 1.2), we have stated the results alternately in terms of totally bounded quasi-uniformities. See the recent work of Smyth (1990) for constructing compactifications directly from quasi-proximities.

Corollary 21. *Let $(X,\mathcal{U})$ be a T_0 quasi-uniform space. If $\mathcal{T}(\mathcal{U}^*)^\sharp = \mathcal{T}(\mathcal{U})$ and $\mathcal{T}(\mathcal{U}^*)^\flat = \mathcal{T}(\mathcal{U}^{-1})$, then the partially ordered space $(X, \mathcal{T}(\mathcal{U}^*), \leqslant)$ determined by $\mathcal{U}$ is a strictly completely regular ordered space.*

Proof. This follows immediately from Theorem 12 and Theorem 18. □

Problem 22. *Does the converse of Corollary 21 hold?*

Quasi-uniform spaces lend themselves very nicely to the consideration of semicontinuous functions taken up in the preceding section. We give one such example.

Theorem 23. *Let $(X,\mathcal{U})$ be a T_0 totally bounded quasi-uniform space such that $\mathcal{U}$ has a countable base. Let L be a continuous lattice which is a connected topological lattice with respect to the Lawson topology and has a countable base with respect to the Scott topology. Then any Scott-continuous function $f\colon (X, \mathcal{T}(\mathcal{U})) \to (L, \text{Scott})$ is the pointwise supremum of an increasing sequence of monotone continuous functions $f_n\colon (X, \mathcal{T}(\mathcal{U}^*), \leqslant) \to (L, \text{Lawson})$.*

Proof. Since $\mathcal{U}$ has a countable base, so does $\mathcal{U}^*$, and hence $\tilde{\mathcal{U}}^*$. Thus $(\tilde{X}, \mathcal{T}(\tilde{\mathcal{U}}^*))$ is metrizable and we have seen previously that it is compact. It then follows from (Gierz et al. 1980, Chapter III.4) that the space $[\tilde{X} \to L]$ of all continuous functions from $(\tilde{X}, \tilde{\mathcal{U}}^*)$ to (L, Scott) has a countable base for the Scott topology.

By Theorem 18 the inclusion $(X, \mathcal{T}(\mathcal{U})) \hookrightarrow (\tilde{X}, \mathcal{T}(\tilde{\mathcal{U}}))$ is an embedding. Since continuous lattices endowed with the Scott topology are injective (Gierz et al. 1980, Chapter II.3), the function $f\colon X \to L$ extends to a continuous function $\tilde{f}\colon \tilde{X} \to L$, which is then continuous and monotone as a function from $(\tilde{X}, \mathcal{T}(\tilde{\mathcal{U}}^*), \leqslant)$ to (L, Scott). Since the continuous lattice $[\tilde{X} \to L]$ has a countable base for the Scott topology, there exists a sequence of functions $h_n \ll \tilde{f}$ with pointwise supremum $\tilde{f}$. By Theorem 14, $\tilde{f}$ is the supremum of monotone functions which are continuous into the Lawson topology, and, as in Corollary 15, we may assume that this collection is directed. We then inductively choose f_n in the collection such that $f_i \leqslant f_n$ for $i < n$ and $h_n \leqslant f_n$. The restriction of these functions f_n to the embedded image of X then gives the desired sequence. □

It is conceivable that one might want to construct models of the lambda calculus or models in denotational semantics consisting of a continuous lattice isomorphic to its space of self-maps, and with the further property that Scott-continuous functions could be approximated by ones that were Lawson-continuous. The preceding theorem shows the kind of models that one would need to build.

Problem 24. Let $I = [0,1]$, and let I^ω denote its countable product. Is $[I^\omega \to I^\omega]$ isomorphic to I^ω?

7.5 Strongly sober compactifications

In this section we wish to consider a type of 'compactification' for a T_0-space X, one that uses compactly ordered spaces for the compactification. (This is an idea that dates back to Fell (1962).) Before we give a precise definition of such compactifications, it will be convenient to give an alternate equivalent way of looking at compactly ordered spaces as a special class of T_0-spaces, namely as locally compact strongly sober spaces. This characterization originated in the exercises to (Gierz et al. 1980, Chapter VII.1); see also (Hofmann and Mislove 1981) for results along this line. Compactifications by means of locally compact strongly sober spaces were considered by Künzi and Brümmer (1987) (although they are not called 'compactifications' there). Smyth (1990) has also considered compactifications by means of such spaces.

A topological space is called *strongly sober* if every ultrafilter has a non-empty set of limit points consisting of the closure of a unique singleton set. Strongly sober spaces, as expected, are sober. (Recall that a space is *sober* if every irreducible closed set is the closure of a unique singleton set.) A space is said to be *locally compact* if it has a basis of (not necessarily open) compact neighbourhoods at each point. A space is said to be *core compact* if, given an open set U and $x \in U$, there exists an open set V containing x which is *compact in U* in the sense that any open cover of U contains a finite subcover of V. A space turns out to be core compact if and only if its lattice of open sets is a continuous lattice. For sober spaces, the notions of local compactness and core compactness coincide.

The *cocompact topology* τ^k on a T_0 space X has as a subbasis for the closed sets all compact upper sets in (X, τ). This topology is a complementary topology for τ.

Locally compact strongly sober spaces are intimately connected with compactly ordered spaces as the next proposition shows. Indeed they are alternate ways of looking at the same objects.

Theorem 25. Let (X,τ) be a T_0-space. The following are equivalent.

1. (X,τ) is locally compact and strongly sober.

2. (X,τ) is locally compact and sober, and the intersection of two compact upper sets is again compact.

3. The patch space of (X,τ,τ^k) is a compactly ordered space.

In these cases, $\tau = \pi^\sharp$ and $\tau^k = \pi^\flat$, where π is the patch topology $\tau \vee \tau^k$.

Conversely, given a compactly ordered space $(X,\pi,\leqslant)$, the space $(X,\pi^\sharp)$ is a locally compact strongly sober space. The two constructions of passing from a locally compact strongly sober space to the compactly ordered space which is the patch of the original space and the cocompact topology and of passing from a compactly ordered space to the locally compact strongly sober space whose open sets are the open upper sets are mutually inverse constructions.

Locally compact strongly sober spaces are also referred to as stably locally compact spaces, because their lattices of open sets are stably continuous lattices (the way-below relation is multiplicative).

Problem 26. Find appropriate conditions for a strongly sober space to be locally compact.

Remark 27. The equivalence of Theorem 25 actually extends to an isomorphism of categories. On the one hand we have the category in which objects are compactly ordered spaces and morphisms are continuous monotone mappings; on the other hand we have the category in which objects are the locally compact strongly sober spaces and morphisms are the continuous proper (or perfect) mappings (where 'proper' means that the inverse image of every compact upper or saturated set is compact). Then these two categories are isomorphic under the object correspondence of Theorem 25 and the morphism correspondence that sends a function to itself (Hoffman 1982, Theorem 6.4).

Definition 28. A strongly sober compactification of a T_0-space X is a topological embedding $j: X \to Y$, where Y is a strongly sober locally compact space and $j(X)$ is dense in the patch topology on Y arising from its topology and the cocompact topology, that is, in its topology making it a compactly ordered space. A strongly sober compactification (Y,j) dominates the compactification (Z,i) if there exists a continuous proper mapping $g: Y \to Z$ such that $g \circ j = i$. The two compactifications are called equivalent if there exists a homeomorphism h from Y onto Z such that $h \circ j = i$.

Let X be a T_0-space. A quasi-uniformity $\mathcal{U}$ on X is said to be *compatible* with the topology of X if the topology $\mathcal{T}(\mathcal{U})$ is the topology of X.

Theorem 29. *Every compatible totally bounded quasi-uniformity on a T_0-space X gives rise to a strongly sober compactification, namely the order compactification of Theorem 18 endowed with the topology of open upper sets. Conversely, up to equivalence of spaces, every strongly sober compactification arises in this way.*

Proof. Let $\mathcal{U}$ be a compatible totally bounded quasi-uniformity of X. That $(\tilde{X}, \mathcal{T}(\mathcal{U}))$ is strongly sober and locally compact is just Theorem 1 of (Künzi and Brümmer 1987). By Theorem 25 $\mathcal{T}(\mathcal{U})$ consists of all open upper sets of the patch topology.

Conversely, let $j: X \to Y$ be a strongly sober compactification. By Theorem 25, the patch space of (Y, τ, τ^k) is a compactly ordered space. It is then a standard result that there is a unique totally bounded quasi-uniformity $\mathcal{U}$ that determines it, namely the collection of all sets in $Y \times Y$ which are open in the product of the patch topologies and contain $\mathrm{Gr}(\leqslant)$ (Fletcher and Lindgren 1982, Theorem 4.21). Now we can give X the structure of a partially ordered topological space which is compactly embeddable by giving it the order and topology it inherits from the compactly ordered patch space Y. It then follows from Theorem 19 and the remarks preceding it that, up to equivalence, Y is the order compactification corresponding to the quasi-uniformity $\mathcal{U}$ restricted to X. □

Theorem 30. *Let $\mathcal{U}$ and $\mathcal{V}$ be compatible quasi-uniformities on the T_0-space X. Then the strongly sober compactification to which $\mathcal{U}$ gives rise dominates the one for $\mathcal{V}$ if $\mathcal{V} \subseteq \mathcal{U}$. The converse holds if the two quasi-uniformities are totally bounded.*

Proof. If $\mathcal{V} \subseteq \mathcal{U}$, then $\mathcal{V}_\omega \subseteq \mathcal{U}_\omega$, and hence every $\mathcal{U}_\omega^*$-Cauchy filter is also one for $\mathcal{V}_\omega^*$. It follows that there is a continuous mapping from the bicompletion for $\mathcal{U}$ to the one for $\mathcal{V}$ which sends a limit point of a $\mathcal{U}_\omega^*$ Cauchy filter to its limit as a $\mathcal{V}_\omega^*$ Cauchy filter. By Remark 27 this mapping is continuous and proper as a mapping of locally compact strongly sober spaces.

The converse is just a restatement of Theorem 4.24 of (Fletcher and Lindgren 1982) with the notion of proximity replaced by the equivalent notion of the unique totally bounded quasi-uniformity that goes along with any proximity. □

Example 31. Let (X, τ) be a T_0-space. There exists a finest compatible totally bounded quasi-uniformity $\mathcal{P}$ on X, the Pervin quasi-uniformity. It is generated by the subbase $\{G \times G \cup (X \setminus G) \times X : G \text{ is open}\}$. The topology $\mathcal{T}(\mathcal{P}^{-1})$ consists of the dual A-discrete sets, that is, all lower sets. The *strongly sober compactification corresponding to this quasi-uniformity can be realized by taking the product of all continuous maps into the two-point Sierpinski space (these are the characteristic functions of all open sets), closing up the image in the product topology for the discrete topology on each two-point space factor, and then giving this closure the relative topology from the product of the Sierpinski topologies. Alternately the strongly sober compactification may be viewed as the Nachbin–Stone–Čech order compactification of $(X, \pi, \leqslant)$, where $\leqslant$ is the specialization order and π is the patch topology of τ and the order dual of the A-discrete topology, and the compactification is endowed with the topology of all open upper sets.*

Proof. That the Pervin quasi-uniformity is the finest totally bounded quasi-uniformity that a topological space admits is shown in (Fletcher and Lindgren 1982, Section 2.1). Let $\leqslant$ denote the specialization order for X. Let $x \in X$. Then $\downarrow x$ is closed, and so $G = X \setminus \downarrow x$ is open. The set $U = G \times G \cup X \times \downarrow x$ then belongs to $\mathcal{P}^{-1}$. For $y \leqslant x$, $U(y) = \downarrow x$, and so $\downarrow x$ is open. Since any lower set is a union of such sets, it follows that they are all open.

For any open set G, let χ_G be the mapping into the two-point Sierpinski space $\{0,1\}$ which is the characteristic function on G. The set $U = G \times G \cup (X \setminus G) \times X$ is then the inverse image of the partial order on $\{0,1\}$, which is a base for its quasi-uniformity as a compact pospace. It then follows that the Pervin quasi-uniformity is the restriction of the product quasi-uniformity on the product of Sierpinski spaces when X is embedded therein via all characteristic functions of open sets. Thus the strongly sober compactification is the closure in the patch topology, which is just the product topology of the discrete two-point spaces.

Finally it follows from Theorem 30 and the fact that the Pervin quasi-uniformity is the finest one that the order compactification corresponding to it dominates all others. Thus it is the Nachbin–Stone–Čech compactification. Theorem 29 then guarantees that, as a strongly sober space, it has the topology consisting of the open upper sets. □

7.6 The Künzi–Brümmer quasi-uniformity

We now consider a special quasi-uniformity on locally compact spaces (or more generally on core compact spaces).

Theorem 32. *If (X, τ) is a T_0-space, then (X, τ, τ^k) is a complemented bitopological space. If X is locally compact, then its patch space is compactly embeddable. If $\mathcal{U}$ is a quasi-uniformity which is compatible with τ, then every τ^k-open set is open in $\mathcal{T}(\mathcal{U}^{-1})$.*

The Künzi–Brümmer quasi-uniformity

Proof. By definition of the cocompact topology τ^k, it is a complementary topology. Suppose that X is locally compact. Then it follows from the spectral theory of continuous distributive lattices and locally compact spaces that X equipped with the patch topology of τ and τ^k embeds by an order-reversing mapping in the spectrum of the lattice of open sets of (X, τ), where that lattice is endowed with the Lawson topology (Hofmann and Lawson 1978, Gierz et al. 1980, Chapter V). Since the lattice of open sets of a locally compact space is a continuous lattice and the Lawson topology makes it a compactly ordered space (Gierz et al. 1980, Chapter VI), it follows that the patch space is compactly embeddable.

Let $\mathcal{U}$ be a quasi-uniformity which is compatible with τ, and let G be a subbasic τ^k-open set such that $A = X \setminus G$ is a compact upper set in the topology $\mathcal{T}(\mathcal{U})$. We wish to show that G is open in $\mathcal{T}(\mathcal{U}^{-1})$. Let $y \in G$. For each $x \in A$, pick $U_x \in \mathcal{U}$ such that $y \notin U_x(x)$. This is possible since $x \in X \setminus \downarrow y$, and the latter is an open set. Pick $V_x \in \mathcal{U}$ such that $V_x \circ V_x \subseteq U_x$. Then there exist finitely many V_i such that A is covered by the collection $\{V_i(x_i)\}$. Let $V = \bigcap V_i(x_i)$.

We claim that $V^{-1}(y) \subseteq G$. If not, there exists $z \in A$ such that $(y, z) \in V^{-1}$, that is, $(z, y) \in V$. Now $z \in A$ implies $z \in V_i(x_i)$ for some i, and so $(x_i, z) \in V_i$, $(z, y) \in V$ implies $(x_i, y) \in V_i \circ V_i \subseteq U_i$, which contradictions the choice of U_i. This completes the proof. □

We recall and slightly expand a result of Künzi and Brümmer (1987).

Theorem 33. *Let X be a core compact T_0-space. Then there is a coarsest quasi-uniformity $\mathcal{Q}$ compatible with the topology, namely the one generated by the sets*

$$(X \setminus G') \times X \cup X \times G \quad \text{where} \quad G' \ll G \text{ in } O(X)$$

If X is locally compact, then $\mathcal{T}(\mathcal{Q}^{-1})$ is the cocompact topology.

Proof. The first part of the proposition is Lemma 5 of (Künzi and Brümmer 1987).

By Theorem 32 the cocompact topology is coarser than $\mathcal{T}(\mathcal{Q}^{-1})$. Let G', G be open sets such that $G' \ll G$, and let $U := (X \setminus G') \times X \cup X \times G$. For each $y \in G$, pick a compact neighbourhood N_y of y such that $N_y \subseteq G$. Since $G' \ll G$, finitely many of the N_i cover G'. The union $N = N_1 \cup \cdots \cup N_m$ is compact, and $G' \subseteq N \subseteq G$. Consider now $U^{-1}(x)$ for $x \in X$. If $x \in G$, then $U^{-1}(x) = X$, and hence we can certainly pick a cocompact neighbourhood of x contained in it. If $x \notin G$, then $U^{-1}(x) = X \setminus G'$ and $X \setminus N$ is a cocompact neighbourhood contained inside it. Since such $U^{-1}(x)$ form a subbasis for the neighbourhoods of x in $\mathcal{T}(\mathcal{U}^{-1})$, we are finished. □

We shall refer to the quasi-uniformity $\mathcal{Q}$ of Theorem 33 as the *Künzi–Brümmer quasi-uniformity*. Spaces that admit a coarsest quasi-uniformity have been considered in some detail by Künzi (1986). They are a generalization of the core compact spaces, but agree with them in certain special cases. The following lemma is often convenient to identify the Künzi–Brümmer quasi-uniformity.

Lemma 34. *Let X be a core compact T_0-space and let $\mathcal{B}$ be a collection of open sets that forms a subbasis for the topology and has the property that, given $x \in B \in \mathcal{B}$, there exists $B' \in \mathcal{B}$ such that $x \in B'$ and $B' \ll B$. Then the Künzi–Brümmer quasi-uniformity is generated by the sets*

$$(X \setminus B') \times X \cup X \times B \quad \text{where} \quad B' \ll B \text{ and } B', B \in \mathcal{B}$$

Proof. Let $U := (X \setminus G') \times X \cup X \times G$ be one of the generating sets of Theorem 33. It suffices to find a V generated by the sets in this lemma such that $V \subseteq U$, because this will guarantee that the filters generated in both cases are the same.

For each $x \in G$, choose a finite subset $\mathcal{B}_x \subseteq \mathcal{B}$ such that $x \in \bigcap \mathcal{B}_x \subseteq G$. For each $B \in \mathcal{B}_x$ pick $B' \in \mathcal{B}$ such that $x \in B'$ and $B' \ll B$. Denote the finite collection of these B' by $\mathcal{B}'_x$, and set $B'_x = \bigcap \mathcal{B}'_x$. Since $G' \ll G$, finitely many of the B'_x cover G', say B'_i for $1 \leqslant i \leqslant m$. Let V be the intersection of all $X \setminus D' \times X \cup X \times D$ where $D' \ll D$, D belongs to one of the collections $\mathcal{B}_i$ and D' belongs to one of the collections $\mathcal{B}'_i$. Then one argues directly that $V \subseteq U$ by taking $(x, y) \in V$ and considering the two cases that $x \notin G'$ and $x \in G'$; in the second case one shows $y \in G$. □

Example 35. Let A be a finite alphabet, and let A^* be the set of finite and infinite words or strings formed from the alphabet A. We partial order the set A^* by the prefix order and define a standard quasi-metric by $d(w_1, w_2) = 0$ if w_1 is a prefix of w_2 and $d(w_1, w_2) = 1/2^n$, where n is the first place in which the two words differ otherwise. Then the quasi-uniformity to which this quasi-metric gives rise is the Künzi–Brümmer quasi-uniformity.

Example 36. Let $(X, \pi, \leqslant)$ be a compactly ordered space and consider the topology consisting of all open upper sets. Then X with this topology is a locally compact strongly sober space. If U and V are open upper sets with $\overline{V}^\pi \subseteq U$, then $V \ll U$ (since any open cover of U has finitely many members covering the compact set $\overline{V}^\pi$ and hence V). Thus the Künzi–Brümmer quasi-uniformity for the open upper sets is generated by all sets of the form $(X \setminus V) \times X \cup X \times U$ where U, V are open upper sets and $\overline{V}^\pi \subseteq U$. A dual result applies to the topology consisting of all lower open sets.

The spectral theory of continuous lattices was developed in (Hofmann and Lawson 1978). It was shown there that a space is core compact if and only if its lattice of open sets $O(X)$ is a continuous lattice. Conversely, the spectrum (the set of meet-primes endowed with the hull–kernel topology) of a distributive continuous lattice is a locally compact sober space. The next result can be viewed as an extension of these results to the setting of quasi-uniformities.

Theorem 37. *Let X be a core compact space embedded in the spectrum of its lattice of open sets $L := O(X)$ via $x \mapsto X \setminus \overline{\{x\}}$. Then the lattice L is a compactly ordered space with respect to the Lawson topology, and the restriction to X of the Künzi–Brümmer quasi-uniformity for the locally compact strongly sober space L endowed with the topology of open lower sets is the Künzi–Brümmer quasi-uniformity on X.*

Proof. It is standard that the Lawson topology on a continuous lattice makes it a compactly ordered space and that the topology consisting of the open lower sets has a subbasis consisting of sets of the form $L \setminus {\uparrow} u$, $u \in L$ (Gierz et al. 1980, Chapters III and VI). If $x \ll y$, then $L \setminus {\uparrow} x \subseteq L \setminus {\Uparrow} x \subseteq L \setminus {\uparrow} y$, and since the middle factor is compact we conclude that $L \setminus {\uparrow} x \ll L \setminus {\uparrow} y$ (see Example 36). By Lemma 34 the sets of the form ${\uparrow} x \times L \cup L \times (L \setminus {\uparrow} y)$, where $x \ll y$, generate the Künzi–Brümmer quasi-uniformity for the topology of open lower sets on L.

We reinterpret x and y in the preceding paragraph as open sets U and V and restrict the quasi-uniformity to the embedded image of X in $O(X)$. Note that $U \subseteq X \setminus \overline{\{x\}}$ (the image of x in $O(X)$) if and only if $x \notin U$. Thus ${\uparrow} U \times L \cup L \times (L \setminus {\uparrow} V)$ restricted to X is just $(X \setminus U) \times X \cup X \times V$, which by Theorem 33 generates the Künzi–Brümmer quasi-uniformity on X. □

Lemma 38. *Let X be a compactly ordered space and let Y be a closed subset. With respect to the relative topology and order, Y is a compactly ordered space. Its Künzi–Brümmer quasi-uniformity with respect to the topology of open upper sets is the restriction to Y of the Künzi–Brümmer quasi-uniformity on X for the topology of open upper sets in X. A dual result holds for the lower open sets.*

Proof. The quasi-uniformities in question are identified in Example 36. Let U be an open upper set in Y. Then $U = U' \cap Y$, where $U' := X \setminus \bigl({\downarrow}(Y \setminus U)\bigr)$ and U' is an open upper set in X (since the lower set of the compact set $Y \setminus U$ is again compact). Also if U_1 and V_1 are open upper sets in X such that $\overline{U}_1 \subseteq V_1$, then for $U := U_1 \cap Y$ and $V := V_1 \cap Y$, $\overline{U} \subseteq V$ (closures are always taken in the patch topology of X). Using these facts, one shows that the restriction to Y of the Künzi–Brümmer quasi-uniformity for the open upper sets on X generates the Künzi–Brümmer quasi-uniformity for the open upper sets on Y. □

7.7 Continuous posets

In this section we briefly illustrate and apply the preceding considerations in the context of continuous dcpo's. These include domains, the objects that one most frequently encounters in models for denotational semantics (see, for example, (Lawson 1988) for background and terminology).

Let P be a continuous dcpo, that is, a continuous partially ordered set in which directed sets have suprema. The closed sets in the *Scott topology* consist of all lower sets that are also closed with respect to taking directed suprema. The *Lawson topology* is the patch topology of the Scott topology and the complementary topology consisting of the weak^d topology, the weak topology on the order dual.

As a rule one considers continuous dcpo's endowed with their Scott topologies. We would like to characterize the spaces that arise in this fashion. Note that condition (3) says that they satisfy a particularly strong form of local compactness.

Theorem 39. *Let X be a T_0- space. The following are equivalent.*

1. *The specialization order on X makes it into a continuous dcpo, and the topology is the Scott topology with respect to this order.*

2. *The space X is sober and its lattice of open sets is a completely distributive lattice.*

3. *The space X is sober and every point of X has a basis of (not necessarily open) neighbourhoods N with the property that any open cover of N has a subcover consisting of one element.*

Proof. The equivalence of (1) and (2) may be found in (Lawson 1979).

$1 \Rightarrow 3$ Let U be a Scott-open set containing x. Then there exists $y \ll x$ such that $y \in U$. Then $\uparrow y \subseteq U$ and contains the Scott-open set $\Uparrow y := \{w : y \ll w\}$, which in turns contains x. Since every open set is an upper set, any neighbourhood of y contains $\uparrow y$. Hence part (3) follows. (Note that X is sober by (2).)

$3 \Rightarrow 1$ We endow X with the specialization order. Since X is sober, each directed set has a supremum to which it converges (Gierz et al. 1980, Exercise 3.17 of Chapter II.3). Thus X is a dcpo.

Let $x \in X$ and let N be a neighbourhood of x such that every open cover of X has a subcover of cardinality 1. Then the open cover $\{X \setminus \downarrow w : w \in N\}$ will be an open cover without a subcover of cardinality 1 unless N has a least element. Let y then be this least element. Then $N \subseteq \uparrow y$. Let U be the interior of N. Then U is an open upper set containing x. Let D be a directed set with supremum $s \geq x$. Then $s \in U$. Since D converges to s, there exists $d \in D$ with $d \in U$. Thus $y \leq d$. This shows that $y \ll x$.

We consider the set of all y_N, where N is a compact neighbourhood of x with the property that every open cover of N has a subcover of cardinality 1, and where y_N is the least element of N. We have just seen that $y_N \ll x$ for each N. Since such neighbourhoods form a basis by hypothesis, we easily conclude that the family $\{y_N\}$ forms a directed set and that it converges to x. Suppose that $y_N \leq z$ for each N, but $x \not\leq z$. Then $X \setminus \downarrow z$ is a neighbourhood of x, and hence contains some y_N. But this would imply $y_N \not\leq z$. It follows that x is the supremum of the y_N. But a dcpo in which point is the directed supremum of a set of elements, each of which is way-below it, is a continuous dcpo.

Since directed sets converge to their suprema in X and since the Scott topology is, essentially by definition, the finest topology with this property, it follows that the Scott topology is finer than the given topology on X. Conversely let U be a Scott-open set containing x. Since the set $\{y_N\}$ previously constructed is directed and has supremum x, it follows from the definition of Scott-open that $y_N \in U$ for some N. But then $x \in N \subseteq \uparrow y_N \subseteq U$. Thus the other inclusion also holds.

□

We examine now how these special spaces behave with respect to notions of previous sections.

Theorem 40. *Let P be a continuous dcpo endowed with the Scott topology.*

1. *The space P is locally compact and sober.*

2. *The Künzi–Brümmer quasi-uniformity $\mathcal{Q}$ is generated by all sets of the form $(P \setminus \Uparrow x) \times P \cup P \times \Uparrow y$, where $y \ll x$ (and $\Uparrow x := \{z \in P : x \ll z\}$).*

3. *The topology $\mathcal{T}(\mathcal{Q}^{-1})$ is the cocompact topology, which in turn is the order dual of the weak topology.*

4. *The topology $\mathcal{T}(\mathcal{Q}^*)$ is the patch topology of these two, which is the Lawson topology λ of P.*

5. *The Scott topology is equal to $\lambda^\sharp = \mathcal{T}(\mathcal{Q}^*)$.*

Proof. Part (1) follows from (3) of Theorem 39. Part (2) follows from Lemma 34 since the sets $\Uparrow x$ form a basis for the Scott topology and if $y \ll x$, then $\Uparrow x \ll \Uparrow y$ since any open cover of $\Uparrow y$ must have some member which contains x and hence $\Uparrow x$.

The first assertion of part (3) follows from Theorem 33; for the second see Section V of (Lawson 1988). The assertion of (4) follows from Theorem 25, and the Lawson topology is by definition the patch topology.

Part (5) is proved for continuous lattices in (Gierz et al. 1980, Theorem 1.6 of Chapter III.1), and the same proof carries over to this setting. □

Problem 41. *Let P be a continuous dcpo. Is P endowed with the Lawson topology a strictly completely regular ordered space? A related question is whether $\lambda^\flat$ is equal to the weak^d topology? More generally, find conditions on a T_0 quasi-uniformity so that $\mathcal{T}(\mathcal{U}^*)^\sharp = \mathcal{T}(\mathcal{U})$ and/or $\mathcal{T}(\mathcal{U}^*)^\flat = \mathcal{T}(\mathcal{U}^{-1})$. Find conditions so that the partially ordered topological space determined by $\mathcal{U}$ is a strictly completely regular ordered space.*

7.8 The Fell compactification

In light of Theorem 33 there exists an order compactification that is dominated by all others. We give a construction for it, and then verify that it is actually the one for the Künzi–Brümmer quasi-uniformity. The construction is an adaptation of the one given originally by Fell (1962). An extensive discussion of the Fell compactification from the perspective of continuous lattice theory has been given by Hoffmann (1982) and from the viewpoint of quasi-uniformities by Künzi (1985), who actually considered a construction for the Fell compactification for a class of spaces somewhat more general than locally compact spaces, namely those admitting a coarsest quasi-uniformity. Much of the material of this section can essentially be found in these references, but we want here to connect it with the earlier portions of the chapter.

Let (X, τ) be a locally compact T_0-space. We consider the set $F(X)$ consisting of all closed sets A such that there exists an ultrafilter in X for which A is the set of limit points for the ultrafilter. We order the set $F(X)$ by inclusion and topologize it by taking as a subbasis of open sets of ν those sets of the form

$$\mathcal{N}(K,V) = \{A \in F(X) : A \cap K = \emptyset,\ A \cap V \neq \emptyset\}$$

where K is compact and V is open.

Theorem 42. *Let X be a locally compact T_0-space. Then the triple $(F(X), \nu, \leqslant)$ is a compactly ordered space and the mapping $x \mapsto \overline{\{x\}} \colon X \to F(X)$ is an embedding into the topology $\nu^\#$ of open upper sets.*

Proof. The assertion that the triple is a compactly ordered space is essentially the result of Fell (1962). □

We follow the lead of Hoffmann and call this compactification the *Fell compactification*.

This construction can be viewed in a slightly alternate way. Let $\Gamma(X)$ be the complete lattice of closed sets. We first of all topologize $\Gamma(X)$ with the weak topology. We embed X into $\Gamma(X)$ be sending x to $\downarrow x$, the closure of singleton x.

To obtain the sobrification $\text{Sob}(X)$ of X, we take all irreducible closed sets (A is *irreducible* if $A = B \cup C$ implies $A = B$ or $A = C$). Note that these include the closures of singleton sets. The topology is just the relative topology for $\Gamma(X)$.

Suppose now that (X, τ) is locally compact. We say that the closed set A is *cocompact with respect to* a closed subset B if there exists a compact set K such that $K \cap B = \emptyset$ and $X = A \cup K$. We say that B is *weakly irreducible* if, whenever the finite union of the closed sets $\{A_1, \ldots, A_n\}$ is cocompact with respect to B, then $B \subseteq A_i$ for some i. The Fell compactification then consists of all weakly irreducible closed sets, endowed with the patch topology from the relative topology and its k-complement (the cocompact topology).

From the viewpoint of continuous lattices, one generally prefers to work in the lattice of open sets. Thus the preceding construction is turned on its head. In this case one again starts with a locally compact T_0-space X, or more generally one that is core compact. Then X can be identified with an order-generating subset of the spectrum of the lattice of open sets $O(X)$ (via $x \mapsto X \setminus \overline{\{x\}}$). The previous compactification can be constructed as the closure in the Lawson topology of $O(X)$ of this embedded image. In this way one picks up the set of weakly prime elements, the order dual notion of the weakly irreducible closed sets. Since $O(X)$ in this case is a continuous lattice, this closure will be a compactly ordered space under the relative Lawson topology. See (Hoffmann 1982) for a detailed treatment of this approach.

Theorem 43. *If X is a core compact T_0-space, then the Fell compactification $F(X)$ is the order compactification constructed from the Künzi–Brümmer quasi-uniformity. The corresponding strongly sober compactification consisting of the open upper sets of the Fell compactification is the smallest strongly sober compactification in the sense that it is the quotient of any other in which (X, τ) embeds.*

Proof. By the work of Hoffmann (1982) (as discussed in the previous remarks), we can obtain $F(X)$ as the closure in the Lawson topology on $O(X)$ of the embedded copy of X in the spectrum. By Theorem 37, the restriction of the Künzi–Brümmer quasi-uniformity on $O(X)$ gives the one for X and by Lemma 38 its restriction to $F(X)$ gives the one for $F(X)$. Hence by Theorem 19, $F(X)$ is the order compactification to which the Künzi–Brümmer quasi-uniformity on X gives rise. The last statement now follows from Theorems 29, 30, and 33. □

In the case that X is locally compact and Hausdorff, the smallest compactification is just the one-point compactification. Hence the Fell compactification in this case is the one-point compactification. In light of this we may view the Fell compactification as a generalization of the one-point compactification for more general locally compact spaces. It is interesting to note that in the Hausdorff case the preceding constructions add the extra point in this case as a ⊥, and that the only neighbourhood of the point is the whole space in the corresponding strongly sober compactification.

We remark that if X is a locally compact sober space with a countable base, then it is shown in (Gierz et al. 1980) that it is a dense G_δ subspace of its closure in the Lawson topology of $O(X)$, where the closure is a compact metrizable space. In light of the preceding, we can say that the patch space of X is a dense G_δ in the Fell compactification. Hence it is a polish space (a second countable space which is metrizable with a complete metric).

Problem 44. *Which polish spaces arise in this way? Which ones arise in this way when one begins with a continuous dcpo?*

It follows from Example 36 and the preceding results that we can always embed any continuous dcpo P into its Fell compactification. The Scott topology embeds in the topology of open upper sets and the Lawson topology embeds in the topology making the Fell compactification a compactly ordered space. Such embeddings might have potential use in some semantic considerations. For example, certain functions on P might extend to the Fell compactification and be contractions with respect to $\mathcal{Q}^*$. Then one might be able to conclude not only the existence of a fixed point (by Tarski's theorem), but use the Banach fixed point theorem to conclude that there was a unique fixed point. In certain instances one might also want to pass to the Fell compactification to do powerdomain constructions, since these are much nicer in situations where the Lawson topology is compact.

References

Beer, G. (1986) Increasing semicontinuous functions and compact topological lattices. *Ricerche di Matematica* **35**, 303–308.

Fell, J. M. G. (1962) A Hausdorff topology for the closed subsets of a locally compact non-Hausdorff space. *Proceedings Amer. Math. Soc.* **13**, 472–476.

References

Fletcher, P., and Lindgren, W. F. (1982) *Quasi-uniform Spaces*, Marcel Dekker, New York.

Gierz, G., Hofmann, K. H., Keimel, K., Lawson, J. D., Mislove, M., and Scott, D. (1980) *A Compendium of Continuous Lattices*, Springer-Verlag, Heidelberg.

Hoffmann, R.-E. (1982) The Fell compactification revisited, *Continuous Lattices and Related Topics*, Proceedings of the Conference on Topological and Categorical Aspects of Continuous Lattices (Workshop V), edited by R.-E. Hoffmann, Mathematik-Arbeitspapiere, Universität Bremen **27**, 68–141.

Hofmann, K. H., and Lawson, J. D. (1978) The spectral theory of distributive continuous lattices. *Trans. Amer. Math. Soc.* **246**, 285–310.

Hofmann, K. H., and Mislove, M. (1981) Local compactness and continuous lattices, *Continuous Lattices*, Lecture Notes in Mathematics 871, edited by B. Banaschewski and R.-E. Hoffmann, Springer-Verlag, 125–158.

Jung, A. (1988) *Cartesian closed categories of domains*, Ph. D. dissertation, Technische Hochschule, Darmstadt, 109 pp.

Künzi, H. P. A. (1985) The Fell compactification and quasi-uniformities. *Top. Proc.* **10**, 305–328.

Künzi, H. P. A. (1986) Topological spaces with a coarsest compatible quasi-proximity. *Quaestiones Math.* **10**, 179–196.

Künzi, H. P. A., and Brümmer, G. C. L. (1987) Sobrification and bicompletion of totally bounded quasi-uniform spaces. *Math. Proc. Camb. Phil. Soc.* **101**, 237–247.

Keimel, K., and Gierz, G. (1982). Halbstetige Funktionen und stetige Verbände, *Continuous Lattices and Related Topics*, Proceedings of the Conference on Topological and Categorical Aspects of Continuous Lattices (Workshop V), edited by R.-E. Hoffmann, Mathematik-Arbeitspapiere, Universität Bremen **27**, 59–67.

Lane, E. P. (1967) Bitopological spaces and quasi-uniform spaces. *Proc. London Math. Soc.* **17**, 241–256.

Lawson, J. D. (1979) The duality of continuous posets. *Houston J. Math.* **5**, 357–386.

Lawson, J. D. (1988) The versatile continuous order, *Mathematical Foundations of Programming Semantics*, Springer Lecture Notes in Computer Science **298**, 134–160.

Main, M., Melton, A., Mislove, M., and Schmidt, D. (Eds.) (1988) *Mathematical Foundations of Programming Language Semantics*, Lecture Notes in Computer Science 298, Springer Verlag.

Nachbin, L. (1965) *Topology and Order*, D. van Nostrand. Princeton.

Scott, D. (1972) Continuous lattices, *Toposes, Algebraic Geometry, and Logic*, Springer LNM **274**, Springer-Verlag, Berlin.

Smyth, M. (1990) *Stable local compactification I* (preprint).

8
Totally bounded spaces and compact ordered spaces as domains of computation

MICHAEL B. SMYTH*

Abstract

We argue that, in searching for 'the' category of domains of computation, the complete totally bounded quasi-metric spaces, along with certain closely related categories, are worthy of consideration. The computational examples help to clarify the basic theory of quasi-metric spaces.

Among the 'closely related' categories considered are the compact ordered spaces. Here we propose a construction of the power space as a free semilattice, generalizing the Plotkin power domain and the Vietoris hyperspace. If compact *pre*-ordered spaces are admitted, the construct gives a refinement of Plotkin's power domain (in the sense that it distinguishes more sets).

Finally, we briefly consider the choice of *morphisms* in relation to questions of computability.

8.1 Introduction

The main aim of this chapter is to contribute to the discussion of 'convenient categories for computation'. We start from the observation that each of the two main types of space used in semantics — namely Scott domains and metric spaces — has its (advantages and) limitations. A metric gives us no means of expressing the ordering of information; however, if all we have is the ordering (as in a domain), only qualitative distinctions can be expressed. We can hardly have both: the only metrizable domain is the one-point space.

In previous works (Smyth 1987a, 1987b), the author has argued that the way out of this dilemma is to move to *quasi*-metric (or quasi-uniform) spaces. In this way we can secure most of the advantages of metric spaces

---
*Michael Smyth was *not* funded by the Office of Naval Research.

while remaining faithful to the information ordering and the Scott topology. But there is one advantage of metric spaces that is not so easily retained when we generalize to quasi-metrics: that of simplicity. The general theory of limits and completeness for quasi-metric and quasi-uniform spaces is rather difficult and not a little controversial. A theme of the previously cited works is that computer science provides an abundance of examples which can be of help in getting the definitions (of completeness etc.) right. But the problem of the difficulty of the theory remains. Our proposal here is to focus on the *totally bounded* spaces. For these, the basic theory poses no difficulty, and hardly any possibility of controversy; moreover, they appear to be of special significance for computation. In addition, they provide the benefit, from the mathematical point of view, of serving as a link between domain theory and the long established theory of compact ordered spaces (Nachbin 1965).

Following the preliminary definitions and examples (Section 8.2), we point out in Section 8.3 some simple but (it seems) hitherto unnoticed facts about Cauchy sequences and filters in totally bounded spaces. In Section 8.4 we recall some more or less well known facts about the equivalence between complete totally bounded spaces and various other notions, especially the compact ordered spaces; we also indicate the computer science motivation for being interested in these concepts.

As an application we point out in Section 8.5 that, using compact pre-ordered spaces, we can formulate a variant notion of power domain (or hyperspace) that enables us to make more distinctions than does the Plotkin (or Vietoris) construction.

In the final section we point out that there are some rather basic questions about computability which need to be addressed if the approach of the present chapter is to be taken further.

The mathematical results will be presented, for the most part, in a fairly sketchy fashion; a fuller account is in preparation.

8.2 Preliminaries

A *complete partial order* (*cpo*) is a poset $(D, \sqsubseteq)$ in which (1) there is a least element $\bot$, and (2) every increasing sequence $x_0 \sqsubseteq x_1 \sqsubseteq \cdots$ has a least upper bound $\bigsqcup x_i$. If only condition (2) is satisfied, we say that D is a *dcpo*. An element e of a dcpo D is *finite* (or *compact*) if, for every increasing sequence (x_i) in D, $e \sqsubseteq \bigsqcup x_i \Rightarrow e \sqsubseteq x_i$ for some i. A dcpo is *algebraic* if every element of D is expressible as the lub of an increasing sequence of finite elements (we shall denote the set of finite elements of D by B_D), and is ω-*algebraic* if in addition B_D is at most countable. A (d)cpo D is *bounded-complete* if every subset of D that is bounded above has a lub (if D is ω-algebraic it is suffices to require that $\bot$ exists, and that each pair of finite elements a, b that is bounded above has the lub $a \sqcup b$). A *Scott domain* is a bounded-complete ω-algebraic cpo.

Example 1. $(\mathcal{P}(\mathbb{N}), \subseteq)$ *is a Scott domain.*

Example 2. *Let Σ^∞ be the set of (finite and infinite) words in an alphabet Σ, and let $\leqslant$ be the prefix order on Σ^∞ ($x \leqslant y$ iff x is a prefix of y). Then $(\Sigma^\infty, \leqslant)$ is an algebraic cpo, which is a Scott domain if Σ is countable.*

There are several important variations and generalizations of the preceding, for which we must refer to the literature: in particular, the SFP domains of Plotkin (1976, 1981), and various flavours of *continuous* cpo's (Gierz et al. 1980). A useful survey paper is (Lawson 1987).

The *Scott topology* of a dcpo $(D, \sqsubseteq)$ has as open sets all those sets O such that:

1. O is an upper set, that is, $O = \uparrow O$;
2. if $\bigsqcup x_i \in O$ $((x_i)$ an increasing sequence), then $x_i \in O$ for some i.

The Scott topology is thus coarser than the Alexandroff topology (for which, of course, the open sets have to satisfy only condition (1)). If D is algebraic, the sets of the form $\uparrow e$ (e finite) constitute a base of the Scott topology; the finite joins of these basic open sets are precisely the compact open sets of D. If D is a Scott domain, the compact open sets form a lattice (a sublattice of $\mathcal{P}(D)$). The Scott topology of a cpo D is T_0, but is T_1 only if D is the one-point space.

A *quasi-metric* on a set X is a map d from $X \times X$ to the non-negative reals (possibly including $+\infty$) satisfying:

1. $d(x, x) = 0$
2. $d(x, y) \leqslant d(x, z) + d(z, y)$
3. if $d(x, y) = d(y, x) = 0$ then $x = y$.

Notice that we do *not* require

3'. if $d(x, y) = 0$ then $x = y$

a condition which forces quasi-metric spaces to be T_1. Note also that it is convenient to admit the possibility that $d(x, y) = \infty$ in the *general* theory of quasi-metrics (see Definition 11 for an illustration), but that this is quite unnecessary in the main body of our work here, which is concerned with totally bounded spaces. Given a quasi-metric d on X, the *conjugate* quasi-metric d^{-1} on X is given by $d^{-1}(x, y) = d(y, x)$, and the *associated metric* d^* by $d^*(x, y) = \max\{d(x, y), d(y, x)\}$. We say that (X, d) is totally bounded if, for every $\varepsilon > 0$, there is a finite subset E of X such that, for every $y \in X$, $d^*(x, y) \leq \varepsilon$ for some $x \in E$; of course this just means that (X, d^*) is a totally bounded metric space in the usual sense. A quasi-metric d is *non-Archimedean* (or is an *ultra*-quasi-metric) if it satisfies the strong triangle law

2'. $d(x, y) \leq \max\{d(x, z), d(z, y)\}$.

The (standard) topology *induced* by a quasi-metric d on X is that in which a set $O \subseteq X$ is considered to be open iff, for every $x \in O$, some ε-ball $B(\varepsilon, x)$ $(= \{y \mid d(x, y) < \varepsilon\})$ is contained in O (we prefer to insert 'standard' here, since we consider that there are topologies other than the standard one which may usefully be associated with a quasi-metric (Smyth 1987b)).

Example 3. For any poset $(P, \leq)$ we have the discrete, or $(0, 1)$-valued, quasi-metric d defined by

$$d(x, y) = \begin{cases} 0 & \text{if } x \leq y \\ 1 & \text{otherwise} \end{cases}$$

which induces the Alexandroff topology, and is totally bounded iff P is finite.

Example 4. On the algebraic cpo Σ^∞, define a quasi-metric by

$$d(x, y) = \inf\{2^{-n} \mid x[n] \leq y[n]\}$$

where $x[n]$ denotes the n-truncation of x, that is, the result of deleting all terms of x after the first n. Notice that $d(x, y) = 0$ iff $x \leq y$. We see that d induces the Scott topology on Σ^∞, and is totally bounded iff Σ is finite.

Example 5. Let D be any Scott domain, and $r : B_D \to \mathbb{N}$ a map (a 'rank function') such that $r^{-1}(n)$ is a finite set for each $n \in \mathbb{N}$. Define a quasi-metric by

$$d(x, y) = \inf\{2^{-n} \mid e \sqsubseteq x \Rightarrow e \sqsubseteq y \text{ for every } e \text{ of rank} \leq n\}$$

Then d is totally bounded and induces the Scott topology of D.

The examples just given are all non-Archimedean. Example 5 can be generalized to any second-countable T_0 space (using a rank function for the countable base of open sets). As our final example we take the following.

Example 6. *Let I be the unit interval $[0, 1]$. Define*

$$d(x, y) = \begin{cases} 0 & \text{if } x \leqslant y \\ x - y & \text{if } y < x \end{cases}$$

Then d is a totally bounded quasi-metric which induces the Scott topology on the (continuous) cpo $(I, \leqslant)$.

Finally a *quasi-uniformity* on a set X is a filter $\mathcal{U}$ of reflexive binary relations over X satisfying

$$\forall U \in \mathcal{U} \,.\, \exists V \in \mathcal{U} \,.\, V^2 (= V \circ V) \subseteq U$$

Any quasi-metric d on X gives rise to a quasi-uniformity $\mathcal{U}_d$ on X by taking as a base of (the filter) $\mathcal{U}_d$ the set of relations $U_\varepsilon (\varepsilon > 0)$, where

$$x \, U_\varepsilon \, y \equiv d(x, y) \leqslant \varepsilon$$

Where necessary (specifically, in Section 8.4), we assume also that $\bigcap \mathcal{U}$ is a partial order; this amounts to saying that the topology is T_0, and corresponds to Axiom 3 for quasi-metric spaces.

The definitions of the conjugate quasi-uniformity $\mathcal{U}^{-1}$, the associated uniformity $\mathcal{U}^*$, total boundedness, and the (standard) induced topology are the obvious analogues of the preceding definitions for quasi-metrics. Corresponding to the non-Archimedean quasi-metrics we have that a quasi-uniformity is *transitive* (or *non-Archimedean*) if it has a base consisting of transitive relations. See (Fletcher and Lindgren 1982) for a fuller account of these notions. Transitive quasi-uniformities are a very common occurrence in computer science. (A base of) such a quasi-uniformity is typically presented as a sequence of finer and finer pre-orders, as for example in the treatment of observational pre-order for CCS processes (Hennessy and Plotkin 1980, Milner 1980). For many purposes it may be convenient to work simply with a transitive quasi-uniformity, rather than with an explicit ultra-quasi-metric.

The reader will have observed that our examples of quasi-metrics and quasi-uniformities feature ordered sets rather prominently. This may be explained by the fact that we have computer science applications in mind. Beyond that, however, we take the view that quasi-metrics and quasi-uniformities may best be seen as elaborated forms of partial (or pre-)orders, so that the ordered set examples are actually the crucial ones to appeal to when trying to develop the appropriate concepts.

8.3 Totally bounded spaces

The definitions in the preceding section are uncontroversial (with the possible exception of Axiom 3 of our definition of a quasi-metric). As soon as the notions of Cauchy sequences (and filters), limits and completeness come into play, however, the situation becomes rather chaotic: many conflicting versions can be found in the mathematical literature. Our approach here will be to focus mainly on the totally bounded quasi-metrics, where the situation is relatively clear, and, of course, to have regard to the computational examples.

Reilly et al. (1982) list seven distinct notions of Cauchy sequence for a quasi-metric space, and others can also be found in the literature (Doitchinov 1988). Several of these definitions require reference to extraneous points of the space. If we ignore these, we are left with just three reasonable 'internal' notions.

Definition 7. *A sequence $x_0, x_1, \ldots$ in the quasi-metric space (X, d) is said to be:*

1. *forward Cauchy if $\forall \varepsilon > 0 . \exists k . k \leqslant l \leqslant m \Rightarrow d(x_l, x_m) \leqslant \varepsilon$;*

2. *backward Cauchy if $\forall \varepsilon > 0 . \exists k . k \leqslant l \leqslant m \Rightarrow d(x_m, x_l) \leqslant \varepsilon$;*

3. *two-way (or bi-)Cauchy if $\forall \varepsilon > 0 . \exists k . k \leqslant l, m \Rightarrow d(x_l, x_m) \leqslant \varepsilon$.*

(NB Our terminology is not standard!)

Since the forward Cauchy sequences seem to be the most significant, at least from a computational point of view, we shall for the most part designate these simply as *Cauchy* sequences. Given a quasi-metric d, the backward and two-way Cauchy sequences can then be characterized as the Cauchy sequences w.r.t. d^{-1} and d^* respectively.

Example 8. *Taking d as the discrete quasi-metric over a poset, the forward, backward, and two-way Cauchy sequences are, respectively, the eventually increasing, eventually decreasing, and eventually constant sequences.*

Example 9. *Now let d be a totally bounded quasi-metric for a Scott domain D, defined in terms of a rank function as above. Then an increasing sequence in D is Cauchy in all three senses.*

The second example just given is an instance of the following general result which, despite its simplicity, appears to be new.

Theorem 10. *In any totally bounded quasi-metric space, all three notions of Cauchy sequence coincide.*

Totally bounded spaces

Proof. Let (X, d) be totally bounded. Clearly it suffices to show that (forward) Cauchy $\Rightarrow$ bi-Cauchy. Let (x_i) be a Cauchy sequence, and let $\varepsilon > 0$ be given. Choose (by total boundedness) a finite subset E of X together with a map $h : \mathbb{N} \to E$ such that $d^*(x_n, h(n)) \leqslant \varepsilon/3$ for all $n \in \mathbb{N}$. Next, choose an index k large enough that

$$k \leqslant m \leqslant n \Rightarrow d(x_m, x_n) \leqslant \varepsilon/3$$

$h^{-1}(e) \cap \{n \mid n \geqslant k\}$ is infinite or empty for each $e \in E$

Now, given any $m, n \geqslant k$, we have $d(x_m, x_n) \leqslant \varepsilon$. For, let p be the least index $\geqslant m$ such that $h(p) = h(n)$. Then

$$d(x_m, x_n) \leqslant d(x_m, x_p) + d(x_p, h(p)) + d(h(p), x_n)$$
$$\leqslant \varepsilon/3 + \varepsilon/3 + \varepsilon/3$$

This shows that (x_i) is bi-Cauchy. □

We shall now briefly discuss limits and completeness. The following seems appropriate, in the context that quasi-metrics are being compared with partial orders, and limits with least upper bounds (see (Rowlands-Hughes 1987), where closely related definitions appear).

Definition 11. *Let (x_i) be a Cauchy sequence in the quasi-metric space (X, d). A point $x \in X$ is a limit of (x_i), written*

$$x = \lim_{i \to \infty} x_i$$

if, for every $y \in X$,

$$d(x, y) = \lim_{i \to \infty} d(x_i, y)$$

The existence in the extended reals of the last-mentioned limit is guaranteed by the Cauchy property of (x_i). The space X is complete if every Cauchy sequence has a limit.

Notice that limits of Cauchy sequences are unique, when they exist.

As pointed out by W. Hunsacker (personal communication), it can easily happen that $d(x_i, y) \to \infty$ if (x_i) is a Cauchy sequence. For this reason it is convenient, in the general theory of limits and completeness, to admit ∞ as a possible distance.

Example 12. *Quasi-metrics for Scott domains based on rank functions (Example 5) are complete; the corresponding statement is not true for general second-countable T_0-spaces.*

214   Totally bounded spaces

A rationale for this definition may be provided as follows. In defining the completion (by Cauchy sequences) of (X, d), we shall have to introduce a distance d^+ over the Cauchy sequences, and d^+ will surely be defined by

$$d^+(\xi, \eta) = \lim_{i \to \infty} \lim_{j \to \infty} d(x_i, y_j)$$

where $\xi = (x_i)$, $\eta = (y_j)$ are Cauchy. Now we regard a point x as a limit of the Cauchy sequence ξ iff the distance from x to any point y is the same as $d^+(\xi, y)$, where we are by abuse treating y as a constant Cauchy sequence.

As to the adequacy of Definition 11 and its associated completion construction, we may check to see what happens in various important special cases. Of course they reduce to the usual constructions in metric spaces: this is the minimum criterion of adequacy. Next, suppose that P is a poset with the discrete quasi-metric. Then a Cauchy sequence is, as we have seen, an (eventually) increasing sequence, while a limit in the sense of Definition 11 is simply a least upper bound. Moreover the completion by Cauchy sequences reduces to the standard (in domain theory) completion of a poset to an algebraic dcpo. Finally, we may consider the case that (X, d) is totally bounded. By Theorem 10, a Cauchy sequence ξ is necessarily bi-Cauchy. Now it is hardly possible to doubt that, if ξ is bi-Cauchy, its limit (if any) is simply the limit of ξ w.r.t. d^* in the ordinary metric sense, and it is readily checked that this is what Definition 11 gives us.

It may be objected that the limit of a Cauchy sequence, according to Definition 11, need not be a limit in the topological sense. Our reply is that, while this is true if the standard topology is understood, it only means that the standard topology of a quasi-metric space is not always the most suitable one. What is wanted is a coarser topology than the standard one, for which limits of Cauchy sequences are topological limits, just as, in studying cpo's, one replaces the Alexandroff topology by the (coarser) Scott topology, for which lubs of increasing sequences are topological limits. We have developed this theme at length in previous work (Smyth 1987a, 1987b). Fortunately, that (somewhat complex) development can be bypassed in the present context.

The point is that, by restricting attention to the totally bounded case, we avoid becoming embroiled in alternative topologies for quasi-metric spaces. The relevant result is Proposition 13. For this, we associate with any quasi-metric space (X, d) the partial order $\leqslant_d$ defined by $x \leqslant_d y \Leftrightarrow d(x, y) = 0$. It is nothing other than the specialization order of X, viewed as a topological space.

Proposition 13. *Let (X, d) be totally bounded, and let $\xi - (x_i)$ be a Cauchy sequence in X. Then if $x = \lim \xi$, x is a (standard) topological limit of ξ, and moreover is the greatest such limit.*

Totally bounded spaces

Proof. Assume $x = \lim \xi$. Let $\varepsilon > 0$ be given. For any sufficiently large index k, we have $\lim_{i \to \infty} d(x_i, x_k) \leqslant \varepsilon$, and so $d(x, x_k) \leqslant \varepsilon$. Hence x is a topological limit. Let y be any topological limit of ξ. Again, let $\varepsilon > 0$ be given. Then, for k sufficiently large, we have both $d(y, x_k) \leqslant \varepsilon$ and $d(x_k, x) \leqslant \varepsilon$, so that $d(y, x) \leqslant 2\varepsilon$. This proves that $y \leqslant_d x$. □

An analogous development can be given for quasi-uniform spaces. We conclude the section with a few remarks on this topic. A suitable notion of Cauchy filter is required. We can arrive at this by considering the filters to which Cauchy sequences give rise. The following notation will be adopted. Let $(X, \mathcal{U})$ be a quasi-uniform space. If $A \subseteq X$ and $U \in \mathcal{U}$, then we write $U(S)$ for $\{y \mid \exists x \in S . xUy\}$, and $U[x]$ for $U(\{x\})$ where $x \in X$. Further, $A \ll B$ (where $A, B \subseteq X$) means that, for some $U \in \mathcal{U}$, $U(A) \subseteq B$. Finally, if $\mathcal{G}$ is a collection of subsets of X, then $\mathcal{U}(\mathcal{G})$ is $\{B \mid \exists A \in \mathcal{G} . A \ll B\}$.

If, as is normally the case in this context, $\mathcal{G}$ is a filter, then $\mathcal{U}(\mathcal{G})$ is the *envelope* of $\mathcal{G}$ (Samuel 1948). We evidently have $\mathcal{U}(\mathcal{G}) = \mathcal{U}(\mathcal{U}(\mathcal{G}))$, and (if $\mathcal{G}$ is a filter) $\mathcal{U}(\mathcal{G}) \subseteq \mathcal{G}$. Again following Samuel, we consider that two filters are *equivalent* if they have the same envelope. We shall also say that a filter $\mathcal{F}$ is *round* if $\mathcal{F} = \mathcal{U}(\mathcal{F})$. For example, the neighbourhood filter of any point is round, and, in a uniform space, a filter $\mathcal{F}$ converges to a point x if and only if $\mathcal{F}$ is equivalent to the neighbourhood filter of x. Each equivalence class of filters contains exactly one round filter (namely, the envelope of any member of the class); the round filters can thus serve as canonical representatives. We observe that, if $\mathcal{G}$ is any filter base, $\mathcal{U}(\mathcal{G})$ is a round filter.

Suppose now that (X, d) is a quasi-metric space. Recall that $\mathcal{U}_d$ is the associated quasi-uniformity. Let a Cauchy sequence $\xi = (x_k)$ be given. For each index k, let $\xi(\uparrow k)$ be the set $\{x_n \mid n \geqslant k\}$. The collection $\mathcal{B}$ of sets $\xi(\uparrow k)$ ($k \in \mathbb{N}$) is a filter base. Define $\mathrm{Fil}(\xi) = \mathcal{U}_d(\mathcal{B})$. This makes Fil a map from the Cauchy sequences to the round filters. We observe that the round filter $\mathcal{F} = \mathrm{Fil}(\xi)$ enjoys the following properties:

$$\forall U \in \mathcal{U} . \forall A \in \mathcal{F} . \exists x \in A . U[x] \in \mathcal{F} \tag{1}$$

$$\mathcal{F} \text{ has a countable base} \tag{2}$$

For 1, the point x can be chosen to be any term x_k for k sufficiently large, by the Cauchy property of ξ. For 2, we have the filter base consisting of the sets of the form $U_r(\xi(\uparrow k))$ ($r \in \mathbb{Q}$). We consider a filter to be *Cauchy* if it satisfies condition 1. It should be noted that 1, as the condition for a filter to be Cauchy, is essentially stronger than what is usually seen in the literature, namely

$$\forall U \in \mathcal{U} . \exists x \in X . U[x] \in \mathcal{F} \tag{3}$$

Condition 3 is much too weak for our purposes: in the spaces with which we most typically deal, for example Scott domains, taking 3 as the condition for Cauchyness would mean that *every* filter is Cauchy (by choosing x to be $\bot$). If our definition of *Cauchy filter* is accepted (see (Smyth 1987a) for an extended discussion), then we can say that each Cauchy sequence gives rise to a countably based round Cauchy filter. Moreover, it is easy to show that two Cauchy sequences ξ, η give rise to the same filter iff $d^+(\xi,\eta) = d^+(\eta,\xi) = 0$.

In the converse direction, suppose that $\mathcal{F}$ is a round Cauchy filter (with respect to $\mathcal{U}_d$), having a countable base which without losing generality we may take to be given as a strongly decreasing sequence $A_0 \gg A_1 \gg \cdots$. Using the Cauchy condition 1 we can construct from the sequence (A_i) a sequence of the form

$$U_1[x_0] \supseteq U_{1/2}[x_1] \supseteq U_{1/4}[x_2] \supseteq \cdots$$

which also constitutes a base of $\mathcal{F}$. Then it is clear that $\xi = (x_i)$ is a Cauchy sequence such that $\mathrm{Fil}(\xi) = \mathcal{F}$. The upshot of these arguments is the following proposition.

Proposition 14. *Let (X,d) be a quasi-metric space. Then the correspondence* Fil, *defined above, determines a bijection between the equivalence classes (w.r.t. d^+) of Cauchy sequences over X and the countably-based round Cauchy filters w.r.t. $\mathcal{U}_d$.*

Having thus shown that the notions of Cauchy sequence (over a quasi-metric) and Cauchy filter (over a quasi-uniformity) are in agreement, we turn now to consider the totally bounded quasi-uniformities. By analogy with Theorem 10 we find that, if $(X,\mathcal{U})$ is totally bounded, there is an equivalence between $\mathcal{U}$-Cauchy filters and $\mathcal{U}^*$-Cauchy filters. More precisely, if $\mathcal{G}$ is a $\mathcal{U}^*$-Cauchy filter, then its envelope $\mathcal{U}(\mathcal{G})$ is a (round) $\mathcal{U}$-Cauchy filter, and every round $\mathcal{U}$-Cauchy filter is of the form $\mathcal{U}(\mathcal{G})$ for a unique round $\mathcal{U}^*$-Cauchy filter $\mathcal{G}$. In summary, and omitting the detailed verification, we have the following proposition.

Proposition 15. *Let $(X,\mathcal{U})$ be totally bounded. Then the assignment $\mathcal{G} \mapsto \mathcal{U}(\mathcal{G})$ gives a bijection from the round $\mathcal{U}^*$-Cauchy filters to the round $\mathcal{U}$-Cauchy filters.*

Next, a word on limits. In previous work (Smyth 1987b, 1987a) we have proposed in effect the following definition.

Definition 16. *Let $\mathcal{F}$ be a Cauchy filter over the quasi-uniform space $(X,\mathcal{U})$. We say that $x \in X$ is a limit of $\mathcal{F}$ ($\mathcal{F} \to x$) if the envelope of $\mathcal{F}$ is the neighbourhood filter of x, that is, $\mathcal{U}(\mathcal{F}) = \mathcal{N}(x)$.* This means that, if $\mathcal{F}$ is a round Cauchy filter, $\mathcal{F} \to x$ iff $\mathcal{F} = \mathcal{N}(x)$.

Notice that Definition 16, unlike Definition 11, refers to the topology of the space. In fact, to accommodate our view that variant (non-standard) topologies need to be considered in the general case, we had (Smyth 1987a) to use a slightly more general definition of 'Cauchy filter' than the one given above. In the totally bounded case, however, the variant topologies are irrelevant. We argue as follows. Let $(X, \mathcal{U})$ be totally bounded, and let $\mathcal{F}$ be a Cauchy filter over X. By Proposition 15, $\mathcal{F}$ has an equivalent $\mathcal{U}^*$-Cauchy filter $\mathcal{G}$. Then the limit of $\mathcal{F}$, if any, should be the same as the limit (in the usual sense) of $\mathcal{G}$ w.r.t. the *uniformity* $\mathcal{U}^*$. This is indeed the case if the above definitions are adopted.

Proposition 17. *Let $(X, \mathcal{U})$ be a totally bounded quasi-uniform space. Let $\mathcal{F}$ be any $\mathcal{U}$-Cauchy filter, and $\mathcal{G}$ a $\mathcal{U}^*$-Cauchy filter such that $\mathcal{F}$ and $\mathcal{G}$ have the same $\mathcal{U}$-envelope. Then, for $x \in X$, $\mathcal{F} \to x$ (w.r.t. $\mathcal{U}$) if and only if $\mathcal{G} \to x$ (w.r.t. $\mathcal{U}^*$).*

In conclusion, we would say that the significance of the totally bounded spaces is twofold. Mathematically, the theory of Cauchy sequences, limits, etc. is particularly simple for these spaces, and should be uncontroversial. This means that, besides the ordinary metric and uniform spaces, totally bounded spaces can serve as a test case with which to examine the adequacy of proposed general definitions of the basic notions: many existing proposals in fact fail this test. (We would add that the definitional proposals should make sense also for cpo's and domains, providing a third type of test case.)

But the totally bounded spaces are, we claim, also significant computationally, although we have done little to substantiate this beyond pointing out that certain types of domain can naturally be viewed as (complete) totally bounded spaces. A *general* justification for total boundedness can be provided as follows. The computational significance of distance is that it provides a measure of difficulty (in terms of computational resources) of distinguishing between points. For a given $\varepsilon > 0$ there will be a particular level of resources that is sufficient to distinguish (to find an observable difference between) any two points x, y such that $d(x, y) \geqslant \varepsilon$. Now, with a given level of resources, there is a bound on the number of points which we can distinguish by observation. This immediately implies that the domain of computation is totally bounded.

8.4 Compact ordered spaces

We begin this section by recalling some of the other guises in which the complete totally bounded spaces appear: notably the stably locally compact spaces and the compact ordered spaces. The results in question are fairly well-known, but it is difficult to find them gathered together in any one place. They may be pieced together from such texts as (Johnstone 1982, Gierz et al. 1980, Fletcher and Lindgren 1982), while the most comprehensive single reference is probably (Künzi and Brümmer 1987). The reader having no previous acquaintance with these ideas may safely skip to the main part of the section (following Theorem 20) in which we consider, in a more leisurely fashion, a restricted case of the equivalences where the computational significance is particularly clear.

Definition 18. *A sober space* (X, T) *is said to be* stably locally compact *if:*

1. *X is locally (quasi-)compact;*

2. *the meet of any finite number of compact upper (with respect to the specialization order $\leqslant_T$) sets is compact (it readily follows that arbitrary meets of compact upper sets are compact).*

It is illuminating to view these spaces in terms of their topologies, or frames (Johnstone 1982): the sober space (X, T) is stably locally compact iff T is a stably continuous lattice, that is, T is continuous and the implication

$$U \ll V, U \ll W \Rightarrow U \ll V \cap W$$

holds in T, where $\ll$ is the way-below relation. The spaces are also known, under a slightly different description, as *compact supersober* (Lawson 1987).

A stably locally compact space (X, T) is, of course, not necessarily Hausdorff. But suppose that we take X with its *patch* topology Π, that is, the least topology on X which refines T and also has the complement of each compact upper set (w.r.t. T) as an open set, and let us also take the specialization order $\leqslant_T$. Then the result $((X, \Pi), \leqslant_T)$ is a compact partially ordered space.

Definition 19. *A* compact pospace *is a pair* $(Z, \leqslant)$ *in which Z is a compact Hausdorff space and $\leqslant$ is a closed partial ordering of Z, that is, a partial order which is closed as a subset of $Z \times Z$. (In Section 8.5, we shall allow the case that $\leqslant$ is only a preorder.)*

Conversely, a compact pospace $(Z, \leqslant)$ can be considered as a stably locally compact space by restricting to the *upper* open sets of Z: the operations by which we have gone from each type of space to the other are inverse.

Next, a compact pospace $(Z, \leqslant)$ can be viewed as a complete totally bounded quasi-uniform space: the quasi-uniformity in question is the set of neighbourhoods of $\leqslant$ in $Z \times Z$. In the other direction, given a complete totally bounded quasi-uniform space $(X, \mathcal{U})$, we recover a compact pospace by taking X with the topology of $\mathcal{U}^*$ as the compact space, and $\bigcap \mathcal{U}$ as the partial order. As before, the two constructions are inverse to each other.

Thus we have three types of space, which may as well be regarded as three descriptions of one type of space. But what of the morphisms? Taking the appropriate morphisms for compact pospaces to be (as usual) the continuous order-preserving maps, these may be identified on the one side with the (quasi-)uniformly continuous maps, and on the other — where the spaces are viewed as stably locally compact — with the continuous compactness-reflecting maps. (A map $f : X \to Y$ reflects compactness if $f^{-1}(Q)$ is compact for any compact $Q \subseteq Y$; it is equivalent, in the presence of continuity, to require that f reflects upper compact sets.) Assuming that the categories are taken with these classes of morphisms, we can summarize as follows.

Theorem 20. *The categories of stably locally compact spaces, of compact pospaces and of complete totally bounded quasi-uniform spaces are isomorphic.*

We shall now look more closely at a special case of the preceding equivalences, in which the computational motivation is easier to discern than in the general situation. Depending on which of the three views we are taking of our class of spaces, we can say that it is the case in which frames are required to be algebraic (not just continuous) lattices, or that it is the zero-dimensional case, or that it is the non-Archimedean case.

The stably locally compact spaces with algebraic frames are spectral (or *coherent* (Johnstone 1982)) spaces. We shall not adopt this as our definition, however, but shall instead follow the original approach of Stone (1937), which is easier to appreciate in computational terms. Schematically, this approach is as follows. We start with a lattice, L; we think of this (in the intended applications) as representing a 'logic' of observable propositions (or properties). Meet and join in L will represent conjunction and disjunction; thus it is appropriate to assume that L is distributive. The order in L represents logical implication. Next, we choose a set X of filters of L. We can think of a filter as representing a deductively closed consistent theory (or specification) in the logic. Then, we topologize X by taking as basic open subsets all those of the form

$$X_a = \{\mathcal{F} \mid a \in \mathcal{F}\} \quad a \in L$$

The open set X_a collects together, as it were, the specifications in which (at least) the property a is satisfied.

Many interesting classes of spaces can be introduced in accordance with this schema, depending on the choice of L and X. In the present context, we want to choose for X the set of prime filters of L.

Definition 21. *A filter $\mathcal{F}$ of a lattice L is prime if, for any $a, b \in L$,*

$$a \vee b \in \mathcal{F} \Rightarrow a \in \mathcal{F} \text{ or } b \in \mathcal{F}$$

We could think of prime filters as unambiguous, or (perhaps) deterministic, specifications. Denote by $\operatorname{Spec} L$ the collection of prime filters of a lattice L, topologized as above. We now have the main definition.

Definition 22. *A space X is spectral if it can be represented as (more precisely, is homeomorphic with) $\operatorname{Spec} L$, where L is a distributive lattice.*

Example 23. *Any Scott domain D is spectral: choose for L the lattice of compact open subsets of D, that is, finite joins of sets of the form $\uparrow a$ (a finite). In fact, the spectral domains are exactly the '2/3-SFP' domains of Plotkin (1981).*

Example 24. *What are usually called Stone spaces are exactly the spaces representable as $\operatorname{Spec} L$, where L is a Boolean algebra. They are the same as the compact ultra-metrizable spaces.*

Spectral spaces thus include the two main types of space used in denotational semantics. As already stated, they are the same as the stably locally compact spaces having algebraic frame, although we must omit the proof. Textbooks which emphasize spectral spaces (under various names) are (Rasiowa and Sikorski 1963, Johnstone 1982, Vickers 1989). A detailed account along the lines of the present chapter may be found in (Smyth 1990).

We now consider the patch topology, by means of which we transformed spaces of the first kind into those of the second kind (compact pospaces). In fact, this has a very simple interpretation in the present context. The 'logic' L from which we build a spectral space as $\operatorname{Spec} L$ has, in general, no provision for negation. We typically work with 'positive' properties, specifically with properties based on the detection of (positive) features in finite approximations. The criterion of a positive property is that, once it holds for an approximate or partial element a, it also holds for all extensions of a. This is of course what lies behind the Scott topology.

In certain contexts it may be appropriate to consider negative properties (negations of positive properties) also. This will be the case in particular if we want to study intermediate results in their own right, and not only as approximations to final results. The difference is that between regarding the value of 'The output sequence begins with 010' as $\bot$, if only one character (a zero) has been output so far, and regarding it as *false*. Technically, the introduction of negative properties may be viewed in two slightly different ways. In the first way, we first extend L to a Boolean algebra, say $\overline{L}$, and then construct $\operatorname{Spec} \overline{L}$; this is of course a Stone space.

Interestingly, in going from Spec L to Spec $\overline{L}$ it turns out that we do not change the points (more precisely, the assignment $\mathcal{F} \mapsto \mathcal{F} \cap L$ is a bijection from Spec $\overline{L}$ to Spec L), but only the topology. In fact (this is the second way of describing the introduction of negative properties) we have replaced the topology of Spec L by its patch. The *patch* of a topological space has been defined above, but in the present context it takes the following form: if $(X, \mathcal{T})$ is spectral, then the patch topology $\mathcal{T}^*$ of X is the topology obtained by taking as sub-basic open sets both the compact open sets of X (which are just the basic open sets X_a, $a \in L$, if X is Spec(L)) and their complements.

In the special case that $(X, \mathcal{T})$ is a domain with Scott topology, the patch topology $\mathcal{T}^*$, explained in terms of positive and negative information, was used by Plotkin (1976) under the name 'Cantor topology'. It is also an instance of the *Lawson* topology, and it is now usual to refer to it as such. For domains, the Lawson and patch topologies do indeed coincide, but for more general spaces, the patch topology seems to be the one which captures the idea of extension by negative properties.

By taking the patch, we reduce a spectral space X to a Stone space, but at the cost (in general) of losing the information ordering, that is, the specialization order of X. To avoid losing the information order, it should be retained separately. The relevant concept is given by the following definition.

Definition 25. *An ordered Stone space, or Priestley space, is a pair $(Z, \leqslant)$, where Z is a Stone space and $\leqslant$ is a partial order on Z satisfying the folloiwng: if $x \not\leqslant y$, then there exists a clopen upper set which contains x but not y. (Note that for the purposes of Section 8.5, we are willing to allow the case that $\leqslant$ is only a preorder).*

It is easy to see that, if $(Z, \leqslant)$ is an ordered Stone space, then $\leqslant$ is closed as a subset of $Z \times Z$; thus ordered Stone spaces are a species of compact ordered spaces.

Spectral spaces can be identified with ordered Stone spaces.

Proposition 26. *Let $(X, \mathcal{T})$ be a spectral space. Then X taken with its patch topology $\mathcal{T}^*$ and the specialization order $\leqslant_\mathcal{T}$ is an ordered Stone space. In the other direction, if $(Z, \leqslant)$ is an ordered Stone space, the open upper sets of Z form a spectral topology on (the carrier of) Z. The two constructions are inverse to each other.*

For further details on ordered Stone spaces and the proof of Proposition 26, see (Priestley 1984) or (Johnstone 1982).

Our third characterization of the spaces is in terms of quasi-metrics. Here we come to the only result of this section that may to some extent be new. We suppose now, as is computationally reasonable, that all our spaces are second-countable. For a spectral space X, this means that the base $\mathcal{K}\Omega(X)$ of compact open sets can be enumerated; let $E_0, E_1, \ldots$ be an enumeration of $\mathcal{K}\Omega(X)$. As with the rank functions of Example 5, we can now introduce a quasi-metric by

$$d(x,y) = \inf\{2^{-n} \mid x \in E_i \Rightarrow y \in E_i \text{ for } i = 0, 1, \ldots, n\}$$

This quasi-metric is non-Archimedean and totally bounded, and induces the topology of X. We shall denote it by d_X (a particular enumeration being understood).

Lemma 27. *Let (X, d) be a totally bounded ultra-quasi-metric space. Then any 'closed' ε-ball $\bar{B}(x, \varepsilon)$ $(= \{y \mid d(x,y) \leqslant \varepsilon\})$, where $\varepsilon > 0$, is clopen w.r.t. the topology of d^*.*

Proof. Given $x \in X$, $\varepsilon > 0$, let Q be a partition of (X, d^*) into closed (therefore clopen) ε-balls. By total boundedness, Q is finite. Moreover, by the non-Archimedean property, if a member B of Q meets $\bar{B}(x, \varepsilon)$ (this ball being taken in (X, d)), then $B \subseteq \bar{B}(x, \varepsilon)$. Hence $\bar{B}(x, \varepsilon)$ is the join of finitely many d^*-clopen sets. □

Notation. Given a quasi-metric space (X, d), we denote by $B^*(x, \varepsilon)$ ($\bar{B}^*(x, \varepsilon)$) the open (closed) ε-ball at x taken w.r.t. d^*. Also, we denote by $\leqslant_d$ the partial order on X induced by d:

$$\leqslant_d (x, y) \Leftrightarrow d(x, y) = 0$$

Lemma 28. *Let (Y, d) be a complete totally bounded quasi-metric space. Let $C \subseteq Y$ be a set which is (1) open w.r.t. d^*, and (2) an upper set w.r.t. $\leqslant_d$. Then C is an open set of (Y, d).*

Proof. Assume $x \in C$. Suppose, if possible, that there does not exist $\delta > 0$ with $B(x, \delta) \subseteq C$. Then we can find a sequence of points $y_0, y_1, \ldots$ such that, for all n, $y_n \notin C$ and $d(x, y_n) \leqslant 1/n$. Since (Y, d^*) is compact, we have a $(d^*\text{-})$convergent subsequence $y_{p(0)}, y_{p(1)}, \ldots$ with limit, say, y. Now for any $\eta > 0$ there exists n such that both $d(x, y_{p(n)}) \leqslant \eta$ and $d(y_{p(n)}, y) \leqslant \eta$, which implies that $d(x, y) \leqslant 2\eta$. Since η is arbitrary, we know that $d(x, y) = 0$, so that $y \in C$. Since C is d^*-open, and $y_{p(n)} \notin C$ for all n, we have a contradiction. □

Proposition 29. *Let $(X, \mathcal{T})$ be a second-countable topological space. Then $(X, \mathcal{T})$ is spectral if and only if $\mathcal{T}$ is induced by a complete totally bounded ultra-quasi-metric.*

Proof.
1. Suppose that X is spectral. It suffices to show that d_X is complete. Now, we claim that d_X^* induces the patch topology $\mathcal{T}^*$. To see this, note that the compact open sets E_i and their complements E_i' ($i = 0, 1, \ldots$) constitute a subbase of $\mathcal{T}^*$, that the d_X^*-ball $B(x, \varepsilon)$ can be expressed as the (finite) intersection

$$\bigcap \{E_i \mid x \in E_i \ \& \ \varepsilon < 2^{-i}\} \cap \bigcap \{\bar{E}_i \mid x \notin E_i \ \& \ \varepsilon < 2^{-i}\}$$

and that E_i can be expressed as $\bigcup \{B(x, 2^{-i}) \mid x \in E_i\}$, with $\bar{E}_i$ likewise. Since $\mathcal{T}^*$ is compact, d_X^* is complete; by the results of Section 8.3, d_X is also complete.

2. Assume that $\mathcal{T}$ is induced by the complete totally bounded ultra-quasi-metric d. Let $Z = (X, \mathcal{T}')$ be the Stone space induced by d^*. Then $(Z, \leqslant_d)$ is an ordered Stone space; indeed, if $x \not\leqslant y$, then $\bar{B}(x, d(x,y)/2)$ is by Lemma 27 a $\mathcal{T}^*$-clopen (and evidently upper) set which separates x from y. Moreover, any $\mathcal{T}$-open set is an upper set which is $\mathcal{T}'$-open (since $\mathcal{T}'$ is finer than $\mathcal{T}$), while any $\mathcal{T}'$-open upper set is open by Lemma 28. It follows that $\mathcal{T}$ is the spectral topology derived from the ordered Stone space $(Z, \leqslant)$ in accordance with Proposition 26. □

The preceding analysis may be summarized in the following way. A domain of computation can be thought of as being built from a lattice of basic positive properties, or propositions, by taking prime filters (consistent unambiguous theories) as the points. The idea is essentially the same as that behind Scott's information systems (Scott 1982), although a little more general. Alternatively, we may admit negative properties in addition to the positive ones, provided that we keep separate account of the information order. Finally we have the description in terms of quasi-uniformities or quasi-metrics: we saw already in Section 8.3 that total boundedness is a natural requirement, and completeness (which we require especially for recursion and fixed points) is automatic on the view that points are to be constructed from properties.

This is not to say that the three approaches are equivalent in all respects. We shall see in the next section that by fully exploiting negative properties we can obtain results that are apparently not achievable with the other approaches. Again, there is the possibility of exploiting the numerical content of a specific quasi-metric (as opposed to a merely 'qualitative' quasi-uniformity), although we shall not pursue this theme here.

8.5 A generalized power domain

We shall not be able to make a survey of the principal domain constructions in this short chapter. We shall just consider one that is of special interest, namely the *power space* — a term that may serve as a compromise between the names by which constructs of this type are known in computer science and in mathematics respectively, namely power *domain* and *hyperspace*. It will be seen that the construction provides a common generalization of the Plotkin power domain and of the Vietoris hyperspace; what may be of some interest is that in one version it strictly refines the Plotkin power domain, in the sense that it enables more subsets to be distinguished.

As vehicle for the construction we shall use compact ordered spaces $(X, \leqslant)$; initially, we assume only that $\leqslant$ is a (closed) *preorder*. In the theory of computation it is usual to emphasize that the (finitary) power construct over an object X should be the free semilattice over X, as being the least extension of X which allows for (bounded) non-deterministic choice. According to this, what we are looking for is the free compact preordered semilattice $(Z, \vee, \leqslant')$ over a given compact preordered space $(X, \leqslant)$; here, the join $\vee$ has to be continuous w.r.t. Z ($(Z, \vee)$ is a topological semilattice) and monotonic w.r.t. $\leqslant'$. The required definition will be given in a moment, but we cannot here provide the full proof that it is satisfactory: that will appear in a separate paper (with W. de Oliveira).

In fact, since $(Z, \vee)$ has to be the free (unordered) compact semilattice over X, we can already identify $(Z, \vee)$ as $(\mathcal{K}(X), \cup)$, where $\mathcal{K}(X)$ is the collection of nonempty compact subsets of X with the finite (or Vietoris) topology: a base of this topology is given by the sets of the form $F_\mathcal{C}$, where $\mathcal{C}$ is a finite collection of open sets of X, and

$$F_\mathcal{C} = \{K \in \mathcal{K}(X) \mid K \subseteq \bigcup \mathcal{C} \ \& \ \forall O \in \mathcal{C} . K \cap O \neq \emptyset\}$$

The space $\mathcal{K}(X)$ so defined is the best-known hyperspace construction, and is the same as that obtained by using a Hausdorff metric. The characterization of $(\mathcal{K}(X), \cup)$ as the free semilattice may be extracted from (Wyler 1981), for example, although a direct proof is also easily given. (Note that Wyler admits the empty set into Z, since in effect his semilattices have zero.)

It remains to identify the ordering $\leqslant'$. We claim that the appropriate choice is the so-called Egli–Milner order, which is as it happens the best-known power domain ordering; that is, we assert

$$K \leqslant' L \Leftrightarrow (\forall x \in K . \exists y \in L . x \leqslant y) \ \& \ (\forall y \in L . \exists x \in K . x \leqslant y) \quad (4)$$

where $\leqslant$ is the given (pre)order of X. In order to justify this, we shall give the outline of an argument to the effect that the right-hand side of Equation (4) is the *least* order on $\mathcal{K}(X)$ which (1) extends the given order of X, (2) renders the join ($\cup$) monotonic, and (3) is closed. This argument can then easily be developed into a full proof of the (external) universal characterization, modulo certain technicalities (notably the 'small subsemilattice' condition (Gierz et al. 1980)).

We begin with a combinatorial lemma on the 'Egli–Milner extension'. If S is any set and R a binary relation on S, we denote by R_{EM} the binary relation on $\mathcal{P}(S)$ defined by

$$A\ R_{EM}\ B \equiv (\forall a \in A\,.\,\exists b \in B\,.\,a\ R\ b)\ \&\ (\forall b \in B\,.\,\exists a \in A\,.\,a\ R\ b)$$

Lemma 30. *Let R be a binary relation on the set S, and A, B finite subsets of S. Then $A\ R_{EM}\ B$ if and only if there exist listings (possibly with repetitions) $(a_1,\ldots,a_k)$, $(b_1,\ldots,b_k)$ (where $k \geqslant 0$) of A, B respectively, such that $a_i R b_i$ for $i = 1,\ldots,k$.*

It follows from Lemma 30 that, if we want to extend the given relation $\leqslant$ to the collection $F^+(X)$ of non-empty finite subsets of X, monotonicity (of $\cup$) requires that we have (at least) $\leqslant_{EM}$ over $F^+(X)$. That is, we know that $\leqslant^F_{EM} \subseteq \leqslant'$, where $\leqslant^F_{EM} = \leqslant_{EM} \cap (F^+(X) \times F^+(X))$.

The next step is to prove that the restriction of $\leqslant_{EM}$ to $\mathcal{K}(X)$, that is, $\leqslant^{\mathcal{K}}_{EM} = \leqslant_{EM} \cap (\mathcal{K}(X) \times \mathcal{K}(X))$, is the *closure* of $\leqslant^F_{EM}$ (in the space $\mathcal{K}(X) \times \mathcal{K}(X)$). Let us at least prove one half of this.

Proposition 31. *With the preceding notation, $\leqslant^F_{EM}$ is dense in $\leqslant^{\mathcal{K}(X)}_{EM}$.*

Proof. Let $F_{\mathcal{C}} \times F_{\mathcal{C}'}$ be a basic neighbourhood of $(K, L) \in \leqslant^{\mathcal{K}(X)}_{EM}$, where $\mathcal{C} = \{O_0,\ldots,O_k\}$ and $\mathcal{C}' = \{O'_0,\ldots,O'_l\}$. We have to find finite sets $A, B \in F^+(X)$ such that $A \in F_{\mathcal{C}}$, $B \in F_{\mathcal{C}'}$, and $A \leqslant_{EM} B$. Begin by putting $A_0 = \{x_0,\ldots,x_k\}$ where, for $i = 0,\ldots,k$, x_k is an arbitrarily chosen element of $K \cap O_i$, and $B_0 = \{y_0,\ldots,y_l\}$ similarly. Then, for $i = 0,\ldots,k$, choose $y'_i \in \bigcup \mathcal{C}'$ such that $x_i \leqslant y'_i$, and, for $j = 0,\ldots,l$, choose $x'_j \in \bigcup \mathcal{C}$ such that $x'_j \leqslant y_j$ (these choices are possible since $K \leqslant_{EM} L$). Then $A = A_0 \cup \{x'_0,\ldots,x'_l\}$, $B = B_0 \cup \{y'_0,\ldots,y'_k\}$ satisfy the required conditions. $\square$

The other half, whose proof we omit, is that $\leqslant^{\mathcal{K}}_{EM}$ is actually closed. We can then deduce that $\leqslant^{\mathcal{K}}_{EM} \subseteq \leqslant'$.

The final step is to show that, in extending to $\leqslant^{\mathcal{K}}_{EM}$, we have not lost monotonicity: the join $\cup: \mathcal{K}(X) \times \mathcal{K}(X) \to \mathcal{K}(X)$ is monotonic with respect to $\leqslant^{\mathcal{K}}_{EM}$. With the proof of this (also omitted) we complete the identification of $\leqslant^{\mathcal{K}}_{EM}$ with $\leqslant'$, as the least order satisfying the requirements (1)–(3) (preceding Lemma 30).

It will be seen that, if an SFP domain is identified with a compact ordered space in the manner indicated in the preceding section, the power space construction just given is, except for one feature, a generalization of Plotkin's construction. The missing feature is *convexity*: Plotkin requires the members of the power domain to be convex as well as compact. The discrepancy is easily explained. What we have constructed is the free preordered compact semilattice. In this (Egli–Milner) preorder, any set is equivalent to its convex closure. With only a little further work, we can show that by restricting to the convex compact subsets of X, we obtain the free *partially* ordered construct (which is what one usually asks for).

The discrepancy may, for some applications, be to the advantage of the construct presented above, for the flexibility of the compact ordered spaces means that more distinctions can be made by means of the compact topology of $\mathcal{K}(X)$, rather than (only) by the information ordering $\leqslant_{EM}^{\mathcal{K}}$. As a rather trivial example, let X be a finite domain, that is, a finite poset with $\bot$. Then the power space of X consists of the non-empty subsets of X with the discrete topology, and with the Egli–Milner preorder. This can be explained by saying that, not only have we admitted 'negative' properties into our topologies in addition to the positive ones (that is, moved from the Scott to the Lawson topology), but we have allowed negative properties at the level of X to be used in forming the basic properties at the level of $\mathcal{P}(X)$. Presumably, if we take negative properties seriously enough, we should be willing to countenance this.

8.6 Discussion

The class of spaces which we have been studying is very robust mathematically, as well as being natural computationally. But now we have to admit that there are one or two features which, at the least, require further thought. We conclude by mentioning two in particular, namely morphisms and computable elements.

We have deliberately said little about morphisms, apart from making what seemed the mathematically inevitable choice (preceding Theorem 20). Now this 'inevitable' choice — viz. the continuous order-preserving maps — reduces, in the case of Scott domains, to the maps which are both Scott- and Lawson-continuous. Thus we have a more restricted class of maps than is usual in domain theory. We have in fact on a previous occasion (Smyth 1989) put forward a 'thesis' to the effect that the properly computable functions do belong to the restricted class: they are *uniformly* continuous. In the case of computing over (possibly infinite) binary sequences, for example, the restriction means that, in order to (try to) obtain k bits of output, no more than $m(k)$ bits of input are required, where m is the modulus of continuity. (Notice that we do not require that $m(k)$ inputs are *sufficient* to produce k outputs, which would mean that only total functions were allowed. The idea is that, even if $m(k)$ inputs do not suffice, nothing is gained by processing more than $m(k)$ inputs. We are dealing with *quasi*-uniformity.) A modulus of continuity can be considered as a bound on the complexity of computation of a function (over possibly infinite data structures), or at least as a necessary condition for the existence of such a bound. Thus there are grounds for attributing priority to the restricted class of morphisms. The argument in terms of complexity bounds is hardly conclusive, however. Although our discussion accords with what is probably the best-known approach to complexity of computation over infinite sequences (Ko and Friedman 1982), we should mention now the approach of Bjerner and Holmström (1989) which requires stability of functions rather than uniform continuity.

While the discussion of uniform continuity bears on the realistic (or feasible) computability of functions, there is a further question about the effective computability of elements. The problem is that some of the equivalences presented above hold only classically, not constructively. In particular, this is true of the equivalence between Cauchy and bi-Cauchy sequences in a totally bounded space. If we try to characterize the computable elements as those which can be represented as limits of effective Cauchy sequences (with recursive modulus of convergence) over a chosen recursively presented dense subset of the space, we shall obtain different results according to which notion of Cauchy sequence is used. It is not difficult to see, for example, that if $\mathcal{P}(\mathbb{N})$ is presented as an effective quasi-metric space in accordance with Example 5 in the natural way, we shall obtain the r.e. or the *recursive* sets as computable elements of $\mathcal{P}(\mathbb{N})$, depending on which notion of effective Cauchy sequence is used. The further consequences of this bifurcation have yet to be explored.

References

Bjerner, B., and Holmström, H. (1989). A compositional approach to time analysis of first order lazy functional programs. In *Proc. 4th. Int. Conf. on Functional Programming Languages and Computer Architecture*.

Doitchinov, D. (1988). On completeness of quasi-metric spaces. *Topology and its Applications*, 30:127–148.

Fletcher, P., and Lindgren, W. (1982). *Quasi-uniform spaces*, Vol. 77 of *Lecture Notes in Pure and Applied Mathematics*. Marcel-Dekker.

Gierz, G., Hofmann, K. H., Keimel, K., Lawson, J. D., Mislove, M., and Scott, D. S. (1980). *A Compendium of Continuous Lattices*. Springer-Verlag, Berlin.

Hennessy, M. C. B., and Plotkin, G. (1980). A term model for CCS. In *Proceedings of the 9th. Mathematical Foundations of Computer Science*, pages 281–274. Springer-Verlag. Lecture Notes in Computer Science Vol. 88.

Johnstone, P. T. (1982). *Stone Spaces*, Vol. 3 of *Cambridge Studies in Advanced Mathematics*. Cambridge University Press, Cambridge.

Ko, K. I., and Friedman, H. F. (1982). Computational complexity of real functions. *Theoretical Computer Science*, 20.

Künzi, H. P. A., and Brümmer, G. C. L. (1987). Sobrification and bicompletion of totally bounded quasi-uniform spaces *Math. Proc. Camb. Phil. Soc.*, 101:237–47.

Lawson, J. D. (1987). The versatile continuous order. In M. Main, A. Melton, M. Mislove, and D. Schmidt (Eds.) *Third Workshop on Mathematical Foundations of Programming Languages Semantics*, pp. 134–160, Berlin. Springer-Verlag. Lecture Notes in Computer Science Vol. 298.

Milner, R. (1980). *A Calculus for Communicating Systems*, Vol. 92 of *Lecture Notes in Computer Science*. Springer-Verlag, Berlin.

Nachbin, L. (1965). *Topology and Order*, Vol. 4 of *New York Math. Studies*. Van Nostrand. Princeton, N.J.

Plotkin, G. D. (1976). A powerdomain construction. *SIAM Journal on Computing*, 5:452–487.

Plotkin, G. D. (1981). Post-graduate lecture notes in advanced domain theory (incorporating the 'Pisa Notes'). Dept. of Computer Science, Univ. of Edinburgh.

Priestley, H. A. (1984). Ordered sets and duality for distributive lattices. *Annals of Discrete Math.*, 23:39–60.

Rasiowa, H. and Sikorski, R. (1963). *The Mathematics of Metamathematics*. PWN — Polish Scientific Publishers, Warszawa. Monografie Matematyczne tom **41**.

Reilly, I. L., Subrahmanyam, P. V., and Vamanamurthy, M. K. (1982). Cauchy sequences in quasi-pseudo-metric spaces. *Monatshefte für Mathematik*, 93:127–120.

Rowlands-Hughes, D. M. (1987). Domains versus metric spaces. M.Sc. thesis, Department of Computing, Imperial College.

Samuel, P. (1948). *Transactions of The Amer. Math. Soc.*, 64:100–132.

Scott, D. S. (1982). Domains for denotational semantics. In M. Nielson and E. M. Schmidt (Eds.) *Automata, Languages and Programming: Proceedings 1982*. Springer-Verlag, Berlin. Lecture Notes in Computer Science **140**.

Smyth, M. B. (1987a). Completion of quasi-uniform spaces in terms of filters. Department of Computing, Imperial College, 1987.

Smyth, M. B. (1987b). Quasi-uniformities: reconciling domains with metric spaces. In *Third Workshop on Mathematical Foundations of Programming Language Semantics*, Lecture Notes in Computer Science Vol. 298, Berlin. Springer Verlag.

Smyth, M. B. (1989). The thesis that computable functions are uniformly continuous. British Theor. Comp. Sci. Symp. 5. Abstract in EATCS Bulletin, 39:211.

Smyth, M. B. (1990). Topology. In S. Abramsky, D. Gabbay, and T. S. E. Maibaum (Eds.) *Handbook of Logic in Computer Science*. Oxford University Press. To appear.

Stone, M. H. (1937). Topological representation of distributive lattices and brouwerian logics. *Časopis pešt. mat. fys.*, 67:1–25.

Vickers, S. J. (1989). *Topology Via Logic*. Cambridge Tracts in Theoretical Computer Science. Cambridge University Press.

Wyler, O. (1981). Algebraic theories of continuous lattices. In *Continuous lattices, Proceedings Bremen 1979*, pp. 390–413. Lecture Notes in Mathematics 871, Springer-Verlag.

9
A characterization of effective topological spaces II

DIETER SPREEN

Abstract

Starting with Scott's work on the mathematical foundations of programming language semantics, interest in topology has grown up in theoretical computer science, under the slogan 'open sets are semidecidable properties'. But whereas on Scott domains all such properties are also open, this is no longer true in general. In order to see which semidecidable sets are open, in (Spreen 199?) we considered countable topological T_0-spaces which satisfy certain additional topological and computational requirements and showed that the given topology is the recursively finest topology generated by semidecidable sets which is compatible with it. In this chapter we present some sufficient conditions for the compatibility of two topologies. By this means we obtain characterizations for some general types of topologies including the Scott topology and the metric topology. From these characterizations we then derive some general results on the effective continuity of effective operators. Such continuity results are basic in computer science, for example, in studies of programming language semantics and in complexity theory.

9.1 Introduction

Topological spaces that satisfy certain natural effectivity requirements have been studied by various authors (Dyment 1984, Hauck 1980, 1981, Hingston 1988, Kalantari 1982, Kalantari and Leggett 1982, 1983, Kalantari and Remmel 1983, Kalantari and Retzlaff 1979, Kalantari and Weitkamp 1985a, 1985b, 1987, Nogina 1966, 1969, 1978a, 1978b, 1981, Vaĭnberg and Nogina 1974, 1976, Xiang 1988). Among these investigations two main directions of interest can be observed, which are related to two different schools, the Russian and the North American. Both schools are interested in studying which topological constructions can be done effectively, but while in the Russian school the main interest is to find out which important results of classical analysis hold effectively (constructively), the North American approach is a more recursion theoretical one and follows Nerode's program for studying recursively enumerable (r.e.) substructures of a recursively presented structure, in this case r.e. classes of open sets and the open sets generated by them.

The present chapter contains results of an ongoing research project that takes its motivation from computer science. In theoretical computer science interest in topology started with Scott's pioneering work on models of the λ-calculus and the semantics of programming languages (Scott 1970, 1971, 1972, 1973, 1976, 1981, 1982, Scott and Strachey 1971). Many important notions in this theory can be interpreted in terms of the canonical topology of a Scott domain, that is, the Scott topology. A similar approach has been developed independently by Eršov (1972, 1973a, 1975, 1977). Scott domains are not the only kind of spaces that are successfully applied in computer science. Several authors used metric spaces to give a mathematical meaning to concurrent programs (America and Rutten 1988, de Bakker and Zucker 1982, Golson and Rounds 1983, Nivat 1979, Reed and Roscoe 1988, Roscoe 1982, Rounds 1985). Smyth (1987, 1988) showed that both types of spaces are special quasi-uniform spaces. He also showed that there are interesting computational examples of the latter kind of spaces arising in studies of 'observational preorder' for CCS and related systems (Hennessy and Plotkin 1980, Milner 1980) which are neither domains nor metric spaces. As he pointed out (Smyth 1983), the use of topology in computing theory is not merely a technical trick. The topology captures an essential computational notion, under the slogan 'open sets are semidecidable properties' (for an introduction to this program see (Vickers 1989)).

In the case of Scott domains the (effectively) open sets are the only semidecidable properties. This follows from a generalization of the Rice–Shapiro theorem on index sets of classes of r.e. sets (Rogers 1967), saying that under certain natural effectivity requirements the restriction of the Scott topology to the computable domain elements is equivalent to the topology generated by all completely enumerable subsets of this set. The latter topology is usually called Eršov topology.

Topologies that are generated by a basis of completely enumerable subsets can be defined on each indexed set. They were first considered by Mal'cev (1971). Therefore we call any such topology a Mal'cev topology. Since the Eršov topology is generated by all completely enumerable subsets of a given indexed set, it is the finest Mal'cev topology on this set. For recursive metric spaces it has been proved in (Spreen 1985) that in general the Eršov topology on such a space is strictly finer than the metric topology. This shows that for some topological spaces there are semidecidable properties which are incompatible with the topological structure.

In (Spreen 199?) we dealt with the question of which of these properties are compatible with the given topology, that is, which generate a Mal'cev topology that is equivalent to the given topology. We considered countable topological T_0-spaces (T, τ) with a countable basis, on which a relation of strong inclusion is defined such that for any two basic open sets and any point belonging to both of them, one can effectively find a further basic open set which contains the given point and is strongly included in each of the given basic open sets. Under some additional effectivity assumptions, which among others imply that every basic open set is completely enumerable, that is, semidecidable, it was then shown that each such topology is the recursively finest Mal'cev topology, that is, compatible with it. Here, saying that a topology η is compatible with the topology τ roughly means that, if some basic open set of topology τ is *not* included in a given basic open set of topology η, then we must be able effectively to find a witness for this, that is, an element of the basic open set of topology τ not contained in the basic open set of topology η. Moreover, topology τ is recursively finer than topology η, if given a basic open set in τ and a point contained in it, we can effectively find a basic open set in η which contains the given point and is contained in the given basic open set of τ. If the space is also recursively separable with respect to τ, which means that it has an effectively enumerable dense base, this implies that given a basic open set in τ one can compute its representation as a union of basic open sets in η.

In the present chapter we give some sufficient conditions for a Mal'cev topology η to be compatible with the given topology τ. As we shall see, this holds if with respect to the specialization order each basic open set of topology τ has a lower bound in a certain neighbourhood of it. It follows that in this case τ is recursively equivalent to the Eršov topology on T. Moreover, η is compatible with τ, if η effectively satisfies the T_3-axiom and the exterior of each of its basic open sets is also completely enumerable. If a Mal'cev topology has the latter property, we say that it is complemented. A similar condition for being compatible with τ is that each basic open set of η is a regular element in the lattice of all completely enumerable subsets of the given space T. Mal'cev topologies with this property are called regular based. Since, under the effectivity assumptions of our general characterization theorem and the additional requirement that (T, τ) is recursively separable, each basic open set of topology τ with completely enumerable exterior is already a regular lattice element, it follows that, if (T, τ) is recursively separable and τ is complemented, then τ is the recursively finest regular based Mal'cev topology on T. As it will turn out, this means that it is recursively equivalent to the effectively semi-regular topology associated with the Eršov topology on T, that is, the topology generated by all regular open sets in the basis of the Eršov topology with completely enumerable exterior.

These results are presented in Section 9.4. In Section 9.2 the general framework is set up and the effectivity assumptions that have to be made are discussed. Then, in Section 9.3, Mal'cev topologies are studied.

The remaining sections contain applications of the results in Section 9.4. First, in Sections 9.5 and 9.6, they are applied to the special cases mentioned in the beginning. In Section 9.5 topological spaces consisting of the computable elements of an effective complete partial order are considered. Beside the above mentioned result that the Scott topology on such a general domain can be generated by the completely enumerable subsets of the domain, we that it can already be generated by the subclass of all completely creative subsets. As a special case we obtain an improvement of the above mentioned Rice–Shapiro theorem.

In Section 9.6 the results of Section 9.4 are applied to recursively separable recursive metric spaces. It follows that the metric topology is the recursively finest complemented Mal'cev topology which effectively satisfies the T_3-axiom. Moreover, it is recursively equivalent to the effectively semi-regular topology associated with the Eršov topology.

Finally, in Section 9.7, the effective continuity of effective operators is studied. Such operators are the canonical homomorphisms of indexed sets (Eršov 1973b). Moreover, they are effectively continuous with respect to suitable Mal'cev topologies such as the Eršov topology. As is well known, this also holds with respect to some other natural topologies (Myhill and Shepherdson 1955, Kreisel et al. 1959, Ceĭtin 1962, Moschovakis 1963, 1964). By considering the inverse image of the range topology under an effective operator we obtain some general results on the effective continuity of such operators including the result presented in (Spreen and Young 1984), which generalizes the theorems due to Myhill and Shepherdson, Kreisel, Lacombe and Shoenfield, Ceĭtin, and Moschovakis cited above. There is a long history of interest in continuity theorems in recursive function theory, and such continuity results are basic in computer science, for example, in programming language semantics studies and in complexity theory (see also (Egli and Constable 1976, Eršov 1975, 1977, Friedberg 1958, Helm 1971, Lachlan 1964, Pour-El 1960, Sciore and Tang 1978b, 1978c, Weihrauch 1981, 1987, Weihrauch and Deil 1980, Young 1968, Young and Collins 1983)).

9.2 Strongly effective spaces

In what follows, let $\langle\ ,\ \rangle : \omega^2 \to \omega$ be a recursive pairing function with corresponding projections π_1 and π_2 such that $\pi_i(\langle a_1, a_2 \rangle) = a_i$, let $P^{(n)}$ ($R^{(n)}$) denote the set of all n-ary partial (total) recursive functions, and let W_i be the domain of the ith partial recursive function φ_i with respect to some Gödel numbering φ. We let $\varphi_i(a)\downarrow$ mean that the computation of $\varphi_i(a)$ stops, and $\varphi_i(a)\downarrow_n$ mean that it stops within n steps. In the opposite cases we write $\varphi_i(a)\uparrow$ and $\varphi_i(a)\uparrow_n$ respectively.

Now, let $\mathcal{T} = (T, \tau)$ be a countable topological T_0-space with a countable basis $\mathcal{B}$. If η is any topology on T, then we also write $\eta = \langle \mathcal{C} \rangle$ to express that $\mathcal{C}$ is a countable basis of η. Moreover, for any subset X of T, $\text{int}_\tau(X)$, $\text{cl}_\tau(X)$, $\text{ext}_\tau(X)$, and $\text{bnd}_\tau(X)$ respectively are the interior, the closure, the exterior, and the boundary of X. In the special cases we have in mind, a relation between the basic open sets can be defined which is stronger than the usual set inclusion, and one has to use this relation in order to derive the characterization result we talked about in the introduction. We call a relation $\prec$ on $\mathcal{B}$ *strong inclusion*, if for all $X, Y \in \mathcal{B}$, from $X \prec Y$ it follows that $X \subseteq Y$. Furthermore, we say that $\mathcal{B}$ is a *strong basis*, if for all $z \in T$ and $X, Y \in \mathcal{B}$ with $z \in X \cap Y$ there is a $V \in \mathcal{B}$ such that $z \in V$, $V \prec X$ and $V \prec Y$.

If one considers basic open sets as vague descriptions, then strong inclusion relations can be considered as 'definite refinement' relations. Strong inclusion relations that satisfy much stronger requirements appear very naturally in the study of quasi-proximities (Fletcher and Lindgren 1982). Moreover, such relations have been used in Császár's approach to general topology (Császár 1963) and in Smyth's work on topological foundations of programming language semantics (Smyth 1987, 1988). Compared with the conditions used in these papers, the above requirements seem to be rather weak, but as we go along, we shall meet a further requirement, and it is this condition which in applications prevents us from choosing $\prec$ to be ordinary set inclusion. For what follows we assume that $\prec$ is a strong inclusion on $\mathcal{B}$ and $\mathcal{B}$ is a strong basis.

Let $x : \omega \rightarrowtail T$ (onto) and $B : \omega \rightarrowtail \mathcal{B}$ (onto) respectively be (partial) indexings of T and $\mathcal{B}$ with domains $\text{dom}(x)$ and $\text{dom}(B)$. The value of x at $i \in \text{dom}(x)$ is denoted, interchangeably, by x_i or $x(i)$. The same holds for the indexing B. A subset X of T is *completely enumerable*, if there is an r.e. set A such that $x_i \in X$ iff $i \in A$, for all $i \in \text{dom}(x)$. We say that B is *computable*, if there is some r.e. set L such that for all $i \in \text{dom}(x)$ and $n \in \text{dom}(B)$, $\langle i, n \rangle \in L$ iff $x_i \in B_n$. Furthermore, the space $\mathcal{T}$ is called *strongly effective*, if B is a total indexing and the property of being a strong basis holds effectively, which means that there exists a function $b \in P^{(3)}$ such that for $i \in \text{dom}(x)$ and $n, m \in \omega$ with $x_i \in B_m \cap B_n$, $b(i, m, n)\downarrow$, $x_i \in B_{b(i,m,n)}$, $B_{b(i,m,n)} \prec B_m$, and $B_{b(i,m,n)} \prec B_n$. Note that, if $\prec$ is only set inclusion and B is not required to be total, then the space is effective in the sense of Nogina (1969). The following lemma presents a natural sufficient condition for a space to be strongly effective (Spreen 199?).

Lemma 1. *Let B be computable and total such that $\{\langle m, n \rangle \mid B_m \prec B_n\}$ is r.e. Then $\mathcal{T}$ is strongly effective.*

As is well known, each point of a T_0-space is uniquely determined by its neighbourhood filter and/or a base of it. If B is computable, a base of basic open sets can effectively be enumerated for each such filter. The next result shows that for strongly effective spaces this can be done in a normed way. An enumeration $(B_{f(a)})_{a\in\omega}$ with $f : \omega \to \omega$ such that $\operatorname{range}(f) \subseteq \operatorname{dom}(B)$ is said to be *normed*, if it is decreasing with respect to $\prec$. If f is recursive, it is also called *recursive* and any Gödel number of f is said to be an *index* of it. In the case which $(B_{f(a)})$ enumerates a base of the neighbourhood filter of some point, we say it *converges* to that point.

Lemma 2. (Spreen 199?) *Let $\mathcal{T}$ be strongly effective and B be computable. Then there are functions $q \in P^{(1)}$ and $p \in P^{(2)}$ such that for all $i \in \operatorname{dom}(x)$ and all $n \in \omega$ with $x_i \in B_n$, $q(i)$ and $p(i,n)$ are indices of normed recursive enumerations of basic open sets which converge to x_i. Moreover, $B_{\varphi_{p(i,n)}(a)} \prec B_n$, for all $a \in \omega$.*

In what follows, we want not only to be able to generate normed recursive enumerations of basic open sets that converge to a given point, but conversely we also need to be able to pass effectively from such convergent enumerations to the point they converge to. We say that B is *acceptable*, if it is computable and there is a function $pt \in P^{(1)}$ such that, if m is an index of a normed recursive enumeration of basic open sets which converges to some point $y \in T$, then $pt(m)\downarrow$, $pt(m) \in \operatorname{dom}(x)$ and $x_{pt(m)} = y$.

In (Spreen and Young 1984) an approach to effective topological spaces has been presented in which similar conditions are used. In that approach most of the concepts are based on effective point sequences. Working with filters instead now seems to us to be more appropriate and natural.

As it is well known, each open set is the union of certain basic open sets. In the context of effective topology one is only interested in such open sets where the union is taken over an enumerable class of basic open sets. We call an open set $O \in \tau$ a *Lacombe set*, if there is an r.e. set $A \subseteq \operatorname{dom}(B)$ such that $O = \bigcup\{B_a \mid a \in A\}$. Set $L_n^\tau = \bigcup\{B_a \mid a \in W_n\}$, if $W_n \subseteq \operatorname{dom}(B)$, and let L_n^τ be undefined otherwise. Then L^τ is an indexing of the Lacombe sets of τ. Obviously, $B \leqslant_m L^\tau$, that is, there is some function $f \in P^{(1)}$ such that $f(n)\downarrow$, $f(n) \in \operatorname{dom}(L^\tau)$ and $B_n = L^\tau_{f(n)}$, for all $n \in \operatorname{dom}(B)$.

We can now effectively compare given topologies effectively. Let $\eta = \langle \mathcal{C} \rangle$ be a further topology on T and $C : \omega \rightarrowtail \mathcal{C}$ (onto) an indexing of $\mathcal{C}$. Then τ is *Lacombe finer* than η ($\eta \subseteq_L \tau$) and η *Lacombe coarser* than τ, if $C \leqslant_m L^\tau$. If both $\eta \subseteq_L \tau$ and $\tau \subseteq_L \eta$, then η and τ are called *Lacombe equivalent*.

There is also another possibility to compare η and τ effectively. τ is said to be *recursively finer* than η ($\eta \subseteq_r \tau$) and η *recursively coarser* than τ if there is some function $g \in P^{(2)}$ such that $g(i,m)\downarrow$, $g(i,m) \in \operatorname{dom}(B)$ and $x_i \in B_{g(i,m)} \subseteq C_m$, for all $i \in \operatorname{dom}(x)$ and $m \in \operatorname{dom}(C)$ with $x_i \in C_m$. If both $\eta \subseteq_r \tau$ and $\tau \subseteq_r \eta$, then η and τ are called *recursively equivalent*.

Lemma 3. (Spreen 199?) *Let B be computable. Then, if η is Lacombe coarser than τ, it is also recursively coarser than τ.*

9.3 Mal'cev topologies

A topology η on T is a *Mal'cev topology*, if it has a basis $\mathcal{C}$ of completely enumerable subsets of T. Any such basis is called a *Mal'cev basis*. Let CE be the class of all completely enumerable subsets of T and $\mathcal{E} = \langle CE \rangle$. $\mathcal{E}$ is called *Eršov topology*. All Mal'cev bases on T can be indexed in a uniform canonical way. Let M_n^η be the set of all x_i with $i \in W_n \cap \text{dom}(x)$, if this set is contained in $\mathcal{C}$, and let M_n^η be undefined otherwise. Then M^η is a computable numbering of $\mathcal{C}$, and for any other numbering C of $\mathcal{C}$, C is computable iff $C \leqslant_m M^\eta$. We shall assume in this chapter that any Mal'cev basis is indexed in a computable way and that CE is indexed by $M^\mathcal{E}$. Then $\mathcal{E}$ is the Lacombe finest Mal'cev topology on T.

For Mal'cev topologies the converse of Lemma 3 is also true, provided our space $\mathcal{T}$ is *recursively separable*, which means that there is some r.e. set $D \subseteq \text{dom}(x)$ such that $\{x_i \mid i \in D\}$ is dense in $\mathcal{T}$, that is, it intersects every basic open set.

Lemma 4. (Spreen 199?) *Let $\mathcal{T}$ be recursively separable. Then any Mal'cev topology on T that is recursively coarser than τ is also Lacombe coarser than τ.*

There are some important classes of Mal'cev topologies which we shall consider now. As is easy to see, the class CE of all completely enumerable subsets of T is a distributive lattice with respect to union and intersection. For $U \in CE$, let U^* denote its *pseudocomplement*, that is, the greatest completely enumerable subset of $T \setminus U$, if it exists. U is called *regular*, if U^* and U^{**} both exist and $U^{**} = U$. We say that a Mal'cev topology is *regular based*, if it has a basis of regular sets. Any such basis is called a *regular basis*. Since the class REG of all regular subsets of T is closed under intersection, it also generates a regular based Mal'cev topology on T, which we denote by $\mathcal{R}$.

Let η be a regular based Mal'cev topology on T with regular basis $\mathcal{C}$ and $C : \omega \rightarrowtail \mathcal{C}$ (onto) be a numbering of $\mathcal{C}$. We say that C is *$*$-computable*, if there is some r.e. set L' such that for all $i \in \text{dom}(x)$ and $m \in \text{dom}(C)$, $\langle i, m \rangle \in L'$ iff $x_i \in C_m^*$. Just as in the general case of all Mal'cev bases, all regular bases on T can also be indexed in a uniform way. Let to this end $R_{\langle m,n \rangle}^\eta = M_m^\eta$, if $m \in \text{dom}(M^\eta)$, $n \in \text{dom}(M^\mathcal{E})$, and $M_m^{\eta\,*} = M_n^\mathcal{E}$, and let $R_{\langle m,n \rangle}^\eta$ be undefined otherwise. Then R^η is a computable and $*$-computable indexing of $\mathcal{C}$, and for any other numbering C of $\mathcal{C}$, C is both computable and $*$-computable, iff $C \leqslant_m R^\eta$. We assume in this chapter that regular bases are always indexed, both computably and ∗-computably, and that REG is indexed by $R^{\mathcal{R}}$. Then $\mathcal{R}$ is the Lacombe finest regular based Mal'cev topology on T.

If η is a topology on T, then a subset X of T is called *weakly decidable*, if its interior and its exterior are both completely enumerable. η is called *effectively semi-regular*, if it is generated by those of its regular open sets which are also weakly decidable. Recall that an open set X is regular open, if $X = \text{int}_\eta(\text{cl}_\eta(X))$. Since the exterior of a regular open set is also regular open, any class of such sets which with each set also contains its exterior generates a topology in which these sets are again regular open. Moreover, the exterior of any of them is the same in both topologies. Hence, the class of all weakly decidable regular open sets of a topology η on T generates a topology η^* which is coarser than η and effectively semi-regular; it is said to be the effectively semi-regular topology *associated* with η.

Since every regular subset of T is regular open and weakly decidable with respect to the Eršov topology on T, and these are the only such sets, the following lemma holds.

Lemma 5. $\mathcal{R}$ *is the effectively semi-regular topology associated with the Eršov topology on T, that is, $\mathcal{R} = \mathcal{E}^*$.*

The reason for the introduction of regular based Mal'cev topologies was that in certain cases one needs to be able to enumerate not only each basic open set, but to a certain extent also its complement. In general, one cannot expect that the whole complement of a basic open set is completely enumerable. So, one has to decide for which part of it this should be the case. As we have just mentioned, every regular set is weakly decidable with respect to the Eršov topology. This leads to another choice of which part of the complements of the basic open sets should be completely enumerable. We say that a Mal'cev topology η with Mal'cev basis $\mathcal{C}$ is *complemented*, if all of its basic open sets are weakly decidable. It follows that every regular based Mal'cev topology is complemented. On the other hand, each complemented Mal'cev topology with a Mal'cev basis of clopen sets is regular based.

Let $C : \omega \rightarrowtail \mathcal{C}$ (onto) be a numbering of $\mathcal{C}$. C is called *co-computable*, if there is an r.e. set L' such that for all $i \in \text{dom}(x)$ and $m \in \text{dom}(C)$, $\langle i, m \rangle \in L'$ iff $x_i \in \text{ext}_\eta(C_m)$. Just as in the case of a regular basis one can construct an indexing P^η of $\mathcal{C}$ that is both computable and co-computable. Set $P^\eta_{\langle m,n \rangle} = M^\eta_m$, if $m \in \text{dom}(M^\eta)$, $n \in \text{dom}(M^\mathcal{E})$, and $\text{ext}_\eta(M^\eta_m) = M^\mathcal{E}_n$, and let $P^\eta_{\langle m,n \rangle}$ be undefined otherwise. As is readily verified, any other indexing $\tilde{C}$ of $\mathcal{C}$ is both computable and co-computable, iff $\tilde{C} \leqslant_m P^\eta$. In what follows we assume that any Mal'cev basis with all of its elements being weakly decidable is indexed not only computably, but also co-computably.

9.4 On compatibility

Let B be computable. Then all basic open sets are completely enumerable, which means that τ is a Mal'cev topology. In (Spreen 199?) we proved that under certain conditions any Mal'cev topology on T which is compatible with τ is recursively coarser than τ. From this result we obtained a characterization of τ.

For $X \subseteq T$, let

$$\mathrm{hl}(X) = \bigcap \{O \in \mathcal{B} \mid \exists O' \in \mathcal{T} \,.\, X \subseteq O' \prec O\}$$

Moreover, let $\eta = \langle \mathcal{C} \rangle$ be some further topology on T, and $C : \omega \rightarrowtail \mathcal{C}$ (onto) be an indexing of $\mathcal{C}$. Then η is said to be *compatible* with τ, if there are functions $s \in P^{(2)}$ and $r \in P^{(3)}$ such that for all $i \in \mathrm{dom}(x)$, $n \in \mathrm{dom}(B)$, and $m \in \mathrm{dom}(C)$ the following hold.

1. If $x_i \in C_m$, then $s(i,m)\!\downarrow$, $s(i,m) \in \mathrm{dom}(C)$ and $x_i \in C_{s(i,m)} \subseteq C_m$.

2. If moreover $B_n \not\subseteq C_m$, then also $r(i,n,m)\!\downarrow$, $r(i,n,m) \in \mathrm{dom}(x)$ and $x_{r(i,n,m)} \in \mathrm{hl}(B_n) \setminus C_{s(i,m)}$.

Theorem 6. *Let $\mathcal{T}$ be strongly effective and B be acceptable. Then any Mal'cev topology that is compatible with τ is recursively coarser than τ. If $\mathcal{T}$ is also recursively separable, than any such topology is even Lacombe coarser than τ.*

Theorem 7. *Let $\mathcal{T}$ be strongly effective, B be acceptable, and τ be compatible with itself. Then τ is the recursively finest Mal'cev topology on T that is compatible with τ. If, in addition, $\mathcal{T}$ is recursively separable, then τ is even the Lacombe finest Mal'cev topology on T which is compatible with τ.*

In what follows we present some conditions for a Mal'cev topology η to be compatible with τ. First, we study certain restrictions on τ, and then we consider requirements for η.

For $y, z \in T$, let $y \leqslant_\tau z$ iff $y \in \mathrm{cl}_\tau(\{z\})$. Since (T, τ) is a T_0-space, it follows that $\leqslant_\tau$ is a partial order. It is called the *specialization order*. We say that a subset X of T is *pointed*, if there is some $y \in \mathrm{hl}(X)$ such that $y \leqslant_\tau z$, for all $z \in X$. This generalizes Eršov's notion of an f-set (Eršov 1972, 1975). $\mathcal{T}$ is said to be *effectively pointed*, if there is some function $h \in P^{(1)}$ such that for all $n \in \mathrm{dom}(B)$, $h(n)\!\downarrow$, $h(n) \in \mathrm{dom}(x)$, $x_{h(n)} \in \mathrm{hl}(B_n)$, and $x_{h(n)} \leqslant_\tau z$, for $z \in B_n$.

Let $(y_a)_{a \in \omega}$ be some sequence of points of T. It is *recursive*, if there is some function $f \in R^{(1)}$ with $\mathrm{range}(f) \subseteq \mathrm{dom}(x)$ such that $y_a = x_{f(a)}$

for all $a \in \omega$. Any Gödel number of f is called an *index* of (y_a). The following lemma states a relationship between recursive sequences (y_a) that are strictly increasing with respect to $\leqslant_\tau$ and have least upper bounds, and convergent normed recursive enumerations of basic open sets.

Lemma 8. *Let $\mathcal{T}$ be effectively pointed. Then, if there is a function $sp \in P^{(1)}$ such that for every index n of a recursive sequence (y_a) which is strictly increasing with respect to $\leqslant_\tau$ and has a least upper bound $sp(n)\!\downarrow$, $sp(n) \in \mathrm{dom}(x)$ and $x_{sp(n)}$ is the least upper bound of (y_a), there is also a function $pt \in P^{(1)}$ with the property that, if m is an index of a normed recursive enumeration of basic open sets which converges to some point $z \in T$, then $pt(m)\!\downarrow$, $pt(m) \in \mathrm{dom}(x)$ and $x_{pt(m)} = z$.*

Proof. Let $h \in P^{(1)}$ witness that $\mathcal{T}$ is effectively pointed, and let m be an index of a normed recursive enumeration $(X_a)_{a \in \omega}$ of basic open sets which converges to some point $z \in T$. Then set $y_a = x(h(\varphi_m(a)))$. Since $y_{a+1} \in \mathrm{hl}(X_{a+1})$ and $X_{a+1} \prec X_a$, it follows that $y_{a+1} \in X_a$. Hence $y_a \leqslant_\tau y_{a+1}$. Because $z \in X_a$, for all $a \in \omega$, we moreover have that z is an upper bound of (y_a). Let $z' \in T$ be a further upper bound of (y_a). Since $y_{a+1} \in X_a$, for all $a \in \omega$, we obtain by the definition of the specialization order that $z' \in X_a$ also, for all $a \in \omega$. Then $z \leqslant_\tau z'$, because (X_a) enumerates an open base of the neighbourhood filter of z. Thus z is the least upper bound of (y_a), which means that for $g \in R^{(1)}$ with $\varphi_{g(i)} = h \circ \varphi_i$, $sp(g(m))\!\downarrow$ and $z = x_{sp(g(m))}$. □

We shall show that every Mal'cev topology on T is compatible with τ, provided B is acceptable, and $\mathcal{T}$ is strongly effective and, in addition, effectively pointed. The next lemma is an essential step in the proof of this result.

Lemma 9. *Let $\mathcal{T}$ be strongly effective and B be acceptable. Then each completely enumerable subset of T is upwards closed under $\leqslant_\tau$.*

Proof. Let X be some completely enumerable subset of T with $x_i \in X$, and let $x_j \in T$ with $x_i \leqslant_\tau x_j$. Moreover, let $pt \in P^{(1)}$, $p \in P^{(2)}$, and $q \in P^{(1)}$ respectively be as in the definition of B being acceptable and Lemma 2, and let $b \in \omega$ such that for $a \in \mathrm{dom}(x)$, $a \in W_b$ iff $x_a \in X$. Finally, let $g(m) = \mu c : \varphi_b(pt(m))\!\downarrow_c$. By the recursion theorem there is then some $n \in \omega$ with

$$\varphi_n(a) = \begin{cases} \varphi_{q(i)}(a) & \text{if } a \leqslant g(n) \\ \varphi_{p(j,\varphi_{q(i)}(g(n)))}(a - g(n)) & \text{otherwise} \end{cases}$$

Suppose that $g(n)\!\uparrow$. Then n is an index of a normed recursive enumeration of basic open sets which converges to x_j. By the acceptability of B it follows that $pt(n)\downarrow$ and $x_{pt(n)} = x_i$. Hence $x_{pt(n)} \in X$, which implies that $\varphi_b(pt(n))\downarrow$. This contradicts our assumption. Therefore $g(n)\downarrow$.

Let $\hat{c} = g(n)$. Since $x_i \in B(\varphi_{q(i)}(\hat{c}))$ and $x_i \leqslant_\tau x_j$, it follows that $x_j \in B(\varphi_{q(i)}(\hat{c}))$ also. By Lemma 2 we obtain that $p(j, \varphi_{q(i)}(\hat{c}))$ is an index of a normed recursive enumeration of basic open sets which converges to x_j. Because

$$B_{\varphi_{p(j,\varphi_{q(i)}(\hat{c}))}(0)} \prec B_{\varphi_{q(i)}(\hat{c})}$$

we have that n is also an index of a normed recursive enumeration of basic open sets which converges to x_i. Hence $pt(n)\downarrow$ and $x_{pt(n)} = x_j$. Since $\varphi_b(pt(n))\downarrow$, it follows that $x_j \in X$. □

Theorem 10. *Let $\mathcal{T}$ be strongly effective and effectively pointed. Moreover, let B be acceptable. Then every Mal'cev topology on T is compatible with τ.*

Proof. Let η be a Mal'cev topology on T with Mal'cev basis $\mathcal{C}$, and let $C : \omega \rightarrowtail \mathcal{C}$ (onto) be a computable indexing of $\mathcal{C}$. Moreover, let $h \in P^{(1)}$ witness that $\mathcal{T}$ is effectively pointed. Then define $s \in P^{(2)}$ and $r \in P^{(3)}$ by $s(i,m) = m$ and $r(i,n,m) = h(n)$. In order to see that (s,r) is a witness for η being compatible with τ, assume that $B_n \not\subseteq C_m$. Then $x_{h(n)} \not\in C_m$, by the above lemma. Hence $x_{h(n)} \in \mathrm{hl}(B_n) \setminus C_{s(i,m)}$. □

With the Characterization Theorem 7 we now obtain the following theorem.

Theorem 11. *Let $\mathcal{T}$ be strongly effective and effectively pointed. Moreover, let B be acceptable. Then τ is recursively equivalent to the Eršov topology on T. If, in addition, $\mathcal{T}$ is recursively separable, then it is even Lacombe equivalent to this topology.*

As we shall see, the assumptions of this theorem are satisfied by spaces consisting of the computable elements of an effective complete partial order. Another example are complete f_0-spaces that satisfy some additional effectivity requirements. Such spaces have been introduced by Eršov (1972, 1975, 1977).

In general one cannot expect that every Mal'cev topology is compatible with the given topology τ. (Spreen 1985) has proved the following.

Theorem 12. *There exists a recursively separable strongly effective space $\mathcal{T}$ with acceptable indexing of the basic open sets such that the Eršov topology on this space is strictly finer than the given topology.*

Note that the space exhibited in (Spreen 1985) is a recursively separable recursive metric space. For what follows, we need the following lemma.

Lemma 13. Let T be recursively separable with dense base DB and strongly effective. Moreover, let B be acceptable. Then for any completely enumerable subset X of T and any basic open set B_n, if B_n intersects X, it also intersects $X \cap DB$.

Proof. Let X be a completely enumerable subset of T and let this be witnessed by $A \subseteq \omega$. Let $L \subseteq \omega$ and $pt \in P^{(1)}$ witness that B is acceptable, and let $p \in P^{(2)}$ and $q \in P^{(1)}$ be as in Lemma 2. Moreover, let $W_b = \{i \in A \mid \langle i, n \rangle \in L\}$, and set $g(m) = \mu a : \varphi_b(pt(m))\downarrow_a$. Finally, let $k \in R^{(1)}$ with range$(k) \subseteq \text{dom}(x)$ such that $(x_{k(a)})_{a\in\omega}$ enumerates DB. Now, assume that $x_j \in B_n \cap X$, and for $m \in \omega$ let $\langle i', m' \rangle$ be the first element in a fixed enumeration of $\{\langle \bar{\imath}, \bar{m} \rangle \mid \langle k(\bar{\imath}), \varphi_{q(j)}(g(\bar{m})) \rangle \in L\}$ with $m' = m$. Define $f(m) = k(i')$. By the recursion theorem there is then some $c \in \omega$ with

$$\varphi_c(a) = \begin{cases} \varphi_{q(j)}(a) & \text{if } a \leqslant g(c) \\ \varphi_{p(f(c), \varphi_{q(j)}(g(c)))}(a - g(c)) & \text{otherwise} \end{cases}$$

Since $x_j \in B_n \cap X$, the assumption that $g(c)\uparrow$ leads to a contradiction. Hence, $g(c)\downarrow$. Set $\hat{a} = g(c)$. Since DB is dense in T, we have that $B(\varphi_{q(j)}(\hat{a})) \cap DB \neq \emptyset$. Thus $f(c)$ is also defined. Set $\hat{\imath} = f(c)$. Then it follows that $pt(c)\downarrow$ and $x_{pt(c)} = x_{\hat{\imath}}$. Because $x_{\hat{\imath}} \in B(\varphi_{q(j)}(\hat{a})) \cap DB$ and $x_{pt(c)} \in B_n \cap X$, we obtain that $B_n \cap X \cap DB \neq \emptyset$. □

Theorem 14. Let T be recursively separable and B be computable. Then every regular based Mal'cev topology on T is compatible with τ.

Proof. Let η be a regular based Mal'cev topology on T with regular basis $\mathcal{C}$, and let $C : \omega \rightarrowtail \mathcal{C}$ (onto) be a computable and $*$-computable indexing of $\mathcal{C}$. First we show for $n \in \text{dom}(B)$ and $m \in \text{dom}(C)$ that, if B_n intersects $T \setminus C_m$, then B_n also intersects C_m^*. Assume to this end that $B_n \cap C_m^* = \emptyset$, but $B_n \cap (T \setminus C_m) \neq \emptyset$. Then $B_n \subseteq T \setminus C_m^*$ and $B_n \cap ((T \setminus C_m^*) \setminus C_m) \neq \emptyset$. It follows that $C_m \subsetneq C_m \cup B_n \subseteq T \setminus C_m^*$. This contradicts the regularity of C_m.

Now, let $L, L' \subseteq \omega$ respectively witness that B is computable and C is $*$-computable, and let $k \in R^{(1)}$ with range$(k) \subseteq \text{dom}(x)$ such that $(x_{k(a)})_{a\in\omega}$ enumerates a dense base of T. For $n, m \in \omega$ let $\langle a', n', m' \rangle$ be the first element in some fixed enumeration of $\{\langle a, b, c \rangle \mid \langle k(a), b \rangle \in L \wedge \langle k(a), c \rangle \in L'\}$ with $n' = n$ and $m' = m$. As we have just seen, there is always such a number a', if $n \in \text{dom}(B)$ and $m \in \text{dom}(C)$ such that $B_n \not\subseteq C_m$. Define $s(i, m) = m$ and $r(i, n, m) = k(a')$. Then (s, r) is a witness that η is compatible with τ. □

Thus for recursively separable T and computable B, τ is compatible with itself, if it is regular based. As we shall see next, for spaces which in addition are strongly effective this is already the case, if τ is complemented.

Lemma 15. *Let T be recursively separable and strongly effective. Moreover, let B be acceptable. Then every weakly decidable basic open set B_n is regular and $B_n^* = \mathrm{ext}_\tau(B_n)$.*

Proof. Let B_n be weakly decidable. First, we show that B_n^* exists and $B_n^* = \mathrm{ext}_\tau(B_n)$. Assume to this end that there is some $X \in CE$ with $X \subseteq T \setminus B_n$ and $X \not\supseteq \mathrm{ext}_\tau(B_n)$. Let $x_j \in X \setminus \mathrm{ext}_\tau(B_n)$. Then $x_j \in \mathrm{bnd}_\tau(B_n)$. Now, let $k \in R^{(1)}$ with $\mathrm{range}(k) \subseteq \mathrm{dom}(x)$ such that $(x_{k(a)})_{a \in \omega}$ enumerates a dense base DB of T. Moreover, let $p \in P^{(2)}$ and $q \in P^{(1)}$ be as in Lemma 2, and let $L \subseteq \omega$ and $pt \in P^{(1)}$ witness that B is acceptable. Then $q(j)$ is an index of a normed recursive enumeration of basic open sets which converges to x_j. Since $x_j \in \mathrm{bnd}_\tau(B_n)$ and DB is dense in T, $B(\varphi_{q(j)}(a)) \cap B_n \cap DB \neq \emptyset$, for all $a \in \omega$. Let $b \in \omega$ such that for $i \in \mathrm{dom}(x)$, $x_i \in X$ iff $i \in W_b$, and set $g(m) = \mu a : \varphi_b(pt(m))\downarrow_a$. Moreover, for $m \in \omega$ let $\langle i', m' \rangle$ be the first element in some fixed enumeration of $\{\langle \bar{i}, \bar{m} \rangle \mid \langle k(\bar{i}), n \rangle, \langle k(\bar{i}), \varphi_{q(j)}(g(\bar{m})) \rangle \in L\}$ with $m' = m$, and define $f(m) = k(i')$. By the recursion theorem there is some $c \in \omega$ with

$$\varphi_c(a) = \begin{cases} \varphi_{q(j)}(a) & \text{if } a \leqslant g(c) \\ \varphi_{p(f(c),\varphi_{q(j)}(g(c)))}(a - g(c)) & \text{otherwise} \end{cases}$$

As is easy to see, the assumption that $g(c)\uparrow$ leads to a contradiction. Thus $g(c)\downarrow$. From what we have shown above it follows that $f(c)\downarrow$ also. Let $\hat{a} = g(c)$ and $\hat{i} = f(c)$. By Lemma 2 we obtain that c is an index of a normed recursive enumeration of basic open sets which converges to $x_{\hat{i}}$. According to our construction we have that $x_{\hat{i}} \in B_n$. Hence $x_{pt(c)} \notin X$, from which it follows that $g(c)\uparrow$, contradicting what we have seen before.

This proves that $B_n^* = \mathrm{ext}_\tau(B_n)$. Next, we show that B_n^{**} exists and $B_n^{**} = B_n$. The proof is very similar to that above. Suppose that there is some $X \in CE$ with $X \subseteq \mathrm{cl}_\tau(B_n)$ but $X \not\supseteq B_n$, and let $x_j \in X \setminus B_n$. Then $x_j \in \mathrm{bnd}_\tau(B_n)$. Hence $B(\varphi_{q(j)}(a)) \cap \mathrm{ext}_\tau(B_n) \cap DB \neq \emptyset$, for all $a \in \omega$. Since B_n is weakly decidable, there is some r.e. set A such that for all $i \in \mathrm{dom}(x)$, $x_i \in \mathrm{ext}_\tau(B_n)$ iff $i \in A$. For $m \in \omega$ let $\langle i', m' \rangle$ be the first element in some fixed enumeration of $\{\langle \bar{i}, \bar{m} \rangle \mid k(\bar{i}) \in A \land \langle k(\bar{i}), \varphi_{q(j)}(g(\bar{m})) \rangle \in L\}$ with $m' = m$. Set $f'(m) = k(i')$ and define φ_c just as above with f' instead of f. Then we can again derive a contradiction to our assumption on X. Hence $B_n^{**} = B_n$. □

With the Characterization Theorem 7 and Lemma 5 we now have the following theorem.

Theorem 16. *Let T be recursively separable and strongly effective. Moreover, let B be acceptable and co-computable. Then τ is Lacombe equivalent to the effectively semi-regular topology associated with the Eršov topology on T.*

This result extends Theorem 6.4 in (Spreen 199?). The requirement for a complemented Mal'cev topology in Theorem 14 to be regular based seems to be very strong. We shall therefore consider another condition such that a complemented Mal'cev topology is compatible with τ.

Let $\eta = \langle \mathcal{C} \rangle$ be a topology on T and $C : \omega \rightarrowtail \mathcal{C}$ (onto) be an indexing of $\mathcal{C}$. We say that η *effectively satisfies the T_3-axiom*, if there is some function $s \in P^{(2)}$ such that for all $i \in \mathrm{dom}(x)$ and $m \in \mathrm{dom}(C)$ with $x_i \in C_m$, $s(i,m)\downarrow$, $s(i,m) \in \mathrm{dom}(C)$, $x_i \in C_{s(i,m)}$, and $\mathrm{cl}_\eta(C_{s(i,m)}) \subseteq C_m$.

Theorem 17. *Let T be recursively separable and strongly effective. Moreover, let B be acceptable. Then every complemented Mal'cev topology on T which effectively satisfies the T_3-axiom is compatible with τ.*

Proof. Let η be a complemented Mal'cev topology with Mal'cev basis $\mathcal{C}$ which effectively satisfies the T_3-axiom. Let this be witnessed by $s \in P^{(2)}$, and let $C : \omega \rightarrowtail \mathcal{C}$ (onto) be a computable and co-computable indexing of $\mathcal{C}$. Moreover, let $k \in R^{(1)}$ with $\mathrm{range}(k) \subseteq \mathrm{dom}(x)$ such that $DB = \{x_{k(a)} \mid a \in \omega\}$ is dense in T. Now, suppose that $B_n \not\subseteq C_m$, for some $n \in \mathrm{dom}(B)$ and $m \in \mathrm{dom}(C)$. Then, for all $i \in \mathrm{dom}(x)$ with $x_i \in C_m$ we have that $B_n \cap \mathrm{ext}_\eta(C_{s(i,m)}) \neq \emptyset$. By Lemma 13 it follows that also $B_n \cap \mathrm{ext}_\eta(C_{s(i,m)}) \cap DB \neq \emptyset$. Let L and L' respectively witness that B is computable and C is co-computable, and for $i, m, n \in \omega$ let $\langle i', m', n', a' \rangle$ be the first element in a fixed enumeration of $\{\langle i, m, n, a \rangle \mid \langle k(a), n \rangle \in L \wedge \langle k(a), s(i,m) \rangle \in L'\}$ with $i' = i$, $m' = m$, and $n' = n$. Define $r(i, n, m) = k(a')$. Then $r \in P^{(3)}$. Moreover, for $i \in \mathrm{dom}(x)$, $n \in \mathrm{dom}(B)$, and $m \in \mathrm{dom}(C)$ with $x_i \in C_m$ and $B_n \not\subseteq C_m$ it follows from what has just been shown that $r(i,n,m)\downarrow$ and $x_{r(i,n,m)} \in B_n \setminus C_{s(i,m)}$. Thus, η is compatible with τ. □

Again, by applying the Characterization Theorem 7 we obtain the following theorem.

Theorem 18. *Let T be recursively separable and strongly effective. Let τ effectively satisfy the T_3-axiom and B be acceptable and co-computable. Then τ is the Lacombe finest complemented Mal'cev topology on T that effectively satisfies the T_3-axiom.*

In the next two sections we study two important types of spaces. As we shall see, they fulfil the requirements of the theorems in this section. First we consider spaces that consist of computable elements of an effective complete partial order, and then we consider recursive metric spaces.

9.5 The domain case

Let $Q = (Q, \sqsubseteq)$ be a partial order. A subset S of Q is *directed*, if for all $y_1, y_2 \in S$ there is some $u \in S$ with $y_1, y_2 \sqsubseteq u$. Q is a *complete* partial order (cpo), if every directed subset S of Q has a least upper bound $\sup S$ in Q. Let $\bot = \sup \emptyset$. Moreover, let $\ll$ denote the *way-below relation* on Q, that is, let $y_1 \ll y_2$ iff for directed subsets S of Q the relation $y_2 \sqsubseteq \sup S$ always implies the existence of a $u \in S$ with $y_1 \sqsubseteq u$. Note that $\ll$ is transitive.

A subset Z of Q is a *basis* of Q, if for any $y \in Q$ the set $Z_y = \{z \in Z \mid z \ll y\}$ is directed and $y = \sup Z_y$. If the cpo Q has a basis, then it is said to be *continuous*. For such cpo's it is shown in (Weihrauch and Deil 1980, Lemma 2.3) that for all $y, y_1, y_2, y_3 \in Q$ the following lemma holds.

Lemma 19.

1. $y_1 \ll y_2 \Rightarrow y_1 \sqsubseteq y_2$.

2. $y_1 \ll y_2 \sqsubseteq y_3 \Rightarrow y_1 \ll y_3$.

3. $y_1 \ll y_2 \Rightarrow \exists z \in Z . y_1 \ll z \ll y_2$.

4. Z_y is directed with respect to $\ll$.

Moreover, for continuous cpo's a canonical topology can be defined (Gierz et al. 1980, Scott 1971). A subset X of Q is open, if $(O1)$ and $(O2)$ hold:

$(O1)$ $\forall u, y \in Q . (u \in X \land u \sqsubseteq y \Rightarrow y \in X)$;

$(O2)$ $\forall u \in X . \exists y \in X . y \ll u$.

The topology thus defined is called the *Scott topology* of Q. If Z is a basis of Q, then $\{O_z \mid z \in Z\}$ with $O_z = \{y \in Q \mid z \ll y\}$ is a basis for this topology. As follows from Lemma 19, the partial order $\leq_Q$ induced by this topology is identical with $\sqsubseteq$.

There have been many suggestions in the literature as to which effectivity requirements a cpo should satisfy in order to develop a sufficiently rich computability theory for these structures (Egli and Constable 1976, Kanda 1980, Kanda and Park 1979, Kreitz 1982, Sciore and Tang 1978a, 1978b, Scott 1971, Smyth 1977). Here we use a very general approach which is due to (Weihrauch and Deil 1980).

Let Z be a basis of Q. If there exists an indexing $e : \omega \to Z$ (onto) of Z such that $\{\langle i, j \rangle \mid e_i \ll e_j\}$ is r.e., then Q is called an *effective* cpo. An element $y \in Q$ is said to be *computable*, if $\{i \mid e_i \ll y\}$ $[= e^{-1}(Z_y)]$ is r.e. Let Q_c denote the set of all computable elements of Q. Then $(Q_c, \sqsubseteq, Z, e)$ is called the *constructive domain*. Q_c can be characterized as that subset D of Q for which, for every directed subset S of Z, $\sup S \in D$ iff $\{i \mid e_i \in S\}$

is r.e. Let σ be the relativization of the Scott topology to Q_c. Then $\mathcal{Q}_c = (Q_c, \sigma)$ is a countable T_0-space with basic open sets $B_n = O_{e(n)} \cap Q_c$ ($n \in \omega$). Moreover, Z is dense in $\mathcal{Q}_c$.

An indexing $x : \omega \to Q_c$ (onto) of the computable cpo elements is called *admissible*, if it satisfies the following axioms.

- $\{\langle i, j\rangle \mid e_i \ll x_j\}$ is r.e.

- There is a function $d \in R^{(1)}$ with $x_{d(i)} = \sup e(W_i)$, for all indices $i \in \omega$ such that $e(W_i)$ is directed.

As is shown in (Weihrauch and Deil 1980, Satz 5) such indexings exist. Moreover, it is easy to verify that B is computable, if x satisfies Axiom 1. If x satisfies Axiom 2, then there is some $g \in R^{(1)}$ with $e = x \circ g$. Since Z is dense in $\mathcal{Q}_c$, it thus follows that $\mathcal{Q}_c$ is recursively separable in this case. In what follows, let x be admissible. Then Axiom 2 also holds if $e(W_i)$ is substituted by $x(W_i)$ (Weihrauch and Deil 1980, Lemma 11). From this it follows that there is a function $sp \in P^{(1)}$ such that, if n is an index of a recursive sequence which is strictly increasing with respect to $\sqsubseteq$, then $sp(n)\downarrow$ and $x_{sp(n)}$ is the least upper bound of this sequence. With Lemma 8 we obtain that B is acceptable.

Now define

$$B_m \prec B_n \Leftrightarrow e_n \ll e_m$$

Then $\ll$ is a strong inclusion and $\{B_n \mid n \in \omega\}$ is a strong basis. Moreover, $\mathcal{Q}$ is effectively pointed and, as a consequence of Lemma 1, strongly effective. With Theorem 11 we thus have the following theorem.

Theorem 20. *Let $(Q_c, \sqsubseteq, Z, e)$ be a constructive domain with Q_c being admissibly indexed. Then the relativization of the Scott topology to Q_c is Lacombe equivalent to the Eršov topology on Q_c.*

(Spreen 1984) has shown that all relativized Scott basic open sets B_n are completely creative, where a subset X of Q_c is *completely creative*, if there is some creative set A such that $x_i \in X$ iff $i \in A$. Let us call a Mal'cev topology *creatively generated*, if it has a basis of completely creative sets. Then we obtain the following improvement of the above result.

Theorem 21. *Let $(Q_c, \sqsubseteq, Z, e)$ be a constructive domain with Q_c being admissibly indexed. Then the relativization of the Scott topology to Q_c is the Lacombe finest creatively generated Mal'cev topology on Q_c.*

Let us now consider the special case of the cpo P_ω of all subsets of ω. With FIN being the subclass of all finite sets and $\rho : \omega \to FIN$ (onto) a canonical indexing of FIN, $(P_\omega, \subseteq, FIN, \rho)$ is an effective cpo. Moreover, the r.e. sets are computable elements of this cpo and the numbering W is admissible (Weihrauch and Deil 1980). Since for pairs of sets such that at least one component is finite the way-below relation coincides with the cpo ordering, we have the following corollary.

Corollary 22. *Let I be a subclass of the class RE of all r.e. sets. Then $W^{-1}(I)$ is creative iff there exists some r.e. set A such that*

$$I = \{X \in RE \mid \exists i \in A \,.\, \rho(i) \subseteq X\}$$

This improves the well-known Rice–Shapiro theorem on index sets of classes of r.e. sets (Rogers 1967).

9.6 The metric case

Let $\mathbb{R}$ denote the set of all real numbers, and let ν be some canonical indexing of the rational numbers. Then a real number z is said to be *computable* if there is a function $f \in R^{(1)}$ such that for all $m, n \in \omega$ with $m \leq n$, the inequality $|\nu_{f(m)} - \nu_{f(n)}| < 2^{-m}$ holds and $z = \lim \nu_{f(m)}$. Any Gödel number of the function f is called an *index* of z. This defines a partial indexing γ of the set $\mathbb{R}_c$ of all computable real numbers.

Now, let (M, δ) be a countable metric space with range$(\delta) \subseteq \mathbb{R}_c$, and let $x : \omega \rightarrowtail M$ (onto) be an indexing of M. Then (M, δ) is said to be *recursive*, if the distance function δ is effective, that is, if there is some function $d \in P^{(2)}$ such that for all $i, j \in \text{dom}(x)$, $d(i,j)\downarrow$, $d(i,j) \in \text{dom}(\gamma)$, and $\delta(x_i, x_j) = \gamma_{d(i,j)}$. As is well known, there is a canonical Hausdorff topology Δ on M. The collection of sets $H_{\langle i,m \rangle} = \{y \in M \mid \delta(x_i, y) < 2^{-m}\}$ ($i \in \text{dom}(x)$, $m \in \omega$) is a basis of this topology. Since there is an r.e. set A with $\langle i, j \rangle \in A$ iff $\gamma_i < \gamma_j$, for $i, j \in \text{dom}(\gamma)$ (Moschovakis 1963, Lemma 5), it follows that H is computable (Moschovakis 1964, Lemma 1). Moreover, it follows that there is some r.e. set L' with $\langle j, \langle i, m \rangle \rangle \in L'$ iff $\delta(x_i, x_j) > 2^{-m}$, for all $i, j \in \text{dom}(x)$ and $m \in \omega$. Hence, H is also co-computable.

Let $\mathcal{M} = (M, \Delta)$ be recursively separable and let $g \in R^{(1)}$ be such that $DB = \{x_{g(n)} \mid n \in \omega\}$ is dense in $\mathcal{M}$. Then the set of all $H_{\langle g(i), m \rangle}$ ($i, m \in \omega$) is also a basis of the metric topology on M. In the case that $\mathcal{M}$ is recursively separable, we shall always use this basis and the numbering $B_{\langle i,m \rangle} = H_{\langle g(i), m \rangle}$. Then B is computable and co-computable also.

For the remainder of this section we assume that $\mathcal{M}$ is recursively separable. Define

$$B_{\langle i,m \rangle} \prec B_{\langle j,n \rangle} \Leftrightarrow \delta(x_{g(i)}, x_{g(j)}) + 2^{-m} < 2^{-n}$$

Using the triangular inequation it is readily verified that $\prec$ is a strong inclusion and $\{B_a \mid a \in \omega\}$ is a strong basis.

As we have just seen, B is computable. We shall now state a condition on Cauchy sequences which ensures that B is also acceptable. Let $(y_a)_{a \in \omega}$ be a sequence with $y_a \in DB$, for all $a \in \omega$. Then (y_a) is said to be *normed* if $\delta(y_m, y_n) < 2^{-m}$, for all $m, n \in \omega$ with $m \leq n$. Moreover, (y_a) is *recursive* if there is some function $f \in R^{(1)}$ with range$(f) \subseteq$ range(g) such that $y_a = x_{f(a)}$, for all $a \in \omega$. Any Gödel number of f is called an *index* of the sequence (y_a). We say that (y_a) satisfies (A) if the following holds (Moschovakis 1964).

(A) There is a function $li \in P^{(1)}$ such that, if m is an index of a converging normed recursive sequence (y_a) of elements of the dense base of $\mathcal{M}$, then $li(m)\!\downarrow$, $li(m) \in \text{dom}(x)$, and $x_{li(m)} = \lim y_a$.

Lemma 23. *Let x satisfy (A). Then $\mathcal{M}$ is strongly effective and B is acceptable.*

Proof. Since $\{\langle\langle i, m\rangle, \langle j, n\rangle\rangle \mid \delta(x_{g(i)}, x_{g(j)}) + 2^{-m} < 2^{-n}\}$ is r.e., it follows from Lemma 1 that $\mathcal{M}$ is strongly effective.

As we have already seen, B is computable. To see that it is also acceptable, let m be an index of a normed recursive enumeration of basic open sets which converges to some point $z \in M$. Moreover, let $f \in R^{(1)}$ with
$$\varphi_{f(n)}(a) = g(\pi_1(\varphi_n(\mu c : a < \pi_2(\varphi_n(c)))))$$
Since the sequence $(B(\varphi_m(c)))$ is strictly decreasing with respect to $\prec$, it follows that $(\pi_2(\varphi_m(c)))$ is a strictly increasing sequence of natural numbers. Thus, $\varphi_{f(m)}$ is a total function. Now, define $y_a = x(\varphi_{f(m)}(a))$ for $a \in \omega$. Then (y_a) is a normed recursive sequence of elements of the dense base of $\mathcal{M}$ which converges to z and $f(m)$ is one of its indices. Because x satisfies (A), we obtain that $li(f(m))\!\downarrow$ and $x_{li(f(m))} = z$. □

As a consequence of Theorem 16 we now have the following theorem.

Theorem 24. *Let (M, δ) be a recursively separable recursive metric space with indexing x satisfying (A). Then the canonical metric topology is Lacombe equivalent to the effectively semi-regular topology associated with the Eršov topology on M.*

It is well known that the metric topology satisfies the T_3-axiom. We shall see next that this also holds effectively.

Lemma 25. *Δ effectively satisfies the T_3-axiom.*

Proof. First we consider the general case where the set of all $H_{\langle i,m \rangle}$ ($i \in$ dom(x), $m \in \omega$) is used as a basis of Δ. From what has been said above it follows that there is some r.e. set E such that for $i, j \in$ dom(x) and $c, m \in \omega$, $\langle i, j, m, c \rangle \in E$ iff $\delta(x_i, x_j) + 2^{-c} < 2^{-m}$. For $i, j, m \in \omega$ let $\langle i', j', m', c' \rangle$ be the first element in some fixed enumeration of E with $i' = i$, $j' = j$, and $m' = m$. Define $s(i, \langle j, m \rangle) = \langle i, c' \rangle$. Then $s \in P^{(2)}$, and for $i, j \in$ dom(x) and $m \in \omega$ with $x_i \in H_{\langle j,m \rangle}$ we have that $s(i, \langle j, m \rangle) \downarrow$. Let $s(i, \langle j, m \rangle) = \langle i, c \rangle$; then it follows moreover that cl$_\Delta(H_{\langle i,c \rangle}) = \{y \in M \mid \delta(x_i, y) \leqslant 2^{-c}\} \subseteq H_{\langle j,m \rangle}$. Thus, Δ effectively satisfies the T_3-axiom.

Let us now consider the case that $\mathcal{M}$ is recursively separable and the set of all $B_{\langle i,m \rangle}$ ($i, m \in \omega$) is used as a basis of Δ. Again there is some r.e. set E such that for $i \in$ dom(x) and $j, m, a, c \in \omega$, $\langle i, j, m, a, c \rangle \in E$ iff $\delta(x_{g(j)}, x_i) + 2^{1-c} < 2^{-m}$ and $\delta(x_{g(a)}, x_i) < 2^{-c}$. For $i, j, m \in \omega$ let $\langle i', j', m', a', c' \rangle$ be the first element in some fixed enumeration of E with $i' = i$, $j' = j$, and $m' = m$. Set $s(i, \langle j, m \rangle) = \langle a', c' \rangle$ in this case. Then s witnesses that Δ effectively satisfies the T_3-axiom. □

Applying Theorem 18 we obtain a further characterization of the metric topology.

Theorem 26. *Let (M, δ) be a recursively separable recursive metric space with indexing x satisfying (A). Then the canonical metric topology is the Lacombe finest complemented Mal'cev topology on M which effectively satisfies the T_3-axiom.*

9.7 Effective operators

Let $\mathcal{T}' = (T', \tau')$ be a second topological space with a countable basis $\mathcal{B}'$, and let $x' : \omega \twoheadrightarrow T'$ (onto) and $B' : \omega \twoheadrightarrow \mathcal{B}'$ (onto) respectively be indexings of T' and $\mathcal{B}'$. We do not require that $\mathcal{T}'$ is a T_0-space. Let $F : T \to T'$. Then F is *effective*, if there is a function $t \in P^{(1)}$ such that $t(i) \downarrow$, $t(i) \in$ dom(x'), and $Fx_i = x'_{t(i)}$, for all $i \in$ dom(x). For $n \in$ dom(B'), let $C_n = F^{-1}(B'_n)$. As is well known, the class of all these C_n generates a topology on T, the inverse image of topology τ' under F, which we denote by $F^{-1}(\tau')$. Now, let F be effective. Then, if B' is computable, C is also computable, which means that $F^{-1}(\tau')$ is a Mal'cev topology on T. Moreover, if B' is co-computable, the same is true for C.

We shall now define when F is effectively continuous. Again there are at least two possibilities. We say that F is *recursively continuous* if there is a function $f \in P^{(2)}$ such that for all $i \in$ dom(x) and $n \in$ dom(B') with $Fx_i \in B'_n$, $f(i, n) \downarrow$, $f(i, n) \in$ dom(B), $x_i \in B_{f(i,n)}$, and $F(B_{f(i,n)}) \subseteq B'_n$. On the other hand, F is *effectively continuous* if there is a function $g \in P^{(1)}$ such that for all $n \in$ dom(B'), $g(n) \downarrow$, and $F^{-1}(B'_n) = L^\tau_{g(n)}$. Obviously,

F is recursively continuous iff $F^{-1}(\tau')$ is recursively coarser than τ, and F is effectively continuous iff $F^{-1}(\tau')$ is Lacombe coarser than τ. Because of Lemmas 3 and 4 we thus have the following lemma.

Lemma 27.

1. Let B be computable. If F is effectively continuous, then it is also recursively continuous.

2. Let T be recursively separable, B' be computable, and F be effective. Then, if F is recursively continuous, it is also effectively continuous.

In (Spreen and Young 1984) the recursive continuity of effective operators has been studied in a general context. One of the main requirements F must fulfil is that of having a witness for non-inclusion. The following definition slightly differs from that in (Spreen and Young 1984). F has a witness for non-inclusion, if there exist functions $s \in P^{(2)}$ and $r \in P^{(3)}$ such that for all $i \in \text{dom}(x)$, $n \in \text{dom}(B)$, and $m \in \text{dom}(B')$ the following hold.

1. If $Fx_i \in B'_m$, then $s(i,m)\downarrow$, $s(i,m) \in \text{dom}(B')$, and $Fx_i \in B'_{s(i,m)} \subseteq B'_m$.

2. If, in addition, $F(B_n) \not\subseteq B'_m$, then also $r(i,n,m)\downarrow$, $r(i,n,m) \in \text{dom}(x)$ and $x_{r(i,n,m)} \in \text{hl}(B_n) \setminus F^{-1}(B'_{s(i,m)})$.

Obviously, F has a witness for non-inclusion iff $F^{-1}(\tau')$ is compatible with τ. Thus, from Theorem 6 we obtain the following theorem.

Theorem 28. *Let T be strongly effective, B be acceptable, and B' be computable. Moreover, let $F : T \to T'$ be effective, and let F have a witness for non-inclusion. Then F must be recursively continuous. If, in addition, T is recursively separable, then F is even effectively continuous.*

This improves the general continuity result in (Spreen and Young 1984). In the remainder of this section we shall consider the two special cases that T is effectively pointed and/or T is recursively separable and τ' is a complemented Mal'cev topology which effectively satisfies the T_3-axiom. In the first case we have the following with Theorem 11 and/or Theorems 10 and 28.

Theorem 29. *Let T be strongly effective and effectively pointed. Moreover, let B be acceptable and B' be computable. Then every effective operator $F : T \to T'$ must be recursively continuous. If T is also recursively separable, then F is even effectively continuous.*

As we have seen in Section 9.5, the requirements of this theorem are fulfilled, for example, by spaces consisting of the computable elements of an effective complete partial order. Hence, this theorem generalizes the Myhill-Shepherdson theorem for such spaces (Egli and Constable 1976, Sciore and Tang 1978b, 1978c, Weihrauch and Deil 1980, Weihrauch 1987). Assume next that τ' effectively satisfies the T_3-axiom. As is readily verified, then $F^{-1}(\tau')$ also effectively satisfies the T_3-axiom. With Theorems 17 and 28 we therefore obtain the following theorem.

Theorem 30. *Let T be recursively separable and strongly effective. Moreover, let B be acceptable, let B' be computable and co-computable, and let τ' effectively satisfy the T_3-axiom. Then every effective operator $F : T \to T'$ must be effectively continuous.*

In Section 9.6 we have seen that the assumptions of this theorem are satisfied if T is a recursively separable recursive metric space with indexing x satisfying (A) and T' is a recursive metric space. For such spaces the effective continuity of effective operators has independently been shown by Ceĭtin (1962) and Moschovakis (1963, 1964).

References

America, P., and Rutten, J. (1988). Solving reflexive domain equations in a category of complete metric spaces. In *Mathematical Foundations of Programming Language Semantics, 3rd Workshop* (Main, M. et al., Eds.), pp. 252–288. Lec. Notes Comp. Sci. 298. Springer, Berlin.

de Bakker, J. W., and Zucker, J. I. (1982). Processes and the denotational semantics of concurrency. *Inform. and Control* 54, pp. 70–120.

Ceĭtin, G. S. (1962). Algorithmic operators in constructive metric spaces. *Trudy Mat. Inst. Steklov* 67, pp. 295–361; English transl., *Amer. Math. Soc. Transl.* (2) 64, pp. 1–80 (1967).

Czászár, A. (1963). *Foundations of General Topology*. Pergamon, New York.

Dyment, E. Z. (1984). Recursive metrizability of numbered topological spaces and bases of effective linear topological spaces. *Izv. Vyssh. Uchebn. Zaved, Mat.* (Kazan), no. 8, pp. 59–61; English transl., *Sov. Math. (Iz. VUZ)* 28, pp. 74–78 (1984).

Egli, H., and Constable, R. L. (1976). Computability concepts for programming language semantics. *Theoret. Comp. Sci.* 2, pp. 133–145.

Eršov, Ju. L. (1972). Computable functionals of finite type. *Algebra i Logika* 11, pp. 367–437; English transl., *Algebra and Logic* 11, pp. 203–242 (1972).

Eršov, Ju. L. (1973a). The theory of A-spaces. *Algebra i Logika* 12, pp. 369–416; English transl., *Algebra and Logic* 12, pp. 209–232 (1973).

Eršov, Ju. L. (1973b). Theorie der Numerierungen I. *Zeitschr. f. math. Logik Grundl. d. Math.* 19, pp. 289–388.

Eršov, Ju. L. (1975). Theorie der Numerierungen II. *Zeitschr. f. math. Logik Grundl. d. Math.* 21, pp. 473-584.

Eršov, Ju. L. (1977). Model $\mathbb{C}$ of partial continuous functionals. *Logic Colloquium 76* (Gandy, R. et al. Eds.), pp. 455–467. North-Holland, Amsterdam.

Fletcher, P., and Lindgren, W. F. (1982). *Quasi-Uniform Spaces.* Dekker, New York.

Friedberg, R. (1958). Un contre-exemple relatif aux fonctionelles récursives. *Compt. Rend. Acad. Sci. (Paris)*, 247, pp. 852–854.

Gierz, G., Hofmann, K. H., Keimel, K., Lawson, J. D., Mislove, M., and Scott, D. S. (1980). *A Compendium on Continuous Lattices.* Springer, Berlin.

Golson, W. G., and Rounds, W. C. (1983). Connections between two theories of concurrency: Metric spaces and synchronisation trees. *Inform. and Control* 57, pp. 102–124.

Hauck, J. (1980). Konstruktive Darstellungen in topologischen Räumen mit rekursiver Basis. *Zeitschr. f. math. Logik Grundl. d. Math.* 26, pp. 565–576.

Hauck, J. (1981). Berechenbarkeit in topologischen Räumen mit rekursiver Basis. *Zeitschr. f. math. Logik Grundl. d. Math.* 27, pp. 473–480.

Helm, J. (1971). On effectively computable operators. *Zeitschr. f. math. Logik Grundl. d. Math.* 17, pp. 231–244.

Hennessy, M., and Plotkin, G. (1980). A term model for CCS. *Mathematical Foundations of Computer Science* (Dembiński, E. P., Ed.), pp. 261–274. Lec. Notes Comp. Sci. 88. Springer, Berlin.

Hingston, Ph. (1988). Non-complemented open sets in effective topology. *J. Austral. Math. Soc. (Series A)* 44, pp. 129–137.

Kalantari, I. (1982). Major subsets in effective topology. *Patras Logic Symposium* (Metakides, G., Ed.), pp. 77–94. North-Holland, Amsterdam.

Kalantari, I., and Leggett, A. (1982). Simplicity in effective topology. *J. Symbolic Logic* 47, pp. 169–183.

Kalantari, I., and Leggett, A. (1983). Maximality in effective topology. *J. Symbolic Logic* 48, pp. 100–112.

Kalantari, I., and Remmel, J. B. (1983). Degrees of recursively enumerable topological spaces. *J. Symbolic Logic* 48, pp. 610–622.

Kalantari, I., and Retzlaff, A. (1979). Recursive constructions in topological spaces. *J. Symbolic Logic* 44, pp. 609–625.

Kalantari, I., and Weitkamp, G. (1985a). Effective topological spaces I: A definability theory. *Ann. Pure Appl. Logic* 29, pp. 1–27.

Kalantari, I. and Weitkamp, G. (1985b). Effective topological spaces II: A hierarchy. *Ann. Pure Appl. Logic* 29, pp. 207–224.

Kalantari, I., and Weitkamp, G. (1987). Effective topological spaces III: Forcing and definability. *Ann. Pure Appl. Logic* 36, pp. 17–27.

Kanda, A. (1980). Gödel numbering of domain theoretic computable functions. Report no. 138, Dept. of Comp. Sci., Univ. of Leeds.

Kanda, A., and Park, D. (1979). When are two effectively given domains identical? *Theoretical Computer Science, 4th GI Conference* (Weihrauch, K., Ed.), pp. 170–181. Lec. Notes Comp. Sci. 67. Springer, Berlin.

Kreisel, G., Lacombe, D. and Shoenfield, J. (1959). Partial recursive functionals and effective operations. *Constructivity in Mathematics* (Heyting, A., Ed.), pp. 290–297. North-Holland, Amsterdam.

Kreitz, Ch. (1982). Zulässige cpo's, ein Entwurf für ein allgemeines Berechenbarkeitskonzept. Schriften zur Angew. Math. u. Informatik Nr. 76, Rheinisch-Westfälische Technische Hochschule Aachen.

Lachlan, A. (1964). Effective operators in a general setting. *J. Symbolic Logic* 29, pp. 163–178.

Mal'cev, A. I. (1971). *The Metamathematics of Algebraic Systems. Collected Papers: 1936–1967* (Wells III, B.F., Ed.). North-Holland, Amsterdam.

Milner, R. (1980). *A Calculus of Communicating Sequences.* Lec. Notes Comp. Sci. 92. Springer, Berlin.

Moschovakis, Y. N. (1963). Recursive analysis. Ph.D. Thesis, Univ. of Wisconsin, Madison, Wis.

Moschovakis, Y. N. (1964). Recursive metric spaces. *Fund. Math.* 55, pp. 215–238.

Myhill, J., and Shepherdson, J. C. (1955). Effective operators on partial recursive functions. *Zeitschr. f. math. Logik Grundl. d. Math.* 1, pp. 310–317.

Nivat, M. (1979). Infinite words, infinite trees, infinite computations. *Foundations of Computer Science III, Part 2* (de Bakker, J. W. et al., Eds.), pp. 1–52. Math. Centre Tracts 109.

Nogina, E. Ju. (1966). On effectively topological spaces. *Dokl. Akad. Nauk SSSR* 169, pp. 28–31; English transl., *Soviet Math. Dokl.* 7, pp. 865–868 (1966).

Nogina, E. Ju. (1969). Relations between certain classes of effectively topological spaces. *Mat. Zametki* 5, pp. 483–495 (Russian); English transl., *Math. Notes* 5, pp. 288–294 (1969).

Nogina, E. Ju. (1978a). Enumerable topological spaces. *Zeitschr. f. math. Logik Grundl. d. Math.* 24, pp. 141–176 (Russian).

Nogina, E. Ju. (1978b). On completely enumerable subsets of direct products of numbered sets. *Mathematical Linguistics and Theory of Algorithms*, pp. 130–132 (Russian). Interuniv. thematic Collect., Kalinin Univ.

Nogina, E. Ju. (1981). The relation between separability and traceability of sets. *Mathematical Logic and Mathematical Linguistics*, pp. 135–144 (Russian). Kalinin Univ.

Pour-El, M. B. (1960). A comparison of five 'computable' operators. *Zeitschr. f. math. Logik Grundl. d. Math.* 6, pp. 325–340.

Reed, G. M., and Roscoe, A. W. (1988). Metric spaces as models for real-time concurrency. *Mathematical Foundations of Programming Language Semantics, 3rd Workshop* (Main, M. et al., Eds.), pp. 330–343. Lec. Notes Comp. Sci. 298. Springer, Berlin.

Roscoe, A. W. (1982). A mathematical theory of communicating processes. D.Phil. Thesis, Oxford Univ.

Rogers, H., Jr. (1967). *Theory of Recursive Functions and Effective Computability.* McGraw-Hill, New York.

Rounds, W. C. (1985). Applications of topology to the semantics of communicating processes. *Seminar on Concurrency.* (Brookes, S. D. et al., Eds.), pp. 360–327. Lec. Notes Comp. Sci. 197. Springer, Berlin.

Sciore, E., and Tang, A. (1978a). Admissible coherent c.p.o.'s, *Automata, Languages and Programming* (Ausiello, G. et al., Eds.), pp. 440–456. Lec. Notes Comp. Sci. 62. Springer, Berlin.

Sciore, E., and Tang, A. (1978b). Computability theory in admissible domains. *10th Annual ACM Symp. on Theory of Computing*, pp. 95–104. Ass. Comp. Mach, New York.

Sciore, E., and Tang, A. (1978c). Effective domains. Manuscript.

Scott, D. (1970). Outline of a mathematical theory of computation. Techn. Monograph PRG-2, Oxford Univ. Comp. Lab.

Scott, D. (1971). Continuous lattices. *Toposes, Algebraic Geometry and Logic* (Bucur, I. et al., Eds.), pp. 97–136. Lec. Notes Math. 274. Springer, Berlin.

Scott, D. (1972). Lattice theory, data types and semantics. *Formal Semantics of Programming Languages* (Rustin, R., Ed.), pp. 65–106. Prentice-Hall, Englewood Cliffs, N.J.

Scott, D. (1973). Models for various type-free calculi. *Logic, Methodology and Philosophy of Science IV* (Suppes, P. et al., Eds.), pp. 157–187. North-Holland, Amsterdam.

Scott, D. (1976). Data types as lattices. *SIAM J. on Computing* 5, pp. 522–587.

Scott, D. (1981). Lectures on a mathematical theory of computation. Techn. Monograph PRG-19, Oxford Univ. Comp. Lab.

Scott, D. (1982). Domains for denotational semantics. *Automata, Languages and Programming* (Nielsen, M. et al., Eds.), pp. 577–613. Lec. Notes Comp. Sci. 140. Springer, Berlin.

Scott, D., and Strachey, C. (1971). Towards a mathematical semantics for computer languages. *Computers and Automata* (Fox, J., Ed.), pp. 19–46. Polytechnic Press, Brooklyn, N.Y.

Smyth, M. B. (1977). Effectively given domains. *Theoret. Comp. Sci.* 5, pp. 257–274.

Smyth, M. B. (1983). Power domains and predicate transformers. *Automata, Languages and Programming* (Diaz, J., Ed.), pp. 662–675. Lec. Notes Comp. Sci. 154. Springer, Berlin.

Smyth, M. B. (1987). Completeness of quasi-uniform spaces in terms of filters. Manuscript.

Smyth, M. B. (1988). Quasi-uniformities: reconciling domains with metric spaces. *Mathematical Foundations of Programming Language Semantics, 3rd Workshop* (Main, M. et al., Eds.), pp. 236–253. Lec. Notes Comp. Sci. 298. Springer, Berlin.

Spreen, D. (1984). On r.e. inseparability of cpo index sets. *Logic and Machines: Decision Problems and Complexity* (Börger, E. et al., Eds.), pp. 103–117. Lec. Notes Comp. Sci. 171. Springer, Berlin.

Spreen, D. (1985). Rekursionstheorie auf Teilmengen partieller Funktionen. Habilitationsschrift, Rheinisch-Westfälische Technische Hochschule Aachen.

Spreen, D. (199?). A characterization of effective topological spaces. *Proc. 1989 Oberwolfach Conf. on Recursion Theory* (Ambos-Spies, K. et al., Eds.). Lec. Notes Math. Springer, Berlin. Forthcoming.

Spreen, D., and Young, P. (1984). Effective operators in a topological setting. *Computation and Proof Theory, Proc. Logic Colloquium '83* (Richter, M. M. et al., Eds.), pp. 437–451. Lec. Notes Math. 1104. Springer, Berlin.

Vaĭnberg Ju. R., and Nogina, E. Ju. (1974). Categories of effectively topological spaces. *Studies in Formalized Languages and Nonclassical Logics*, pp. 253–273 (Russian). Izdat. 'Nauka', Moscow.

Vaĭnberg, Ju. R., and Nogina, E. Ju. (1976). Two types of continuity of computable mappings of numerated topological spaces. *Studies in the Theory of Algorithms and Mathematical Logic*, Vol. 2, pp. 84–99, 159 (Russian). Vycisl. Centr Akad. Nauk SSSR, Moscow.

Vickers, S. (1989). *Topology via Logic.* Cambridge Univ. Press, Cambridge.

Weihrauch, K., and Deil, Th. (1980). Berechenbarkeit auf cpo-s. Schriften zur Angew. Math. u. Informatik Nr. 63, Rheinisch-Westfälische Technische Hochschule Aachen.

Weihrauch, K. (1981). Computability on metric spaces. Informatik Berichte Nr. 21, Fernuniversität Hagen.

Weihrauch, K. (1987). *Computability.* Springer, Berlin.

Xiang, Li. (1988). Everywhere nonrecursive r.e. sets in recursively presented topological spaces. *J. Austral. Math. Soc. (Series A)* 44, pp. 105–128.

Young, P. (1968). An effective operator, continuous but not partial recursive. *Proc. Amer. Math. Soc.* 19, pp. 103–108.

Young, P., and Collins, W. (1983). Discontinuities of provably correct operators on the provably recursive real numbers. *J. Symbolic Logic* 48.

10
The importance of cardinality, separability, and compactness in computer science with an example from numerical signal analysis

KLAUS E. GRUE

Abstract

This chapter gives an example where topological guidance has been essential in developing a numerical algorithm for solving a problem from signal analysis. The chapter explains the importance of cardinality, separability, and compactness in numerical analysis, and provides examples of spaces that can be made separable or compact by a non-standard choice of topology. Furthermore, the chapter suggests a definition of 'approximate computability' and analyses some immediate consequences of the definition.

10.1 Introduction

Computers can perform accurate computations on some sets like the set of integers, and approximate computations on other sets like the set of real numbers. There are also sets on which computers can perform neither accurate nor approximate computations. One observation to be stated is that computers can only perform approximate computations over separable spaces (separability is necessary but not sufficient). In particular, computers can only perform approximate computations over spaces of cardinality at most $2^{\aleph_0}$.

In the example to be given, the problem is to find a bounded real function with certain properties. As the set of bounded real functions has cardinality exceeding $2^{\aleph_0}$, the problem is reformulated to take place in L_∞ which has cardinality $2^{\aleph_0}$.

We equip L_∞ with the so-called weak*-topology to obtain separability. The unit sphere (w.r.t. $\|\bullet\|_{1_\infty}$) in L_∞ is compact (w.r.t. the weak*-topology), and this result is used as follows:

258 The importance of cardinality, separability, and compactness

1. to prove that solutions to the problem exist;
2. to prove that the sequence of approximations generated by the numerical algorithm has a subsequence that converges towards a solution;
3. to prove a property about the solutions.

From the point of view of numerical analysis, it may be somewhat surprising that it is possible to compute with bounded functions (or, rather, L_∞-functions) without imposing any further restrictions on these functions. Applications of computation on L_∞ are numerous. As one example, a grey tone screen image can be represented as an L_∞-function which gives the intensity of each point as an L_∞-function of time and two spatial coordinates. Colour images can be represented as three L_∞-functions. By representing computer images this way it becomes possible to write programs that produce images independently of screen resolution and refresh rate.

The concept of 'computability' is important in computer science. The chapter ends by giving some suggestions of what 'approximate computability' could mean and states the observation (well known from intuitionistic logic) that merely continuous functions can be 'approximately computable'.

The example is interdisciplinary in nature, but the results needed from each discipline are not particularly deep.

10.2 An optimization problem

The problem which was actually solved by topological guidance is described in (Grue 1985). However, this chapter considers a slightly different problem which is more interesting from the point of view of topology and exhibits less technicalities.

The problem to be considered is the following 'optimization problem'. Let $a_1, \ldots, a_n \in \mathbb{R}$ (the set of real numbers). Let $h_0, \ldots, h_n \in L_1$, that is, let $h_0, \ldots, h_n : \mathbb{R} \to \mathbb{R}$ be real functions such that

$$\|h_i\|_1 = \int_{-\infty}^{+\infty} |h_i(t)|\, dt < +\infty \quad i \in \{0, \ldots, n\}$$

(All integrals are Lebesgue integrals.) Find a function $g : \mathbb{R} \to \mathbb{R}$ such that

$$\forall t \in \mathbb{R} : |g(t)| \leq 1$$

and

$$\langle g, h_i \rangle = \int_{-\infty}^{+\infty} g(t)\, h_i(t)\, dt \leq a_i \quad i \in \{1, \ldots, n\}$$

$$\langle g, h_0 \rangle = \int_{-\infty}^{+\infty} g(t)\, h_0(t)\, dt \quad \text{is maximal}$$

(Also prove the existence of g.) The importance of this problem follows from Golomb and Weinberger (1959), except that Golomb and Weinberger deal mainly with finite energy signals (L_2) rather than bounded signals. Furthermore, an application of the solution to the problem is stated in the Appendix.

The importance of separability

Let $\mathbb{Z}$ denote the set of integers, and let $D = \{m2^n \mid m, n \in \mathbb{Z}\}$ be the set of finite binary expansions. The set D is countable, which means that elements of D can be represented using finitely many bits in a computer. Different elements of D may require different numbers of bits, and elements of D may require arbitrarily many bits, but each element of D merely requires finitely many bits. The set D is a dense subset of $\mathbb{R}$, which means that any element of $\mathbb{R}$ can be approximated arbitrarily well by elements of D. D is suited to numerical computations in $\mathbb{R}$ exactly because D is a countable dense subset of $\mathbb{R}$. A topological space is 'separable' if it has a countable dense subset. Hence, a necessary requirement to make approximate computations in a topological space is that the space is separable.

As any separable space has cardinality at most $2^{\aleph_0}$, another necessary requirement is that the space has cardinality at most $2^{\aleph_0}$.

Satisfaction of the cardinality condition

The optimization problem mentioned above aims at finding a bounded function g. However, the set of bounded functions has cardinality exceeding $2^{\aleph_0}$, and so we formulate the problem to require $g \in L_\infty, \|g\|_{1\infty} \leqslant 1$ instead of $g : \mathbb{R} \to \mathbb{R}, \forall t \in \mathbb{R} : |g(t)| \leqslant 1$. Assuming $g \in L_\infty$ means the following.

1. We identify functions that are equal almost everywhere, that is, we identify functions g_1 and g_2 for which $\int_{-\infty}^{+\infty} |g_1(t) - g_2(t)| \, dt = 0$.

2. We assume that $\int_0^x g(t) \, dt$ is defined and finite for all $x \in \mathbb{R}$.

Both these restrictions are inessential to the optimization problem. The two restrictions together reduce the space to be considered to cardinality $2^{\aleph_0}$.

Satisfaction of the separability condition

The norm $\|\bullet\|_{1\infty}$ induces a topology on L_∞ which, unfortunately, is not separable. Fortunately, however, the so-called weak*-topology on L_∞ makes L_∞ separable. The weak*-topology is uniquely determined by the following property.

Let $g, g_1, g_2, \ldots \in L_\infty$. We have $g_i \to g$ for $i \to +\infty$ w.r.t. the weak*-topology iff $\langle g_i, h \rangle \to \langle g, h \rangle$ for all $h \in L_1$.

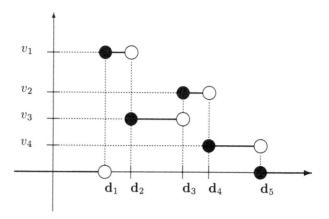

Fig. 10.1 An example step function

One may think of this concept of convergence as an adaptation of pointwise convergence to L_∞.

The choice of the weak*-topology has another benefit. Alaoglu's theorem (Rudin 1973) states that the unit sphere

$$B = \{g \in L_\infty \mid \|g\|_{1_\infty} \leqslant 1\}$$

is compact w.r.t. the weak*-topology, that is, any sequence $g_1, g_2, \ldots \in B$ has a subsequence which converges w.r.t. the weak*-topology (and the limit of any such sequence again belongs to B).

A countable dense subset of L_∞

Let $n \geqslant 1$ and let $v = (v_1, \ldots, v_{n-1}) \in D^{n-1}$. Let $d = \{d_1, \ldots, d_n\} \subseteq D$ be a set of n distinct elements of D (the set of finite binary expansions), and let $(\mathbf{d}_1, \ldots, \mathbf{d}_n)$ be the d_is sorted in ascending order. We define the step function $S_{v,d}(t)$ as

$$S_{v,d}(t) = \begin{cases} 0 & \text{for } t \in (-\infty, \mathbf{d}_1) \\ v_i & \text{for } t \in [\mathbf{d}_i, \mathbf{d}_{i+1}), i \in \{1, \ldots, n-1\} \\ 0 & \text{for } t \in [\mathbf{d}_n, +\infty) \end{cases}$$

In particular, for $n = 1$, we define $S_{v,d}(t) = 0$ for $t \in (-\infty, \infty)$. Figure 10.1 shows $S_{(v_1,\ldots,v_4),\{d_1,\ldots,d_5\}}$ for some $v_1, \ldots, v_4$ and $d_1, \ldots, d_5$.

If $d \subseteq D$ has n elements, then we define

$$V_d = \{S_{v,d} \mid v \in D^{n-1}\}$$

The set V_d is an $(n-1)$-dimensional vector space. Further define V as the union of the V_ds:

$$V = \bigcup \{V_d \mid d \subseteq D, d \text{ finite }\}$$

The set V is a countable dense subset of L_∞ (with respect to the weak*-topology). Hence, V is a candidate for performing approximate computations with elements of L_∞. Any countable superset of V may also serve as a candidate, as any superset of V is also dense in L_∞.

A numerical solution to the optimization problem

Let $d_1, d_2, \ldots$ be an enumeration of D, that is, let $d_1, d_2, \ldots$ be any sequence without repetitions such that $\{d_1, d_2, \ldots\} = D$. Define $e(i) = \{d_1, \ldots, d_i\}$. We can easily verify that

$$V = \bigcup \{V_{e(i)} \mid i \in \mathbb{N}\}$$
$$V_{e(1)} \subseteq V_{e(2)} \subseteq \cdots$$

We may use this observation to outline a numerical algorithm for solving the optimization problem. The algorithm is not known to be present in the literature. Let

$$W = \{g \in L_\infty \mid \|g\|_{1\infty} \leqslant 1 \wedge \langle g, h_1 \rangle \leqslant a_1 \wedge \cdots \wedge \langle g, h_n \rangle \leqslant a_n\}$$

We may state the optimization problem as follows: find $g \in W$ such that $\langle g, h_0 \rangle$ is maximal (and prove the existence of g).

If $W = \emptyset$ then, obviously, the problem has no solutions, and so we rule out this case. In the problem considered in the Appendix, we have $a_i > 0$, $i \in \{1, \ldots, n\}$, and so the zero function is an element of W in that case.

Further, to rule out pathological cases, we assume that W is the closure of $W \cap V$, that is, that any element of W can be approximated arbitrarily well by elements of V. This will normally be easy to verify for practical applications.

As an example of a pathological case, let $n = 2$ and let h_1 be the constant function $h_1(t) = 1$. Let a_1 be any real number which has no finite binary expansion (for example, π) and let $h_2 = -h_1$ and $a_2 = -a_1$. We have $W = \{g \in L_\infty \mid \|g\|_{1\infty} \leqslant 1 \wedge \langle g, h_1 \rangle = a_1\}$ so that $W \neq \emptyset$ and $W \cap V = \emptyset$.

From the assumptions we deduce that $W \cap V \neq \emptyset$. Hence, for any sufficiently large $k \in \mathbb{N}$, we have $W \cap V_{e(k)} \neq \emptyset$. Now, for each $i \geqslant k$ we may solve the following problem: find $\mathbf{g}_i \in W \cap V_{e(i)}$ such that $\langle \mathbf{g}_i, h_0 \rangle$ is maximal. As we shall see later, this may be done using the simplex method (Rockafellar 1970). According to Alaoglu's theorem, as $\|\mathbf{g}_i\|_{1\infty} \leqslant 1$

for $i \geqslant k$, the series $\mathbf{g}_k, \mathbf{g}_{k+1}, \ldots$ has a convergent subseries (w.r.t. the weak*-topology), and the limit g for such a series obviously maximizes $\langle g, h_0 \rangle$ for $g \in W$.

Hence, we have a method for finding a series $\mathbf{g}_k, \mathbf{g}_{k+1}, \ldots$ which can be thinned into a convergent subseries. The process of actual thinning is irrelevant for the problem in the Appendix, where it is the value of $\langle g, h_0 \rangle$ which is needed. The series $\langle \mathbf{g}_k, h_0 \rangle, \langle \mathbf{g}_{k+1}, h_0 \rangle, \ldots$ of real numbers converges monotonically towards $\langle g, h_0 \rangle$.

As the sequence $d_1, d_2, \ldots$ may be any enumeration of D, a clever implementation of the above method may determine this sequence 'on the fly' such that d_i is placed where g is expected to be most dynamic.

The simplex method

The set $V_{e(i)}$ consists of all functions $S_{v,e(i)}$ where $v = (v_1, \ldots, v_{i-1}) \in D^{i-1}$. To find $\mathbf{g}_i \in W \cap V_{e(i)}$ such that $\langle \mathbf{g}_i, h_0 \rangle$ is maximal we need to solve the following problem: find $v = (v_1, \ldots, v_{i-1}) \in D^{n-1}$ such that

$$-1 \leqslant v_j \leqslant 1 \qquad j \in \{1, \ldots, i-1\}$$
$$v_1 c_{1k} + \cdots + v_{i-1} c_{(i-1)k} \leqslant a_j \quad k \in \{1, \ldots, n\}$$
$$v_1 c_{10} + \cdots + v_{i-1} c_{(i-1)0} \quad \text{is maximal}$$

where

$$c_{ij} = \int_{d_i}^{d_{i+1}} h_j(t)\, dt$$

Hence we have $2i - 2 + n$ linear inequalities and one linear form to optimize, which is exactly what the simplex method can solve.

The above problem may have several solutions. Among these solutions one can find solutions for which at least $i - 1$ of the inequalities become equalities (Rockafellar 1970). Hence, one can find solutions for which $|v_j| = 1$ for at least $i - 1 - n$ different j, and $|v_j| \neq 1$ for at most n different j.

A property of the solutions

We have now outlined a numerical algorithm which finds a sequence $\mathbf{g}_k, \mathbf{g}_{k+1}, \ldots$ where each $\mathbf{g}_i$ is found using the simplex method. The sequence is known to have a convergent subsequence, and the limit of any such sequence is a solution to the optimization problem. Each function $\mathbf{g}_i$ is a step function $S_{v,d}$ where $v = (v_1, \ldots, v_{i-1})$.

If we choose each $\mathbf{g}_i$ such that $|v_j| \neq 1$ holds for at most n values, then any limit g of any convergent subsequence must satisfy $|g(t)| = 1$ for almost all t. To see this, proceed as follows: for any $h \in L_1$ we have

$$\int_{-\infty}^{+\infty} (1 - |\mathbf{g}_i(t)|) h(t) \, dt \to 0 \quad i \to +\infty$$

and hence

$$\int_{-\infty}^{+\infty} (1 - |g(t)|) h(t) \, dt = 0$$

As this holds for all $h \in L_1$, $1 - |g(t)| = 0$ for almost all t.

Hence, among the solutions to the optimization problem there are functions that bounce back and forth between -1 and 1. This indicates that it is reasonable to approximate solutions to the optimization problem by step functions. This is an important result seen from the point of view of numerical analysis. For other problems it might be more reasonable to approximate by piecewise linear continuous functions, splines, or other countable dense subsets of L_∞. The choice of a countable dense subset of L_∞ affects the pace of convergence and thereby the efficiency of the algorithm.

10.3 Further work

An interesting issue which remains to be studied is the notion of 'approximate computability'. It is clear that for example, addition of real numbers can be approximated arbitrarily well by computers. It is also possible to verify that the Fourier transform $F : L_2 \to L_2$ can be approximated arbitrarily well if, for example, we choose V defined earlier as a countable dense subset of L_2.

The signum function $f : \mathbb{R} \to \mathbb{R}$ defined by $f(x) = -1$ for $x < 0$, $f(0) = 0$, and $f(x) = 1$ for $x > 0$ is not approximately computable, for if $y = 0$, then regardless of the accuracy with which the computer knows y, the computer cannot decide whether $f(y) = -1$, $f(y) = 0$, or $f(y) = 1$. (This is a standard example from intuitionistic logic (Heyting 1966).) However, the function $g : \mathbb{R} \setminus \{0\} \to \mathbb{R}$ defined by $g(x) = -1$ for $x < 0$ and $g(x) = 1$ for $x > 0$ is computable. As we can see, a necessary condition for a function g to be approximately computable is that g is continuous.

Just as with the concept of 'computability', it is not obvious what 'approximate computability' should mean. For the concept of 'computability', a number of suggestions have been made by Markov, Turing, Herbrand-Gödel, and others, and all these suggestions have been proved to be equivalent (Mendelson 1979). It seems reasonable to follow the same approach for 'approximate computability', that is, to state definitions of the concept, to study consequences of the definitions, and to compare various definitions to see if they are equivalent.

One definition of 'approximate computability' could proceed as follows. We first choose a domain S for performing computations. The choice of domain is somewhat arbitrary. The choice and its impacts are discussed later.

Choice of domain

Let $(S, \leqslant)$ be the 'cpo' (complete partial order) given by the domain equation (Schmidt 1986, Scott 1982):

$$S = (\{nil\} + S \times S)_\perp$$

Further, let S_f be the finite elements of S, that is, let S_f be the least set such that $nil \in S_f$, $\perp \in S_f$, and $\forall x, y \in S_f : (x, y) \in S_f$. Let S_m be the set of maximal elements of S, that is, let $S_m = \{x \in S \mid \forall y \in S : x \not< y\}$. If $x, y \in S$ and $x \leqslant y$, then we say that x 'approximates' y.

One property of S is: for each maximal element y there is a chain $x_1 \leqslant x_2 \leqslant \cdots$ of finite elements such that y is the only element of S for which $x_i \leqslant y$, $i \in \{1, 2, \ldots\}$. In other words, maximal elements may be approximated arbitrarily well by finite elements.

The cpo S is interesting from a computer science point of view because computers can, to some extent, compute with elements of S_f. The qualification 'to some extent' covers the fact that computers cannot do just anything with elements of S_f. In particular, a computer cannot do anything reasonable with the bottom element $\perp$ of the cpo, because $\perp$ represents 'total absence of information'.

Representation of topological spaces

Let T be a topological space, let $S' \subseteq S_m$, and let t' be a surjective (or 'onto' or 'epimorphic') function of type $t' : S' \to T$. For each $x' \in S'$ we say that x' is a 'representation' of $t'(x') \in T$.

For all $x \in S$ define $t(x) = \{t'(x') \mid x' \in S' \wedge x \leqslant x'\}$. We have that t is a function of type $t : S \to \mathbb{P}(T)$ where $\mathbb{P}(T)$ denotes the powerset of T. For each $x \in S$ we say that x is an 'approximate representation' of each $y \in T$ for which $y \in t(x)$. Hence $x \in S$ is an approximate representation of any element of $t(x)$.

We say that $t : S \to \mathbb{P}(T)$ is a 'representation' function for T if it can be formed as above and the following holds: for any $y \in T$ and for any open neighbourhood Y of y, there is an $x \in S_f$ such that $y \in t(x) \subseteq Y$. In other words, t is a representation function if it is possible to approximate any element of T arbitrarily well by finite elements of S.

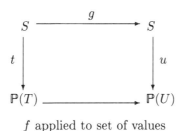

Fig. 10.2 Representing the function f by g

We say that a topological space T is 'representable' if there exists a representation function for T. As a direct consequence, any representable T is separable. To see this, choose (by the axiom of choice) an element $e_x \in t(x)$ for each $x \in S_f$, and form $E = \{e_x \mid x \in S_f\}$. E is a dense subset of T and, as S_f is countable, E is also at most countable.

It is trivial, but somewhat lengthy, to prove that the following topological spaces are representable: the real numbers with the usual topology; the integers with the point topology; any L_p space, $p \in [1, +\infty)$, with the $\|\bullet\|_{1_p}$-norm topology; the space L_∞ with the weak*-topology. Further, if A and B are representable, then the cartesian product $A \times B$ and the direct sum $A \oplus B$ are representable.

Representation of functions between spaces

Now let T and U be representable topological spaces and let t and u be representation functions for T and U respectively (see Figure 10.2). Let f and g be functions of type $f : T \to U$ and $g : S \to S$ respectively. We say that f and g are 'compatible' (w.r.t. t and u) if $f(t(x)) \subseteq u(g(x))$ for all $x \in S$ where $f(t(x))$ is shorthand for $\{f(y) \mid y \in t(x)\}$. Hence, if f and g are compatible, and if $x \in S$ represents $y \in T$, then $g(x)$ represents $f(y)$. Further, we say that g 'represents' f if f and g are compatible and the following holds: for all $y \in T$ and all neighbourhoods Z of $z = f(y) \in U$, there exists a neighbourhood Y of y such that

$$\forall x \in S : (t(x) \subseteq Y \Rightarrow u(g(x)) \subseteq Z)$$

In other words, we can approximate $f(y)$ arbitrarily well by $g(x)$ if only x approximates y sufficiently well.

We say that $f : T \to U$ is 'approximately computable' (w.r.t. t and u) if there exists a computable $g : S \to S$ which represents f. Hence, approximate computability is defined in terms of usual computability. Any of the equivalent definitions of usual computability will do. As an example we may state that the lazy Lisp (Friedman and Wise 1976, Henderson and Morris 1976, Henderson 1980) functions of type $S \to S$ are exactly the computable functions.

As a direct consequence, any approximately computable function is continuous. To see this, consider any element $y \in T$ and any neighbourhoods Z of $z = f(y) \in U$. There exists a neighbourhood Y of y such that $\forall x \in S : (t(x) \subseteq Y \Rightarrow u(g(x)) \subseteq Z)$. Combined with $f(t(x)) \subseteq u(g(x))$ this gives $\forall x \in S : (t(x) \subseteq Y \Rightarrow f(t(x)) \subseteq Z)$. For each $x' \in T$ there exists an $x \in S$ such that $t(x) = \{x'\}$. Hence $\forall x' \in T : (x' \in Y \Rightarrow f(x') \in Z)$. The continuity of f follows directly.

It is trivial but somewhat lengthy to prove that the following functions are approximately computable (w.r.t. suitable representation functions): addition, subtraction, and multiplication of real numbers and integers; division by non-zero real numbers and integers; addition of elements of L_p, $p \in [1, +\infty]$; multiplication of elements of L_p, $p \in [1, +\infty]$ by real numbers and integers; Fourier transformation of elements of L_2. As examples, addition of real numbers is an approximately computable function of type $\mathbb{R} \times \mathbb{R} \to \mathbb{R}$, and division is an approximately computable function of type $\mathbb{R} \times (\mathbb{R} \setminus \{0\}) \to \mathbb{R}$.

Further, the following is trivial to prove: if $f_1 : T_1 \to T_2$ and $f_2 : T_2 \to T_3$ are approximately computable (w.r.t. suitable representation functions), then $f_2 \circ f_1$ is also approximately computable. The identity function on any representable topological space is approximately computable. If $f_1 : T \to T_1$ and $f_2 : T \to T_2$ are computable, then $f : T \to T_1 \times T_2$ given by $f(x) = (f_1(x), f_2(x))$ is computable. Further, the projections $f_1 : T \to T_1$ and $f_2 : T \to T_2$ of any approximately computable $f : T \to T_1 \times T_2$ are approximately computable. Similar results hold for direct sums.

Negative results

If T ($\neq \emptyset$) and U are topological spaces and U is of cardinality $2^{\aleph_0}$, then there exist constant functions $k : T \to U$ that are not computable, for there are $2^{\aleph_0}$ constant functions but merely $\aleph_0$ computable functions. Likewise, if U is a representable topological space of cardinality $2^{\aleph_0}$ with a representation function u, then there exists $x \in U$ such that $\{x\} = u(y)$ holds for no computable constant $y \in S$. Hence, representability of U does not ensure computability of each of its elements.

The domain S itself is not representable. An explanation of this phenomenon is that S contains elements that are only 'partially defined' in computer science terms. From a mathematical point of view, the set S is well defined and each element of S is a distinct object. From a computer science view, however, non-maximal elements are elements that are only partially defined. In particular, the bottom element ⊥ is completely undefined, and any process that does anything with ⊥ except passing it around, is doomed to loop indefinitely (as a consequence of Turing's halting problem (Mendelson 1979)). Hence, from a computer science point of view, ⊥ is not a distinct object, for any attempt to distinguish it from any other object by means of a computer makes the computer loop indefinitely. Intuitively, the definition of 'representation function' rules out spaces with partial objects. The definition of representability has been chosen to fit the normal mathematical view that topological spaces consist of well-defined distinct objects.

Consequences of the choice of domain

The choice of the domain S above was somewhat arbitrary. Many other choices will lead to the same class of 'approximately computable functions'. The choice of one particular domain is useful when proving the approximate computability of compositions of approximately computable functions. The domain S above is the simplest that makes trivial the proof of the representability of cartesian products and direct sums.

Some relations to the literature and representation of real numbers

It is interesting to compare the above framework with the programs in (Boehm et al. 1986). Boehm et al. consider several possible representations of real numbers and provide computer programs for addition, subtraction, multiplication, and division with arbitrary precision for these representations. For simplicity, we merely consider a simplified version of one of the representations.

Expressed in the framework above, the representation of elements of $[-1, 1]$ in (Boehm et al. 1986) is as follows. Consider the cpo $\underline{S}$ given by

$$\underline{S} = (\{-1, 0, 1\} \times \underline{S})_\perp$$

The maximal elements of $\underline{S}$ are infinite sequences $(a_1, a_2, \ldots)$ where $a_i \in \{-1, 0, 1\}$ for $i \in \{1, 2, \ldots\}$. Now, for each maximal element $x = (a_1, a_2, \ldots)$ of $\underline{S}$ define

$$t'(x) = \sum_{i=1}^{+\infty} a_i 2^{-i}$$

Then define $t : \underline{S} \to \mathbb{P}([-1, 1])$ by $t(x) = \{t'(x') \mid x' \in \underline{S}_m \wedge x \leqslant x'\}$, and let $\phi : S \to \underline{S}$ be any computable surjective function that maps maximal elements to maximal elements. We have that $t \circ \phi$ is a representation function for the topological space $[-1, 1]$ (with the usual topology). (Boehm et al. (1986) also state representations of all of $\mathbb{R}$.)

As explained in (Boehm et al. 1986), usual binary expansions, that is, elements of $\underline{S}' = (\{0, 1\} \times \underline{S}')_\perp$ cannot be used to represent elements of $[0, 1]$. One problem with $\underline{S}'$ is that the addition $1/3 + 1/6 = 1/2$ requires infinite carry look ahead and hence cannot be performed in finite time. If $\underline{S}'$ is substituted for $\underline{S}$ in the definition of t above, then $t \circ \phi$ does not become a representation function. Furthermore, as explained in (Boehm et al. 1986), the use of $\underline{S}'' = (\{-2, -1, 0, 1, 2\} \times \underline{S}'')_\perp$ instead of $\underline{S}$ increases the efficiency of algorithms by reducing the need for carry look ahead.

For further inspiration, Clenshaw and Olver (1980, 1986) provide algorithms for computing certain real functions with arbitrary precision, and Collatz (1966) is an exponent for using functional analysis in numerical analysis.

10.4 Conclusion

- Any set of cardinality greater than $2^{\aleph_0}$ is out of reach of computers.
- Approximate computations are only possible in separable spaces.
- In order to keep the cardinality at $2^{\aleph_0}$ when working with functions $f : \mathbb{R}^n \to \mathbb{R}^m$, identify functions that are equal almost everywhere, and require f to satisfy that $\int_0^{x_1} \cdots \int_0^{x_n} f(t_1, \ldots, t_n)\, dt_1 \cdots dt_n$ is defined and finite for all real $x_1, \ldots, x_n$.
- In order to obtain separability when working in L_∞, use the weak*-topology.
- Use Alaoglu's theorem, that is, make use of the fact that any closed subset of L_∞ which is bounded w.r.t. $\|\bullet\|_{1\infty}$ is compact w.r.t. the weak*-topology.
- In numerical analysis, do not hesitate to represent real functions as data structures rather than as computer programs.

In addition to the above conclusions, the chapter has suggested a definition of 'approximate computability' and investigated some immediate consequences.

Acknowledgements

My thanks are due to Nils Andersen, DIKU, for comments on the manuscript, and to Professor Dr. Dieter Spreen and others at the conference for useful comments and corrections.

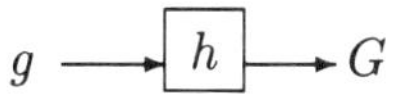

$$g \longrightarrow \boxed{h} \longrightarrow G$$

Fig. 10.3 The filter h

Appendix An example from numerical signal analysis

In signal analysis, a 'filter' is a physical device which takes a signal as input and delivers a signal as output. Such filters may be electrical filters that input and output electrical signals or they may be other kinds of filters such as microphones that input a sound signal and output an electrical signal.

The 'linear filters' are of particular interest because of their well understood properties. A linear filter is characterized by an 'impulse response' H, and its output G depends on the input g as follows:

$$G(t) = (g * H)(t) = \int_{-\infty}^{+\infty} g(\tau) H(t-\tau)\, d\tau$$

We define the 'inverse impulse response' h by $h(\tau) = H(-\tau)$. Hence

$$G(t) = (g * H)(t) = \int_{-\infty}^{+\infty} g(\tau) h(\tau - t)\, d\tau$$

In particular we have $G(0) = \langle g, h \rangle$.

Graphically, we represent the filter as in Figure 10.3. In signal analysis it is customary to assume $g \in L_2$ even though $g \in L_\infty$ is more reasonable for most physical signals. The assumption $g \in L_2$ leads to simpler mathematics. In this Appendix, however, we assume $g \in L_\infty$. Correspondingly, we assume $h \in L_1$.

Now assume that we want to build a filter with inverse impulse response $h \in L_1$ which is going to filter signals $g \in L_\infty$ for which $\|g\|_{1_\infty} \leqslant 1$. However, for economic reasons, we have to use a number of cheap filters with impulse responses $h_1, \ldots, h_n \in L_1$ and to combine these into one filter with approximately the impulse response h as in Figure 10.4. In Figure 10.4, F represents a function of type $F : \mathbb{R}^n \to \mathbb{R}$. We have

$$\underline{G}(t) = F(G_1(t), \ldots, G_n(t))$$
$$G_i(t) = \int_{-\infty}^{+\infty} g(\tau) h_i(\tau - t)\, d\tau + e_i(t)$$

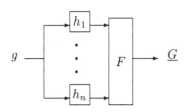

Fig. 10.4 Combined filters $h_1 \ldots h_n$

The functions e_i are unknown error functions which describe the imperfection of the filters $h_1, \ldots, h_n$. We assume $\|e_i\|_{1\infty} \leq \epsilon_i, i \in \{1, \ldots, n\}$, where $\epsilon_1, \ldots, \epsilon_n \in \mathbb{R}_+$ are known constants.

For any choice of F we define the 'error bound' E_F as the maximal difference between the G of Figure 10.3 and the $\underline{G}$ of Figure 10.4:

$$E_F = \sup\{\|G - \underline{G}\|_{1\infty} \mid g \in L_\infty \wedge \|g\|_{1\infty} \leq 1 \wedge \|e_i\|_{1\infty} \leq \epsilon_i, i \in \{1, \ldots, n\}\}$$

Hence

$$E_F = \sup\{|\langle g, h \rangle - F(\langle g, h_1 \rangle + d_1, \ldots, \langle g, h_n \rangle + d_n)| \mid g \in L_\infty \wedge \|g\|_{1\infty} \leq 1 \wedge |d_i| \leq \epsilon_i\}$$

One somewhat surprising result (Golomb and Weinberger 1959) is that among the functions $F : \mathbb{R}^n \to \mathbb{R}$ which minimize E_F, there is at least one which is linear in its arguments. To see this, proceed as follows: define

$$C = \{(\langle g, h \rangle, \langle g, h_1 \rangle + d_1, \ldots, \langle g, h_n \rangle + d_n) \mid g \in L_\infty \wedge \|g\|_{1\infty} \leq 1 \wedge |d_i| \leq \epsilon_i\}$$

Obviously, C is a closed bounded convex subset of $\mathbb{R}^{n+1}$, and C is symmetric around $(0, \ldots, 0)$. Define $A = \sup\{x_0 \in \mathbb{R} \mid (x_0, 0, \ldots, 0) \in C\}$. We have that $(A, 0, \ldots, 0)$ is on the boundary of C. From (Rockafellar 1970) we have that there is at least one tangent to C through $(A, 0, \ldots, 0)$ which is not parallel to $(1, 0, \ldots, 0)$. Let $x_0 = a_1 x_1 + \cdots + a_n x_n + A$ be such a tangent, and let $\underline{F}(x_1, \ldots, x_n) = a_1 x_1 + \cdots + a_n x_n$. We have

$$\forall (x_0, \ldots, x_n) \in C : |x_0 - \underline{F}(x_1, \ldots, x_n)| \leq A$$

and equality is obtained for $(A, 0, \ldots, 0)$ and $(-A, 0, \ldots, 0)$. Hence, $E_{\underline{F}} = A$ for this particular $\underline{F}$. However, as $(A, 0, \ldots, 0)$ and $(-A, 0, \ldots, 0)$ both belong to C, $E_F \geq A$ for any F, and so $\underline{F}$ minimizes E_F.

To find A we have to find $\sup\{x_0 \in \mathbb{R} \mid (x_0, 0, \ldots, 0) \in C\}$, that is, to find the maximal value of

$$x_0 = \langle g, h \rangle$$

where

$$g \in L_\infty \quad \|g\|_{1_\infty} \leqslant 1$$
$$\langle g, h_1 \rangle \leqslant \epsilon_1 \quad \langle g, -h_1 \rangle \leqslant \epsilon_1$$
$$\vdots \qquad\qquad \vdots$$
$$\langle g, h_n \rangle \leqslant \epsilon_n \quad \langle g, -h_n \rangle \leqslant \epsilon_n$$

This is exactly the optimization problem stated in the main text. To find $\underline{F}$, that is, to find $a_1, \ldots, a_n$, one has to perform numerical differentiation of the surface of C at the point $(A, 0, \ldots, 0)$ and hope that C is differentiable. The numerical differentiation requires the optimization problem to be solved n times with small perturbations of $e_1, \ldots, e_n$, and $2n$ times if differentiability has to be verified. We omit the details.

We have now outlined one optimization problem in signal analysis which can be solved by solving the optimization problem in the main text. Many similar optimization problems in signal analysis can be solved the same way.

References

Boehm, H. J., Cartwright, R., Riggle, M., and O'Donell, M. J. (1986) Exact Real Arithmetic: A Case Study in Higher Order Programming. *1986 ACM Symposium on Lisp and Functional Programming.*

Clenshaw, C. W., and Olver, F. W. J. (1980) An unrestricted algorithm for the exponential function. *SIAM Journal on Numerical Analysis.* Vol. 17, pp. 310–331.

Clenshaw, C. W., and Olver, F. W. J. (1986) Unrestricted algorithms for reciprocals and square roots. *BIT,* Vol. 26, pp. 476–492.

Collatz, L. (1966) *Functional Analysis and Numerical Mathematics.* Academic Press.

Friedman, D. P., and Wise, D. S. (1976) CONS Should Not Evaluate Its Arguments. In: S. Michaelson and R. Milner (Ed.), *Automata, Languages and Programming.* Edinburgh University Press, pp. 257–284.

Golomb, M., and Weinberger, H. F. (1959) Optimal Approximation and Error Bounds. In: R. Langer (Ed.), *On Numerical Approximation,* pp. 117–190. The University of Wisconsin Press.

Grue, K. E. (1985) Optimal Reconstruction of Bandlimited Bounded Signals. *IEEE Transactions on Information Theory,* Vol. IT-31, No. 5, pp. 594-601.

Henderson, P., and Morris, J. H. Jr. (1976) A Lazy Evaluator. *Conference Record of the Third ACM Symposium on Principles of Programming Languages.*

Henderson, P. (1980) *Functional Programming, Application and Implementation.* Prentice-Hall.

Heyting, A. (1966) *Intuitionism, An Introduction.* North-Holland Publishing Company.

Mendelson, E. (1979) *Introduction to Mathematical Logic.* D.Van Nostrand Company.

Rockafellar, R. T. (1970) *Convex Analysis.* Princeton University Press.

Rudin, W. (1973) *Functional Analysis.* McGraw-Hill Book Company.

Schmidt, D. A. (1986) *Denotational Semantics.* Allyn and Bacon Inc.

Scott, D. S. (1982) Domains for denotational semantics, in LNCS 140: *Proc. 9th ICALP*, pp. 577–613, Springer.

11
Digital topology: a comparison of the graph-based and topological approaches

T. Y. KONG AND A. ROSENFELD

Abstract

In image processing, the study of the topological properties of binary images has usually been based on adjacency relations defined on the black and white point sets (most often a combination of the so-called 4-, 8-, 6-, 18-, and 26-adjacency relations). We call this the *graph-based* approach to digital topology.

A different approach has been proposed by Khalimsky and other topologists (Kopperman, Meyer, and Wilson). This alternative approach associates binary images with finite or locally finite topological spaces, and so we call it the *topological* approach to digital topology. Kovalevsky has recently (and independently) developed a similar theory and used it in practical applications.

This chapter describes and compares the graph-based and topological approaches to digital topology.

11.1 Introduction

An n-dimensional binary *digital image* is a partition of a set V of grid points in Euclidean n-space (usually $V = \mathbb{Z}^n$) into two sets: the points in one set are called *black* points, while the points in the other set are called *white* points. In practice n is normally 2 or 3, and the set of black points is finite.

The set V is used as a discrete approximation ('digitization') of $\mathbb{R}^n$. The set of black points is a digitization of regions of interest. In the digital image of a printed page the value of n is 2 and the set of black points might be a digitization of the letters printed on the page.

Topological notions such as connectedness can be defined for binary digital images. Digital topology is about these derived notions. Concepts of digital topology are used in the theory of topology-related image processing operations such as object counting and labelling, contour filling, border and boundary following (as in surface tracking), and thinning.

11.2 Graph-based digital pictures; conventional digital pictures

An n-dimensional *graph-based digital picture* is a quadruple (V, β, ω, B) where V is a set of grid points in $\mathbb{R}^n$ (usually $V = \mathbb{Z}^n$), $B \subseteq V$, and each of β and ω is a set of straight line segments joining points in V. If $\mathcal{P} = (V, \beta, \omega, B)$, then points in V are called *points* of $\mathcal{P}$, points in B are called *black points* of $\mathcal{P}$, and points in $V - B$ are called *white points* of $\mathcal{P}$. The *adjacency relation* of $\mathcal{P} = (V, \beta, \omega, B)$ is the irreflexive symmetric relation on V in which two black points are related (and said to be *adjacent*) if they are endpoints of a line segment in β, and two white points or a white point and a black point are related (and said to be *adjacent*) if they are endpoints of a line segment in ω.

We may regard a digital picture $\mathcal{P} = (V, \beta, \omega, B)$ as a graph, with vertex set V, whose adjacency relation is the adjacency relation of $\mathcal{P}$. We identify each subset S of V with the subgraph induced by S, and use graph-theoretic terminology (Bollobás 1979) when talking about S. For example, we say that $S \subseteq V$ is *connected* if the subgraph induced by S is connected. Similarly, if $S \subseteq T \subseteq V$ we say S is a *component* of T if the subgraph induced by S is a component of the subgraph induced by T (that is, if S is a maximal connected subset of T).

A component of the set of all black (white) points is called a *black* (*white*) *component*. Provided that the resolution of the digital image is high enough, each black component should be the digitization of a single object, and each finite white component should be the digitization of a hole or cavity in an object. For this reason component counting and labelling are important operations in image processing.

When $V = \mathbb{Z}^2$ or $\mathbb{Z}^3$, the sets β and ω are often given by one of the basic adjacency relations known as the 4- and 8-adjacency relations ($V = \mathbb{Z}^2$) and the 6-, 26-, and 18-adjacency relations ($V = \mathbb{Z}^3$). However, some authors (for example, Klette and Voss (1987)) have considered other kinds of adjacency relation[1]. For $p \in \mathbb{Z}^n$ we write $N(p)$ for the set of points in $\mathbb{Z}^n$ that differ from p by at most one in each coordinate—that is, the points in $\mathbb{Z}^n$ whose ℓ^∞ distance from p is at most 1. If $n = 2$ or 3, two points $p, q \in \mathbb{Z}^n$ are said to be *8-adjacent* ($n = 2$) or *26-adjacent* ($n = 3$) if they are distinct and $p \in N(q)$. Two points are said to be *4-adjacent* ($n = 2$) or *6-adjacent* ($n = 3$), if they differ in just one coordinate and differ in that coordinate by just 1. When $n = 3$, two points are said to be *18-adjacent* if they are distinct, differ in at most two coordinates, and differ in each of those coordinates by at most 1.

If *just one* of these basic adjacency relations is used as the adjacency relation between all pairs of points in $\mathbb{Z}^n$ (that is, the digital picture is

---
[1] See (Kong and Rosenfeld 1989, Section 11) for further references.

Fig. 11.1 Connectivity paradoxes in (4,4) and (8,8) digital pictures (1s are black points, 0s are white points)

0	0	0	0	0
0	0	1	0	0
0	1	0	1	0
0	0	1	0	0
0	0	0	0	0

($\mathbb{Z}^n, \beta, \omega, B$) where $\beta = \omega$ is the set of all straight line segments joining k-adjacent points, for some fixed k), then the black and white point sets may have paradoxical connectedness properties. For example, in Figure 11.1 if 4-adjacency is used then the black points are totally disconnected but still separate the set of white points into two components, while if 8-adjacency is used then the black points form the discrete analogue of a Jordan curve but the set of white points is nevertheless connected. However, certain combinations of *two* basic adjacency relations are free from such paradoxes.

Thus in $\mathbb{Z}^2$ one possibility is to call two points adjacent if both points are black and the points are 8-adjacent, or if at least one point is white and the points are 4-adjacent. (This is the adjacency relation of (V, β, ω, B) when β is the set of all straight line segments joining 8-adjacent points and ω is the set of all straight line segments joining 4-adjacent points.) We call this the (8, 4)-*adjacency relation*. The (4, 8)-*adjacency relation* is defined in the same way, with the roles of 4 and 8 interchanged. Similarly, in $\mathbb{Z}^3$ we can use the (6, 26)-, (26, 6)-, (6, 18)-, or (18, 6)-adjacency relations.[2]

An (m, n) *digital picture* is a graph-based digital picture (V, β, ω, B) in which β is the set of straight line segments joining every pair of m-adjacent points in V and ω is the set of straight line segments joining every pair of n-adjacent points in V. Thus the adjacency relation of an (m, n) digital picture is the (m, n)-adjacency relation. A (4, 8) or (8, 4) digital picture is two-dimensional ($V = \mathbb{Z}^2$); a (6, 26), (26, 6), (6, 18), or (18, 6) digital picture is three-dimensional ($V = \mathbb{Z}^3$). We shall abuse the (V, β, ω, B) notation by writing (V, m, n, B) for an (m, n) digital picture with black point set B. (Examples are $(\mathbb{Z}^2, 8, 4, B)$ and $(\mathbb{Z}^3, 18, 6, B)$.)

By a *conventional* digital picture we mean an (m, n) digital picture where $(m, n) = (4, 8), (8, 4), (6, 26), (26, 6), (6, 18),$ or $(18, 6)$. Conventional digital pictures are used more often than any other kind of digital picture in image processing.

---

[2] The (18, 18)-, (18, 26)-, (26, 18)-, and (26, 26)-adjacency relations induce the paradoxical 8-adjacency relation for both black and white points in each coordinate plane, while the (6, 6)-adjacency relation induces the paradoxical (4, 4)-adjacency relation in each coordinate plane. Consequently these adjacency relations may produce connectivity paradoxes and so they are usually avoided.

Digital Jordan curve theorems

In conventional digital pictures the connectedness properties of the black and white point sets of digital images are analogous to the connectedness properties of polyhedra and their complements in $\mathbb{R}^n$. The following digital Jordan curve theorem illustrates this.

Proposition 1. (Rosenfeld 1970, 1973, 1979, Proposition 2.4.5)
Suppose $\mathcal{P} = (\mathbb{Z}^2, m, n, B)$ where $(m, n) = (8, 4)$ or $(4, 8)$, B is connected and contains at least five points, and each point in B is adjacent to just two other points in B. Then $\mathcal{P}$ has just two white components, each of which is adjacent to every black point.

Figure 11.1 shows that this result is false for $(8, 8)$ digital pictures, and it is easy to find a counterexample to show that it fails for $(4, 4)$ digital pictures.
 The converse of Proposition 1 is also true.

Proposition 2. (Rosenfeld 1975, 1979, Corollary 2.5.7) *Suppose we have $\mathcal{P} = (\mathbb{Z}^2, m, n, B)$ where $(m, n) = (8, 4)$ or $(4, 8)$, and $\mathcal{P}$ has just two white components, each of which is adjacent to every black point. Then B is connected and each point in B is adjacent to just two other points in B.*

There is also a digital Jordan surface theorem for three-dimensional conventional digital pictures.

Proposition 3. (Morgenthaler and Rosenfeld 1981, Reed and Rosenfeld 1982, Reed 1984, Kong and Roscoe 1985) *Let $\mathcal{P} = (\mathbb{Z}^3, m, n, B)$ where $(m, n) = (6, 26), (26, 6), (6, 18),$ or $(18, 6)$. Suppose that B is connected and that for each $p \in B$ the point p is adjacent to just two components of $N(p) - B$, both of which are adjacent to each neighbour of p in B. Then $\mathcal{P}$ has just two white components, each of which is adjacent to every black point.*

Proposition 3 says (very roughly) that 'local separators are global separators'. It is rather less satisfactory than Proposition 1 because the properties it requires of the black point set B are not intrinsic to B but involve white points as well. To obtain a true analogue of the Jordan surface theorem we would need to have an intrinsic definition of a digital Jordan surface. But it appears that such a definition has to be quite complicated and inelegant, especially if we want a single definition that applies to $(6, 26), (6, 18), (26, 6),$ and $(18, 6)$ digital pictures.
 The converse of Proposition 3 is false: as Morgenthaler and Rosenfeld (1981, p. 237) observed, digital surfaces 'may touch themselves without globally affecting connectivity'.

Graph-based digital pictures; conventional digital pictures

```
1 1 1 1 1 1 1 1 1 0 0 0 0 0 0 0 0
1 0 1 1 1 1 1 0 1 0 1 0 0 0 0 1 0
1 0 1 1 1 1 1 0 1 0 1 0 0 0 0 1 0
1 1 0 1 1 1 0 1 1 0 0 1 0 0 1 0 0
1 1 0 1 1 1 0 1 1 0 0 1 0 0 1 0 0
1 1 1 0 1 0 1 1 1 0 0 0 1 0 1 0 0
1 1 1 0 1 0 1 1 1 0 0 0 1 0 1 0 0
1 1 1 1 0 1 1 1 1 0 0 0 0 1 0 0 0
1 1 1 1 0 1 1 1 1 0 0 0 0 1 0 0 0
1 1 1 1 1 1 1 1 1 0 0 0 0 0 0 0 0
```

Fig. 11.2 Kovalevsky's example shows the need for a more general kind of digital picture space (1s are black points, 0s are white points)

Graph-based digital picture spaces

In a digital picture (V, β, ω, B) the black point set B is determined by the image data; β and ω are chosen by the user to structure and interpret any image data that might be produced in a given application. The data-independent part of the digital picture $\mathcal{P}$ will be referred to as $\mathcal{P}$'s *digital picture space*. More precisely, we shall say that $\mathcal{P} = (V, \beta, \omega, B)$ is a digital picture on the digital picture space $\mathcal{S} = (V, \beta, \omega)$. Thus a digital picture space is a triple (V, β, ω), where V is a set of grid points and each of β and ω is a set of straight line segments joining points in V. Again we may abuse notation and refer to the digital picture spaces $(\mathbb{Z}^3, 26, 6)$, $(\mathbb{Z}^2, 4, 8)$, etc. Image processing operations such as thinning or digital rotation would normally transform a digital picture to another digital picture on the *same* digital picture space.

Not all digital picture spaces (V, β, ω) are well behaved. For example, we observed earlier that the digital Jordan curve theorem fails on the spaces $(\mathbb{Z}^2, 4, 4)$ and $(\mathbb{Z}^2, 8, 8)$. Kong et al. (199?b) and Kong (1989) identify a large class of well-behaved digital picture spaces, which we call *strongly normal*.

As Kovalevsky (1989) points out, we may want to use the black / white point configuration in the vicinity of two points p, q to decide whether or not p and q are adjacent. Thus in Figure 11.2 we may wish to use $(4, 8)$-adjacency on the left half of the digital image and $(8, 4)$-adjacency on the right half to produce a white 'V' and a black 'V'. This suggests that the above notion of a digital picture space (V, β, ω) in which β and ω are independent of the black point set is too restrictive—Figure 11.2 shows that the user may not be sure what β and ω should be until the image data is known. However, we can generalize the above definition by making a digital picture space a triple $(V, \mathcal{B}, \mathcal{W})$ where each of $\mathcal{B}$ and $\mathcal{W}$ is a *function* that maps the black point set to a set of straight line segments joining points in V. A digital picture on that space is then a digital picture

(V, β, ω, B) where $\beta = \mathcal{B}(B)$ and $\omega = \mathcal{W}(B)$. The earlier definition of a digital picture space corresponds to the special case of this definition in which $\mathcal{B}$ and $\mathcal{W}$ are constant functions.

However, Kovalevsky (1989) gave a different solution to this difficulty, recommending the use of topological digital pictures which we now introduce.

11.3 Topological digital pictures

Khalimsky (1977, 1986) and more recently Kovalevsky (1989) have proposed that a digital image be associated with a topological space. Then instead of having to define *analogues* of topological notions (such as connectedness) for digital pictures, standard topological notions can be used almost directly.

Khalimsky n-space

Consider the partition of the real line given by the open intervals $\{(2i, 2i+2) \mid i \in \mathbb{Z}\}$ together with the singleton sets $\{\{2i\} \mid i \in \mathbb{Z}\}$ (the sets contain the endpoints of the open intervals). By taking the cartesian product of n copies of this partition of the real line we get a partition of $\mathbb{R}^n$. Let $\mathcal{D}_n$ be the elements of this partition of $\mathbb{R}^n$. Thus, in the case $n = 2$, $\mathcal{D}_n$ consists of open squares $\{(2i, 2i+2) \times (2j, 2j+2) \mid i, j \in \mathbb{Z}\}$, line segments (without endpoints) that are edges of the squares, and singleton sets containing points (with even integer coordinates) that are corners of the squares.

The topological space used in Khalimsky's and Kovalevsky's topological digital pictures is just the quotient space of this partition of $\mathbb{R}^n$. This is the space whose points are the elements of $\mathcal{D}_n$, in which a set of elements $X \subseteq \mathcal{D}_n$ is an open set iff $\bigcup X$ is open in $\mathbb{R}^n$. Equivalently, $X \subseteq \mathcal{D}_n$ is closed iff $\bigcup X$ is closed in $\mathbb{R}^n$. We call $\mathcal{D}_n$ equipped with this topology *Khalimsky n-space*. Khalimsky n-space is a T_0 Alexandroff space[3] in which every point has a finite neighbourhood and the closure of every one-point set is finite.

Khalimsky originally constructed his n-space as a product of n copies of Khalimsky 1-space. The latter is a special case of the connected ordered topological spaces he investigated in (Khalimsky 1969a, 1969b).

Three important properties of $\mathcal{D}_n$ are listed in Proposition 4. In informal language, the proposition says that the interior, closure, and boundary of any set $X \subseteq \mathcal{D}_n$ consist of just those elements of $\mathcal{D}_n$ that are in the interior, closure, and boundary of $\bigcup X$ in $\mathbb{R}^n$.

---

[3] An Alexandroff space is a space in which every point has a minimal neighbourhood.

Proposition 4. *For all $S \subseteq \mathcal{D}_n$,*

1. $\bigcup \operatorname{int} S = \operatorname{int}(\bigcup S)$
2. $\bigcup \operatorname{cl} S = \operatorname{cl}(\bigcup S)$
3. $\bigcup \operatorname{fr} S = \operatorname{fr}(\bigcup S)$

All three properties are easy consequences of the fact that the membership map $q : \mathbb{R}^n \to \mathcal{D}_n$ (which sends each point x in $\mathbb{R}^n$ to the member of $\mathcal{D}_n$ that contains x) is continuous, open, and onto.

The connectedness properties of $S \subseteq \mathcal{D}_n$ and $\bigcup S$ are also the same.

Proposition 5. *For all $S \subseteq \mathcal{D}_n$,*

1. *S is connected iff $\bigcup S$ is connected*
2. *$X \subseteq S$ is a component of S iff $\bigcup X$ is a component of $\bigcup S$*

In fact much more than this is true.

Proposition 6. *If X is any subspace of $\mathcal{D}_n$, $p' \in p \in X$ are arbitrarily chosen base points, and $q : \bigcup X \to X$ is the membership map (which sends each point x in $\bigcup X$ to the element of X that contains x), then for any based metric space A:*

1. *every continuous base-point-preserving map $f : A \to X$ can be factored as $f = qF$ for some continuous base-point-preserving map $F : A \to \bigcup X$*
2. *if $F_0 : A \to \bigcup X$ and $F_1 : A \to \bigcup X$ are arbitrary continuous base-point-preserving maps such that $qF_0 = qF_1$ then there is a fixed base-point homotopy $H : A \times [0,1] \to \bigcup X$ between F_0 and F_1 such that qH is a constant homotopy.*

In the terminology of (Kong and Khalimsky 1990), properties 1 and 2 make $\bigcup X$ a *metric analogue* of X. It follows that many homotopy properties of X are the same as the corresponding homotopy properties of $\bigcup X$ (Kong and Khalimsky 1990, Section 3).

Topological digital picture spaces; topological digital pictures

Following Khalimsky et al. (1990) call the set of open points in Khalimsky n-space (that is, the open n-cells in $\mathcal{D}_n$) the *open screen*. We denote the open screen by O_n. We now identify $\mathbb{Z}^n$ with O_n. Specifically, we identify each point $(i_1, i_2, \ldots, i_n)$ in $\mathbb{Z}^n$ with the open n-cell in O_n whose centre has coordinates $(2i_1 + 1, 2i_2 + 1, \ldots, 2i_n + 1)$. (This allows us to say that two points in O_2 are 4-connected etc.)

Define an n-dimensional *topological digital picture space* to be a pair $(V,^*)$ where $V = O_n$ and * is a function that maps each subset B of V to a subset B^* of $\mathcal{D}_n$ such that $\text{int}(\text{cl } B) \subseteq B^* \subseteq \text{cl } B$. (We could generalize this definition to allow V to be the set of open points in a suitably well-behaved topological space other than $\mathcal{D}_n$, such as some of the spaces considered in (Kronheimer 199?). However, we shall not consider such generalizations in this chapter.)

Define an n-dimensional *topological digital picture* to be a triple $(V,^*, B)$ where $(V,^*)$ is an n-dimensional topological digital picture space and $B \subseteq V$. If $\mathcal{P} = (V,^*, B)$ then points in V are called *points* of $\mathcal{P}$, points in B are called *black points* of $\mathcal{P}$, and points in $V - B$ are called *white points* of $\mathcal{P}$.

Note that the condition $\text{int}(\text{cl } B) \subseteq B^*$ ensures that when $B = O_n$ we have $B^* = \mathcal{D}_n$. The condition $B^* \subseteq \text{cl } B$ ensures that when B is empty so is B^*. These two conditions also imply that, for all $B \subseteq V$, B^* is a regular set[4] containing B whose (topological) boundary is the boundary of $\text{cl } B$.

A *black component* of a topological digital picture $(O_n,^*, B)$ is a set $C \cap O_n$ where C is a component of B^*; a *white component* is a set $C \cap O_n$ where C is a component of $\mathcal{D}_n - B^*$.

Kronheimer (199?) has pointed out that $S \subseteq O_2$ is 4-connected iff $\text{int}(\text{cl } S)$ is connected, and $S \subseteq O_2$ is 8-connected iff $\text{cl } S$ is connected. It follows that in a two-dimensional topological digital picture $\mathcal{P} = (O_2,^*, B)$ if p and q are in the same 4-component of the black point set or white point set, then p and q are in the same black or white component. Conversely if p and q are in the same black or white component, then p and q are in the same 8-component of the black point set or white point set. Analogous results obtain for higher-dimensional digital pictures.

The role of the * function in a topological digital picture $(V,^*, B)$ is analogous to the role of β and ω in a graph-based digital picture (V, β, ω, B).

Simple examples

In a two-dimensional topological digital picture $(O_2,^*, B)$ where $B^* = \text{cl } B$ the black components are the 8-components of the black point set B and the white components are the 4-components of the white point set $O_2 - B$.

In a topological digital picture $(O_2,^*, B)$ where $B^* = \text{int}(\text{cl } B)$ the black components are the 4-components of the black point set B and the white components are the 8-components of the white point set $O_2 - B$.

Similarly, in a three-dimensional picture $(O_3,^*, B)$, if $B^* = \text{cl } B$ then the black components are the 26-components of B and the white components are the 6-components of $O_3 - B$, while if $B^* = \text{int}(\text{cl } B)$ then the

---

[4] X is said to be a *regular set* if $\text{cl}(\text{int } X) = \text{cl } X$ and $\text{int}(\text{cl } X) = \text{int } X$.

black components are the 6-components of B and the white components are the 26-components of $O_3 - B$.

Multicoloured digital pictures

We have so far considered only binary (that is, two-colour) digital images, in which O_n is partitioned into just two sets (the black and the white points). The notion of a topological digital picture generalizes easily to k-coloured digital pictures in which O_n is partitioned into k sets.

Instead of the * function we have to choose a rule which, given any partition of O_n into sets $(S_1, S_2, \ldots, S_k)$, will produce a partition of $\mathcal{D}_n$ into sets $(\mathcal{S}_1, \mathcal{S}_2, \ldots, \mathcal{S}_k)$ such that $\mathrm{int}(\mathrm{cl}\, \mathcal{S}_i) \subseteq \mathcal{S}_i \subseteq \mathrm{cl}\, \mathcal{S}_i$. Kovalevsky (1989, Section 4) calls such a rule a *global face membership* rule and gives several examples.

11.4 The fundamental group of a digital picture; continuous analogues of graph-based digital pictures

In studying the topology of three-dimensional digital images we need an analogue for digital pictures of the fundamental group $\pi_1(X, p)$ of a topological space X with base point p.

Graph-based digital pictures

If $\mathcal{P} = (V, \beta, \omega, B)$ is a graph-based digital picture and $X \subseteq V$ then let $\mathcal{P}_X$ denote the union of X with all straight line segments that join adjacent points in X. Also, define the *complement of* $\mathcal{P}$, written $\overline{\mathcal{P}}$, to be the digital picture $(V, \omega, \beta, V - B)$.

Now suppose $\mathcal{P}$ is an n-dimensional graph-based digital picture, where $n = 2$ or 3, with black point set B and white point set W, and suppose $p \in B$. Then we define the *fundamental group of* $\mathcal{P}$ *with base point* p, written $\pi(\mathcal{P}, p)$, to be the subgroup of $\pi_1(\mathbb{R}^n - \mathcal{P}_W, p)$ consisting of those equivalence classes of loops[5] which contain a loop in $\mathcal{P}_B$. In other words, $\pi(\mathcal{P}, p)$ is obtained from $\pi_1(\mathbb{R}^n - \mathcal{P}_W, p)$ by omitting those equivalence classes of loops which do not contain a loop in $\mathcal{P}_B$. For alternative definitions of the group $\pi(\mathcal{P}, p)$ (up to a natural group isomorphism), and an application of the group to the theory of three-dimensional image thinning algorithms, see (Kong 1989).

It is shown in (Kong et al. 199?b) that for a wide variety of two- and three-dimensional graph-based digital pictures[6] $\mathcal{P} = (V, \beta, \omega, B)$, including

---

[5] Two loops based at p are equivalent iff there is a fixed base-point homotopy of one loop to the other in $\mathbb{R}^n - \mathcal{P}_W$.

[6] All digital pictures on strongly normal digital picture spaces.

all conventional digital pictures, one can find a set $C(\mathcal{P}) \subseteq \mathbb{R}^n$ with all of the following seven properties, where W is the white point set and n is the dimension of $\mathcal{P}$:

1. $\mathcal{P}_B \subseteq C(\mathcal{P})$.

2. $\mathcal{P}_W \subseteq \mathbb{R}^n - C(\mathcal{P})$.

3. Each component of $C(\mathcal{P})$ meets V in a black component of $\mathcal{P}$.

4. Each component of $\mathbb{R}^n - C(\mathcal{P})$ meets V in a white component of $\mathcal{P}$.

5. The boundary of a component X of $C(\mathcal{P})$ meets the boundary of a component Y of $\mathbb{R}^n - C(\mathcal{P})$ iff there is a black point in X that is adjacent to a white point in Y.

6. For each black point p in $\mathcal{P}$, the inclusion of $\mathcal{P}_B$ in $C(\mathcal{P})$ induces an isomorphism of $\pi(\mathcal{P}, p)$ to $\pi_1(C(\mathcal{P}), p)$.

7. For each white point q in $\mathcal{P}$, the inclusion of $\mathcal{P}_W$ in $\mathbb{R}^n - C(\mathcal{P})$ induces an isomorphism of $\pi(\overline{\mathcal{P}}, q)$ to $\pi_1(\mathbb{R}^n - C(\mathcal{P}), q)$.

Call a set $C(\mathcal{P}) \subseteq \mathbb{R}^n$ with these properties a *continuous analogue*[7] of $\mathcal{P}$. Note that if $\mathcal{P}$ is an n-dimensional conventional digital picture ($n = 2$ or 3) then the complement in $\mathbb{R}^n$ of a continuous analogue of $\mathcal{P}$ is a continuous analogue of $\overline{\mathcal{P}}$.

When $\mathcal{P}$ is a two-dimensional (4, 8) or three-dimensional (6, 26) digital picture, a continuous analogue of $\mathcal{P}$ is given by the union of $\mathcal{P}_B$ with all unit lattice squares and, in the (6, 26) case, all unit lattice cubes whose edges are contained in $\mathcal{P}_B$. When $\mathcal{P}$ is a two-dimensional (8, 4) or three-dimensional (26, 6) digital picture a continuous analogue of $\mathcal{P}$ is given by the union of the convex hulls of all sets of pairwise-adjacent black points. These continuous analogues were used in (Kong and Roscoe 1985) to deduce the (6, 26) and (26, 6) cases of Proposition 3 from the Jordan–Brouwer separation theorem for $\mathbb{R}^3$. Similar but somewhat more elaborate constructions yield continuous analogues of (6, 18) and (18, 6) digital pictures (Kong 1989, Kong et al. 199?b).

When $\mathcal{P}$ is an (8, 4) digital picture, a different continuous analogue of $\mathcal{P}$ is given by the union of all squares with centroid in B and edges of length 1 parallel to the coordinate axes. Similarly, when $\mathcal{P}$ is a (26, 6) digital picture, a continuous analogue of $\mathcal{P}$ is given by the union of all cubes with centroid in B and edges of length 1 parallel to the coordinate axes.

The existence of a fairly simple continuous analogue of each digital picture on a digital picture space $\mathcal{S}$ is evidence that connectedness on $\mathcal{S}$

---

[7]In this chapter we do not require a continuous analogue to be a polyhedral set.

is analogous to connectedness in Euclidean space, and properties 6 and 7 provide confirmation that $\pi(\mathcal{P}, p)$ has been appropriately defined for digital pictures on $\mathcal{S}$.

Topological digital pictures

It is not necessary to give a special definition of the fundamental group for topological digital pictures since the standard definition can be used. More precisely, we define the fundamental group of the topological digital picture $\mathcal{P} = (V,{}^*, B)$ with base point $p \in B$ to be $\pi_1(B^*, p)$. This definition applies to any topological digital picture—unlike the above definition of the fundamental group of a graph-based digital picture, which is only valid in two and three dimensions.

It is a consequence of Proposition 6 that the group $\pi_1(B^*, p)$ is isomorphic to the group $\pi_1(\bigcup B^*, p')$ where p' is any point in p (Kong and Khalimsky 1990, Section 3).

Similarly, we may define the nth *homotopy group* of $\mathcal{P} = (V,{}^*, B)$ with base point $p \in B$ to be $\pi_n(B^*, p) = \pi_n(\bigcup B^*, p')$ where $p' \in p$. In the same vein we might reasonably define a *continuous map* from a topological space A to a topological digital picture $\mathcal{P} = (V,{}^*, B)$ to be a continuous map from A to B^*. Proposition 6 provides useful insight into the properties of such maps in the case when A is a metric space. There seems to be no easy way to define these concepts for general graph-based digital pictures.

11.5 Graph-based equivalents of topological digital pictures

Given a topological digital picture $\mathcal{P} = (O_n,{}^*, B)$, where $n = 2$ or 3, we can often construct a graph-based digital picture $\mathcal{P}'$ such that $\bigcup B^*$ is a continuous analogue of $\mathcal{P}'$.

To do this, let $A_\mathcal{P}$ be the adjacency relation on O_n such that $p \in O_n$ is adjacent to $q \in O_n$ iff $p \neq q$ and one of the following conditions is satisfied:

1. p and q are black points and $\mathrm{cl}\{p\} \cap \mathrm{cl}\{q\} \cap B^*$ is non-empty;

2. p and q are white points and $\mathrm{cl}\{p\} \cap \mathrm{cl}\{q\} - B^*$ is non-empty;

3. p is a black point, q is a white point, and $\mathrm{cl}\{p\} \cap \mathrm{cl}\{q\} - (B - \{p\})^*$ is non-empty;

4. q is a black point, p is a white point, and $\mathrm{cl}\{q\} \cap \mathrm{cl}\{p\} - (B - \{q\})^*$ is non-empty.

Note that conditions 3 and 4 can be restated informally as 'one of p and q is black, the other is white, and the two points would satisfy condition 2 if the black point were replaced by a white point'.

Example 7. When $n = 2$, if the * function is given by $S^* \equiv \mathrm{cl}\, S$ then $A_\mathcal{P}$ is the $(8,4)$-adjacency relation, while if $S^* \equiv \mathrm{int}(\mathrm{cl}\, S)$ then $A_\mathcal{P}$ is the $(4,8)$-adjacency relation.

For $p \neq q \in \mathcal{D}_n$ we say p is a face of q if $p \in \mathrm{cl}\{q\}$. So for $p, q \in \mathcal{D}_n$, $p \neq q$ the set of common faces of p and q is $\mathrm{cl}\{p\} \cap \mathrm{cl}\{q\}$. Each open 3-cell in $\mathcal{D}_3$ has six two-dimensional faces (its geometrical faces, without edges or corners), twelve one-dimensional faces (its edges, without endpoints), and eight zero-dimensional faces (singleton sets each containing one corner). Say that a topological digital picture $(V,^*,B)$ is *face convex* if:

1. whenever two black points p, q have a common face in B^*, the highest-dimensional common face of p and q is in B^*;

2. whenever two white points p, q have a common face that is not in B^*, the highest-dimensional common face of p and q is not in B^*.

Note that all two-dimensional topological digital pictures are face convex. A three-dimensional topological digital picture $(O_3,^*,B)$ is face convex iff for every pair of black points p, q that are 18-adjacent but not 6-adjacent the set $\mathrm{cl}\{p\} \cap \mathrm{cl}\{q\} \cap B^*$ is either empty or contains the common edge of p and q, and for every pair of white points p, q that are 18-adjacent but not 6-adjacent the set $\mathrm{cl}\{p\} \cap \mathrm{cl}\{q\} - B^*$ is either empty or contains the common edge of p and q.

The following result justifies the assertion at the beginning of this section. Recall that O_n is identified with $\mathbb{Z}^n$, as we explained when we introduced O_n in Section 11.3.

Proposition 8. *Let $\mathcal{P} = (V,^*,B)$ be a two-dimensional or face convex three-dimensional topological digital picture. Let $\mathcal{P}'$ be a graph-based digital picture, with point set V and black point set B, whose adjacency relation is $A_\mathcal{P}$. Then $\bigcup B^*$ is a continuous analogue of $\mathcal{P}'$.*

We omit the proof. The most important step is to show that $\bigcup B^*$ has properties 6 and 7 of a continuous analogue of $\mathcal{P}'$.

Graph-based digital pictures for which there is no analogous topological digital picture

It is easy to see that if $\mathcal{P} = (O_n,^*,B)$, where $n = 2$ or 3, is any two- or three-dimensional topological digital picture, then in every unit lattice square or cube in $\mathbb{Z}^n$ (which we identify with O_n) either the black points or the white points are $A_\mathcal{P}$-connected. Also, if $\mathcal{P}$ is face convex, then no straight line segment joining two $A_\mathcal{P}$-adjacent black points can cross a straight line segment joining two $A_\mathcal{P}$-adjacent white points. Consequently some graph-based digital pictures, such as the paradoxical $(4,4)$, $(8,8)$, and three-dimensional $(6,6)$ digital pictures, have no equivalent among topological digital pictures. The $(26,26)$, $(18,18)$, $(18,26)$, and $(26,18)$ digital pictures have no equivalent among face convex topological digital pictures.

Unfortunately, some graph-based digital pictures that have no equivalent among topological digital pictures may have important practical applications. The digital pictures used by Gordon and Udupa (GU) (1989), and which we now define, are good examples.

Say that $p, q \in \mathbb{Z}^3$ are GU-*adjacent* if they are 6-adjacent, or if they are 18-adjacent and have different z-coordinates. Thus each point is GU-adjacent to exactly fourteen other points. Let $\mathcal{G}$ be the set of all straight line segments that join GU-adjacent points in $\mathbb{Z}^3$. A digital picture on the digital picture space $(\mathbb{Z}^3, 6, \mathcal{G})$ will be called a $(6, \mathrm{GU})$ *digital picture*. In a $(6, \mathrm{GU})$ digital picture two black points are adjacent iff they are 6-adjacent, while two white points or a white point and a black point are adjacent iff they are GU-adjacent.

The $(\mathbb{Z}^3, 6, \mathcal{G})$ digital picture space is of practical interest in three-dimensional medical imaging because Gordon and Udupa (1989) published a fast boundary-tracking algorithm for $(6, \mathrm{GU})$ digital pictures based on Proposition 9 below. In their tests this boundary-tracking algorithm was, on the average, at least 35 per cent faster than that of Artzy et al. (1981), which was already an efficient algorithm[8].

A *voxel* is a closed unit cube in $\mathbb{R}^3$ with edges parallel to the co-ordinate axes whose centre has integer coordinates. If $T \subseteq \mathbb{Z}^3$, then we call a voxel with center in T a T-*voxel*. The *boundary* between a black component C and a white component D is the set $\{f \mid f$ is the common face of a C-voxel and a D-voxel$\}$.

In a $(6, \mathrm{GU})$ digital picture, say that two distinct elements f_1, f_2 of the boundary between a black component C and a white component D are *surface-adjacent* if f_1 and f_2 have an edge in common that is parallel to the xy-plane and f_1, f_2 are faces of (1) the same C-voxel, or (2) two 6-adjacent C-voxels, or (3) two 18-adjacent C-voxels that are 6-adjacent to the same third C-voxel. Thus each element of the boundary that is perpendicular to the z-axis is surface-adjacent to exactly four other elements of the boundary, but each element of the boundary that is perpendicular to the x- or the y-axis is surface-adjacent to just two other boundary elements.

We can now state the result that provides the basis for Gordon and Udupa's fast boundary-tracking algorithm.

---

[8]In discussing the merits of their algorithm, Artzy et al. (1981, p. 2) remarked that 'previously reported computer programs [would track boundaries] at a computer cost which is at least an order of magnitude greater than required by our software'.

Proposition 9. In a $(6, \mathrm{GU})$ digital picture with only finitely many black points let S be the boundary between a black component and an adjacent white component. Then given any distinct $s, s' \in S$ there is a sequence $s = s_0, s_1, \ldots, s_n = s'$ of elements of S such that s_i is surface-adjacent to s_{i+1} for $0 \leqslant i < n$.

For a proof of this, see (Kong et al. 199?c). There is no simple way to restate this as a theorem about topological (rather than graph-based) digital pictures, because the $(6, \mathrm{GU})$-adjacency relation induces the paradoxical $(4, 4)$-adjacency relation in planes perpendicular to the z-axis.

11.6 Khalimsky's Jordan curve theorem and a generalization

Khalimsky originally used his n-space in a rather different way: he identified $\mathbb{Z}^n$ with $\mathcal{D}_n$ itself, rather than with the open screen O_n.

In our notation, each point $p \in \mathbb{Z}^n$ may be identified with the member of $\mathcal{D}_n$ whose centroid is at p. Thus if all coordinates of p were odd, then p would be identified with an open n-cell in $\mathcal{D}_n$. As another example, in the case $n = 2$ if p had coordinates $(2i + 1, 2j)$ for integers i, j, then p would be identified with the horizontal line segment $(2i, 2i + 2) \times \{2j\}$.

This identification of $\mathbb{Z}^n$ with $\mathcal{D}_n$ eliminates the need to augment the black point set B with fictitious points to give B^*, as is necessary when $\mathbb{Z}^n$ is identified with O_n. But different points in $\mathbb{Z}^n$ are now dealt with differently and end up having different topological properties—points whose coordinates are all even become closed, points whose coordinates are all odd become open, and the other points become neither open nor closed.

Say that $x, y \in \mathbb{Z}^n$ are *Khalimsky adjacent* if $x \neq y$ and the two-point set $\{x, y\}$ is connected (after making the above identification of $\mathbb{Z}^n$ with $\mathcal{D}_n$). Figure 11.3 shows this adjacency relation in the case $n = 2$. If $\mathbb{Z}^n$ is identified with $\mathcal{D}_n$, then the black point set can be regarded as a graph-based digital picture whose adjacency relation is Khalimsky adjacency. It is readily confirmed that $S \subseteq \mathbb{Z}^n$ is a connected subspace if and only if S is a connected set of points in the graph-based digital picture.

Khalimsky has shown that the digital Jordan curve theorem stated above for $(4, 8)$ and $(8, 4)$ digital pictures is also valid for the Khalimsky adjacency relation. In fact there is a general n-dimensional Jordan surface theorem. For $S \subseteq \mathbb{Z}^k$ write $A(p, S)$ for the set of points in S that are Khalimsky adjacent to p. In $\mathbb{Z}^k$ define a *0-surface* to be a disconnected two-point subspace, and, for all positive integers $n < k$, recursively define an *n-surface* to be a connected subspace S such that for each point p in S the set $A(p, S)$ is an $(n-1)$-surface. Then we have the following theorem.

Proposition 10. (Khalimsky et al. 199?, Kopperman et al. 199?, Kong et al. 199?a) *For any positive integer n, if S is an n-surface in $\mathbb{Z}^{n+1}$ then*

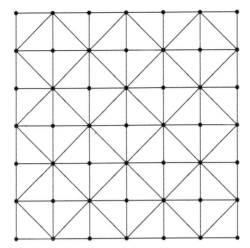

Fig. 11.3 Two-dimensional Khalimsky adjacencies

$\mathbb{Z}^{n+1} - S$ has just two components, each of which is Khalimsky adjacent to every point p in S.

The simple intrinsic definition of an n-surface (due to Kopperman) makes this result much more elegant than the Jordan surface theorem for three-dimensional conventional digital pictures which we stated as Proposition 3.

11.7 Concluding remarks

The topological approach to digital topology has some mathematical advantages over the graph-based approach.

For example, there is no need to define and construct continuous analogues of topological digital pictures, as is necessary for graph-based digital pictures. This is because $\bigcup B^*$ is essentially a 'continuous analogue' of the topological digital picture $\mathcal{P} = (V, *, B)$.

We have to invent a special definition of the fundamental group of a graph-based digital picture, and then verify the intuitive soundness of that definition by constructing continuous analogues. This is unnecesary for topological digital pictures, since we can use the standard definition of the fundamental group almost directly. There are equally natural definitions of the higher homotopy groups of a topological digital picture, and of continuous maps from a topological space to a topological digital picture, whereas there seems to be no easy way to define these concepts for general graph-based digital pictures. Moreover, the present definitions of the fundamental group and continuous analogues of a graph-based digital picture are only valid in two and three dimensions, while the topological approach is valid in any finite number of dimensions.

On the other hand the topological approach, though fairly general, is still not as general as the graph-based approach. Thus there is no topological digital picture space that is analogous to the graph-based digital picture space used by Gordon and Udupa (1989) for their fast boundary-tracking algorithm.

References

Artzy, E., Frieder, G., and Herman, G. T. (1981) The theory, design, implementation and evaluation of a three-dimensional boundary detection algorithm, *Computer Graphics and Image Processing* **15**, 1–24.

Bollobás, B. (1979) *Graph Theory: an introductory course*, Graduate Texts in Mathematics 63, Springer Verlag, New York.

Gordon, D., and Udupa, J. K. (1989), Fast surface tracking in three-dimensional binary images, *Computer Vision, Graphics and Image Processing* **45**, 196–214.

Halimskii, E. (Khalimsky, E.) (1969) The generalized segment and its use in homotopy theory, *Soviet Math. Doklady* **10**, 351–353.

Halimskii, E. (Khalimsky, E.) (1969) On topologies of generalized segments, *Soviet Math. Doklady* **10**, 1508–1511.

Halimskii, E. (Khalimsky, E.) (1977) *Uporiadochenie Topologicheskie Prostranstva (Ordered Topological Spaces)*, Naukova Dumka Press, Kiev.

Khalimsky, E. (1986) Pattern analysis of N-dimensional digital images, *Proc. 1986 IEEE International Conference on Systems, Man and Cybernetics*, CH2364-8/86, 1559–1562.

Khalimsky, E., Kopperman, R. D., and Meyer, P. R. (199?) Computer graphics and connected topologies on finite ordered sets. To appear in *Topology and its Applications*.

Khalimsky, E., Kopperman, R. D., and Meyer, P. R. (1990) Boundaries in digital planes. *Journal of Applied Mathematics and Stochastic Analysis* **3**, pp. 27–55.

Klette, R., and Voss, K. (1987) *The Three Basic Formulae of Oriented Graphs*, CAR-TR-305, Center for Automation Research, University of Maryland, College Park, MD 20742.

Kong, T. Y., and Roscoe, A. W. (1985) Continuous analogues of axiomatized digital surfaces, *Computer Vision, Graphics and Image Processing* **29**, 60–86.

Kong, T. Y. (1989) A digital fundamental group, *Computers and Graphics* **13**(2), 159–166.

Kong, T. Y., and Rosenfeld, A. (1989) Digital topology: introduction and survey, *Computer Vision, Graphics and Image Processing* **48**, 357–

393.

Kong, T. Y., and Khalimsky, E. (1990) Polyhedral analogues of locally finite topological spaces. To appear in R. M. Shortt (Ed.) *General Topology and Applications: Proceedings of the 1988 Northeast Conference*, Marcel Dekker, New York, pp. 153–164.

Kong, T. Y., and Udupa, J. K. (199?) *Justification of a Fast Surface-Tracking Algorithm*. Paper in preparation.

Kong, T. Y., Kopperman, R. D., and Meyer, P. R. (199?) *Metric and Polyhedral Analogues of Topological Spaces*. Paper in preparation.

Kong, T. Y., Roscoe, A. W., and Rosenfeld, A. (199?) *Concepts of Digital Topology*. Submitted.

Kopperman, R. D., Meyer, P. R., and Wilson, R. G. (199?) A Jordan surface theorem for three-dimensional digital spaces. To appear in *Discrete and Computational Geometry*.

Kovalevsky, V. A. (1989) Finite topology as applied to image analysis, *Computer Vision, Graphics and Image Processing* **46**, 141–161.

Kronheimer, E. H. (199?) *The Topological Structure of Digital Images*, Dept. of Mathematics and Statistics, Birkbeck College, University of London, England. Submitted.

Morgenthaler, D. G., and Rosenfeld, A. (1981) Surfaces in three-dimensional digital images, *Information and Control* **51**, 227–247.

Reed, G. M., and Rosenfeld, A. (1982) Recognition of surfaces in three-dimensional digital images, *Information and Control* **53**, 108–120.

Reed, G. M. (1984) On the characterization of simple closed surfaces in three-dimensional digital images, *Computer Vision, Graphics and Image Processing* **25**, 226–235.

Rosenfeld, A. (1970) Connectivity in digital pictures, *J. ACM* **17**, 146–160.

Rosenfeld, A. (1973) Arcs and curves in digital pictures, *J. ACM* **20**, 81–87.

Rosenfeld, A. (1975) A converse to the Jordan Curve Theorem for digital curves, *Information and Control* **29**, 292–293.

Rosenfeld, A. (1979) *Picture Languages*, Academic Press, New York.

12
Tiling the plane with one tile

D. GIRAULT-BEAUQUIER AND M. NIVAT

Abstract

We deal with tilings of the plane obtained by translations of only one tile. The tiles which can tile the plane according to this condition are called *exact* tiles. A very simple characterization of exact tiles is given. Moreover, we prove that every tiling with one exact tile is half-periodic, that is, invariant by at least one translation.

12.1 Introduction

Tiling the plane is a very old activity of mankind. Grunbaum and Shepherd (1986) give many beautiful examples of decorative tilings belonging to nearly all ancient civilizations. Most decorative tilings are those we call *periodic*, which means that they are invariant by two independent translations.

Among periodic tilings some are called *regular*: if the tiling uses just one tile and translated instances of itself, the tiling is said to be *regular* if and only if the surrounding of each instance of the tile in the tiling is the same. Grunbaum and Shephard (1986) give regularity a completely different meaning (the tilings which are shown on p. 34 of their book are not regular in their sense but clearly regular in our sense).

From our definitions it follows that a periodic tiling U by a finite number of different tiles induces a regular tiling with just one tile T: this means that there exists a tile T exactly covered by a finite number of instances of the given tiles such that U contains a regular tiling by T. This is shown in Figure 12.1. The well-known and widely used tiling with one regular octagon and one square contains the regular tiling by one tile exactly covered by one octagon and one adjacent square.

A tile is *exact* if there exists at least one tiling of the plane by translated instances of that tile. Our chapter is devoted to the characterization of exact tiles.

We first investigated exact *polyominoes* (a polyomino is a polygon exactly covered by unit squares whose edges are horizontal and vertical).

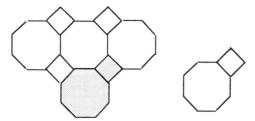

Fig. 12.1

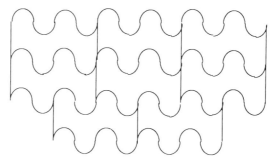

Fig. 12.2

We refined a recent result of Wijshoff and van Leeuwen (1984) about exact polyominoes. We proved that a polyomino is exact if and only if it is a *pseudo-hexagon*, and that every tiling with an exact polyomino is half-periodic (that is, invariant by some translation). It follows from the result of Wijshoff and van Leeuwen that it is decidable whether a polyomino is exact, and this was their aim when studying skewing schemes and data transfer functions in various types of parallel processing machines. The purely geometrical properties of exact polyominoes do not appear in their work. On the contrary we focused on these properties and this led us to extend the results on polyominoes to the widest possible family of tiles, that is, all the subsets of the plane which are homeomorphic to a closed disc and whose boundary is piecewise C^2, with a finite number of inflection points (we do need be able to define the length of the boundary). All the results remain true, namely the fact that a tile is exact if and only if it admits a surrounding by translated images of itself (the number of which can be proved to be at least four and at most eight), this being equivalent to saying that the tile is a pseudo-hexagon (which can be degenerated). Such an exact tile and a tiling of the plane by this tile are shown in Figure 12.2.

The fact that the exact tile shown above is a pseudo-hexagon is shown in Figure 12.3.

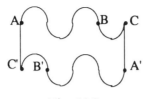

Fig. 12.3

The translation $\vec{AB'}$ maps the edge $[AB]$ onto the edge $[B'A']$, the translation $\vec{BC'}$ maps the edge $[BC]$ onto the edge $[C'B']$, and the translation $\vec{CA}$ maps the edge $[CA']$ onto the edge $[AC']$.

Furthermore the following two results hold:

- there exists a tiling with one (exact) tile if and only if there exists a regular tiling;
- all tilings by an exact tile are half-periodic.

A conclusion of this work is that Wang's (1978) conjecture holds for just one tile. Wang conjectured that for all finite set of tiles there exists a tiling if and only if there exists a periodic tiliong. This conjecture was proved to be false by exhibiting sets of tiles such that the set of tilings is not empty but does not contain a periodic tiling. Such sets have been exhibited in that order by Berger (1966), Robinson (1971) and Penrose (1978), the last one being the simplest. In fact all these authors allow rotations of the tile, but their results can easily be converted into results for tiles which are only translated (the number of possible rotations for their tiles is finite). Penrose's set of tiles (kites and darts) is equivalent to a set of translated tiles containing twenty elements. Whether there exists a tiling of the plane for a given set of tiles containing a small number of elements (between two and twenty) is still an open problem. We may guess that the problem is decidable for two and formulate the following conjecture.

Conjecture 1. *There exists a tiling with two tiles if and only if there exists an exact tile which is exactly covered by a bounded number of instances of each tile (the bound depending on the lengths of the two tiles).*

12.2 Preliminaries

The tiles we deal with are homeomorphic to a closed disc. Their *boundary* is a curve, oriented in the clockwise sense (which will be called the *positive* or *direct sense*), and we assume (it is a reasonable assumption after all) that this boundary is piecewise C^2 and that each component arc of class C^2 admits a finite number of inflection points. One consequence of this

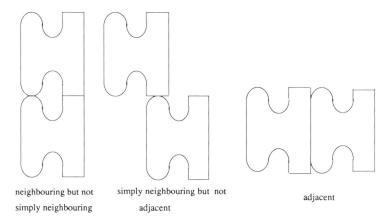

neighbouring but not   simply neighbouring but not
simply neighbouring   adjacent   adjacent

Fig. 12.4

hypothesis is that the boundary has a length. Let q be a tile. Two points A and B on the boundary of q define an oriented Jordan arc which is the path from A to B in the positive sense along the boundary of q. It will be denoted by $[AB]$. The image of q in the translation of vector u will be denoted by $q(u)$, and $q(u)$ is called an *instance* of q. The boundary of a tile q is denoted by $b(q)$ and its length is $|b(q)|$. If A belongs to the boundary of q, we denote by A' the 'symmetric' point of A on this boundary, that is, the point such that $|[AA']| = |[A'A]| = |b(q)|/2$ (it implies that $(A')' = A$). Two instances $q(u)$ and $q(v)$ are said to be:

- *neighbouring* if $q(u) \cap q(v)$ is a non-empty set with an empty interior;
- *simply neighbouring* if $q(u)$ and $q(v)$ are neighbouring and $q(u) \cap q(v)$ is a connected set;
- *adjacent* if $q(u)$ and $q(v)$ are simply neighbouring and $q(u) \cap q(v)$ is not reduced to a point.

A *tiling of the plane* P by the tile q can be represented as a set U of vectors such that $P = \bigcup_{u \in U} q(u)$ and for two distinct vectors u and v of U, $q(u)$ and $q(v)$ are not overlapping, that is, are disjoint or neighbouring (Figure 12.4).

We shall focus on some objects that will play an important role later on, namely the *edges* of a tiling. The edges of a tiling are the common boundaries between two simply neighbouring tiles. We have to give a precise definition of this notion. Let $q(u)$ and $q(v)$ be two simply neighbouring tiles, and let us suppose that A and B are the extremities of the arc $q(u) \cap q(v)$ such that $[BA]$ is the directly oriented arc on $q(u)$ and $[AB]$ is the directly oriented arc on $q(v)$ (Figure 12.5(a)).

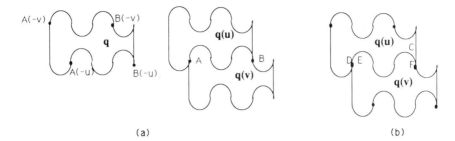

Fig. 12.5

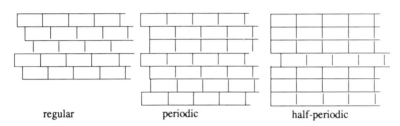

Fig. 12.6 Examples

We denote by $[q(u), q(v)]$ the edge of $q(u)$ related to $q(v)$, that is, the arc $[BA]$ *referred to the reference tile* q, so it is the arc $[B(-u)A(-u)]$ of q. We define in the same way $[q(v), q(u)] = [A(-v)B(-v)]$. And, by abuse of notation, if $[q(u), q(v)] = [CD]$ and $[q(v), q(u)] = [EF]$, we shall represent these edges on $q(u)$ and $q(v)$ as shown in Figure 12.5(b).

A tile q is *exact* if it can tile the plane. A tiling U of the plane with an exact tile q is said to be:

- *periodic* if there exist two independent vectors u and v such that $U = U + u = U + v$;
- *regular* if there exist two independent vectors u and v and a vector u_0 such that $U = u_0 + \{ku + k'v/k, k' \in \mathbb{Z}\}$;
- *half-periodic* if there exists a vector $u \neq 0$ such that $U = U + u$. (Figure 12.6)

Surroundings

Let q be a tile. A *surrounding of* q is a circular sequence $(q(u_0), \ldots, q(u_{k-1}))$ of instances of q such that, for $i = 0, \ldots, k-1$, q and $q(u_i)$ are simply neighbouring, $q(u_i)$ and $q(u_{i+1})$ are simply neighbouring (indices are defined modulo k), and if $[q, q(u_i)] = [A_i B_i]$, then $A_{i+1} = B_i$.

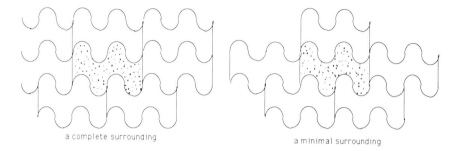

a complete surrounding   a minimal surrounding

Fig. 12.7

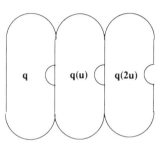

Fig. 12.8

The surrounding is said to be *complete* if for each i, $q(u_i)$ and $q(u_{i+1})$ are adjacent, and *minimal* if for each i, $|[q, q(u_i)]| > 0$ (Figure 12.7).

Let us observe that the concatenation of arcs $[A_i B_i]$ provides the boundary $b(q)$. We now give some results established in (Wijshoff and van Leeuwen 1984) for polyominoes but they hold for our tiles as well (only the fact that the tile is a closed topological disc is needed, but the proof is a little more 'sophisticated'). The proof is exactly the same for the first two lemmas, replacing 'a cell' by 'an interior point'.

Lemma 2. *If the tiles q and $q(u)$ do not overlap then q, $q(u)$, $q(2u)$ do not overlap each other (Figure 12.8).*

Lemma 3. *Let q, $q(u)$, $q(v)$ be three instances of q, pairwise neighbouring. Then q, $q(u)$, $q(v)$, $q(u-v)$, $q(-u)$, $q(-v)$, $q(v-u)$ do not overlap each other (Figure 12.9).*

Lemma 4. *Let q be a tile q with boundary $b(q)$. Suppose there are two or three neighbouring instances of q that form a hole h (homeomorphic to a closed disc) (Figure 12.10). Then the size of the interior boundary I of these instances with respect to h is strictly less than $|b(q)|$.*

Preliminaries

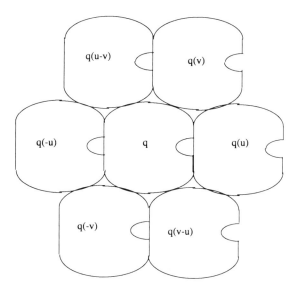

Fig. 12.9

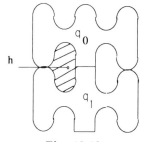

Fig. 12.10

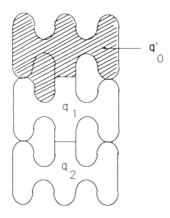

Fig. 12.11

Proof. We give only the proof for two neighbouring instances of q because the modification of the proof of (Wijshoff and van Leeuwen 1984) is the same for two or for three tiles.

Suppose that $q(u)$ and $q(v)$ are neighbouring, and form a hole h. Let S be the set of edges belonging to the boundary of the hole h or to both $q(u)$ and $q(v)$. These edges have positive or null length. Let $l(I)$ be the length of the boundary I of h and $l(S)$ the sum of the lengths of the edges in S. Of course, $l(I) < l(S)$. We want to prove that two edges of S correspond to non-overlapping arcs on the boundary of p. By this we mean that if $[A(u)B(u)]$ and $[C(v)D(v)]$ are two edges of S (possibly the same edge) then $[AB] \cap [CD]$ is empty or reduced to a point (in other words these arcs are not overlapping).

First of all, if $[A(u)B(u)]$ and $[C(v)D(v)]$ are the same edge of S then $[AB]$ and $[CD]$ are disjoined on the boundary of q. Actually, $[DC] = [AB](v - u)$, and if $[AB]$ and $[DC]$ overlap along an arc $[MN]$ then the interior of q would be both on the left and on the right of $[MN]$; if $[AB]$ and $[DC]$ have a common extremity then we would have $B = C = B(v-u)$ or $A = D = A(v-u)$. So, $[AB]$ and $[DC]$ are disjoined. So we can suppose that $[A(u)B(u)]$ and $[C(v)D(v)]$ are distinct edges of S. Let $q_0 = q(u)$ $q_1 = q(v)$ $q_2 = q(2v - u)$. Because of Lemma 2, q_0, q_1, and q_2 do not overlap each other. The relative position of q_0 and q_1 is the same as that of q_1 and q_2. So we can extend both q_0 and q_1 to q'_0 and q'_1 in such a way that q'_0 covers exactly q_0 and the hole h and q'_1 covers q_1 and the corresponding hole (Figure 12.11).

Then again q'_0, q'_1, q_2 do not overlap each other. Suppose that $[A(u)B(u)]$, $[C(v)D(v)]$ are elements of S such that $[AB]$ and $[CD]$ are overlapping. Let $[MN]$ be an arc of strictly positive length included in

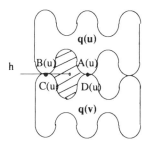

Fig. 12.12

$[AB] \cap [CD]$. If $[A(u)B(u)]$ and $[C(v)D(v)]$ are both parts of q_0 and q_1, then q'_0 and q_2 border q'_1 along the same arc $[M(v)N(v)]$, which is a contradiction with the fact that q'_0 and q_2 do not overlap. The two other possibilities lead to the same kind of arguments.

Now if q_0 and q_1 have a common edge of strictly positive length the proof is achieved because we have

$$l(I) < l(S) \leqslant |b(q)|$$

It remains the case where q_0 and q_1 have no common edge of strictly positive length.

In that case $l(I) = l(S)$. But in that case S contains only two edges of strictly positive length (Figure 12.12) $[C(v)D(v)]$ where $A(u) = D(v)$ and $B(u) = C(v)$. But $[AB]$ and $[DC]$ are disjoined (for the same reason as above) and $l(I) < |b(q)|$. □

Lemma 5. *Given a tile q with boundary $b(q)$. Let E be the exterior boundary of the union of any non-empty collection of instances of q which do not overlap each other and form no hole. Then $|E| \geqslant |b(q)|$.*

Proof. Here again we have to refine the proof of (Wijshoff and van Leeuwen 1984). Let $C = (q(u_1), \ldots, q(u_k))$ be a non-empty finite collection of instances of q which do not overlap each other and whose union has no hole. We subdivide the boundary of q in a finite number of arcs in the following way: let c be the convex hull of q. There exists a finite sequence $\alpha_1, \alpha_2, \ldots, \alpha_n$ of points of $b(q) \cap b(c)$ such that the neighbouring of α_i in $b(q)$ is different to the neighbouring of α_i in $b(c)$ (Figure 12.13). (This is a consequence of the assumption that we made concerning the tiles we deal with.)

If $[\alpha_i \alpha_{i+1}]$ is an arc of $b(q) \cap b(c)$ we insert between α_i and α_{i+1} a large enough number of points $\alpha_{i_1} = \alpha_i, \ldots, \alpha_{i_{n_i}} = \alpha_{i+1}$ in such a way that if β and γ are two consecutive points in this sequence, and if there

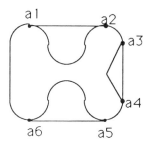

Fig. 12.13

exists an integer j such that $[\beta(u_j)\gamma(u_j)]$ overlaps an edge between $q(u_j)$ and another instance of q, $q(u'_j)$, then $[\beta(u_j)\gamma(u_j)]$ is entirely contained in this edge. This is possible because the number of edges is finite. We still call the new sequence $(\alpha_1,\ldots,\alpha_n)$.

Now we want to prove (and it is sufficient to prove that $l(E) \geqslant l(b(q))$ that for every i there exists a j such that $[\alpha_i\alpha_{i+1}](u_j)$ is included in E. Let $s = [\alpha_i\alpha_{i+1}]$. We define a straight line l in the following way.

1. If s is an arc of c, then l is a line perpendicular to the line $\delta = (\alpha_i\alpha_{i+1})$.

2. If s is not an arc of c, then l is a line perpendicular to the line δ cutting s in a single point m (s is a convex curve, so this is possible).

The lines δ and l are oriented in such a way that q is on the left of δ and the measure of the angle (l,δ) is $+\pi/2$. Now we project all the arcs $s(u_1),\ldots,s(u_n)$ on l (oriented as on the figure). Let p be the rightmost point of these projections; p belongs to the projection of an arc $s(u_{k_0})$. Suppose that $s(u_{k_0}) \not\in E$. There exists $q(u_k)$ such that $b(q(u_k)) \cap s(u_{k_0}) \neq \emptyset$.

1. Then $c(u_k)$ whose area is strictly larger than the area of $c(u_{k_0})-q(u_{k_0})$ has to project out of $c(u_{k_0})-q(u_{k_0})$ and $s(u_k)$ has a projection beyond p; this contradicts the fact that p is the rightmost point on l.

2. Then $b(q(u_k)) \supset s(u_{k_0})$, and again $c(u_k)$ lies beyond the line $\delta(u_{k_0})$ and this is also a contradiction.

□

12.3 Combinatorics and curves

We consider the set $\mathbf{C}_0$ of oriented curves which are a finite union of geometric arcs of class C^2 [1] (some parts of the curve can be 'multiple' arcs).

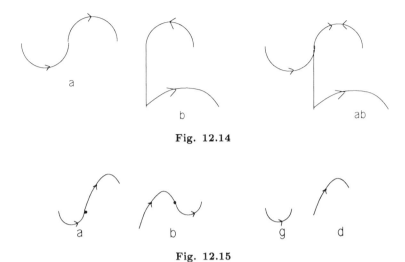

Fig. 12.14

Fig. 12.15

We can define in $\mathbf{C}_0$ a product (partially defined) which is a natural operation: if a and b are curves such that the extremity of a is the origin of b then ab is the curve obtained as the 'union' of a and b. Its origin is the origin of a and its extremity is the one of b. Clearly ab belongs to $\mathbf{C}_0$.

If we define in $\mathbf{C}_0$ the relation $a \sim b$ if there exists a translation t such that $t(a) = b$, then clearly the above defined product is compatible with this equivalence relation and induces a product in the quotient set $\mathbf{C} = \mathbf{C}_0/\sim$ which has in this way a monoid structure. If α and β are two classes, the product $\alpha\beta$ is the equivalence class of the product of an element a of α with an element b of β such that the extremity of a is equal to the origin of b (Figure 12.14).

This monoid $\mathbf{C}$ has combinatorial properties which look like the combinatorial properties of words on some alphabet; we shall now establish them.

Elements of $\mathbf{C}$ will be called *curves* (for simplicity). The empty curve is denoted by 1.

A curve $\alpha \neq 1$ is said to be *primitive* if there exists no curve β such that $\alpha \in \beta\beta^+$.

Two curves α and β are *conjugate* if there exist two curves γ and δ such that $\alpha = \gamma\delta$ and $\beta = \delta\gamma$ (Figure 12.15).

Let α be a curve. The *mirror image* of α is the curve obtained by changing the orientation of α. The origin of α is the extremity of $\tilde{\alpha}$ and conversely.

Let $\alpha = \beta\gamma\delta$. Then γ is a *factor* of α; it is a *left factor* if $\beta = 1$ and a *right factor* if $\delta = 1$.

Fig. 12.16

Fig. 12.17

Lemma 6. *If α is a factor of β^+ for a curve β of arbitrary small length, then α is a line segment.*

Proof. This is a clear consequence of the fact that a real continuous and periodic function of a real argument which has an arbitrarily small period is a constant one. □

Lemma 7. *If $\alpha\beta = \gamma\alpha$ there exist a curve $\delta = \delta_1\delta_2$ and a conjugate $\delta_c = \delta_2\delta_1$ such that $\alpha \in \delta^*\delta_1$, $\beta \in \delta_c^+$ and $\gamma \in \delta^*$.*

Proof. If $|\alpha| \leqslant |\gamma|$, we have

$$\gamma = \alpha\lambda$$
$$\beta = \lambda\alpha$$

So the result is proved, if we write $\delta = \alpha\lambda$ and $\delta_c = \lambda\alpha$ (Figure 12.16).

If $|\alpha| \geqslant |\gamma|$ we have

$$\alpha = \gamma^k\gamma_1$$

with γ_1 a strict left factor of γ (Figure 12.17).

So the equality $\alpha\beta = \gamma\alpha$ becomes

$$\gamma_1\beta = \gamma\gamma_1 \quad |\gamma_1| < |\gamma|$$

Then β and γ are conjugate and we have

$$\gamma = \gamma_1\gamma_2$$
$$\beta = \gamma_2\gamma_1$$
$$\alpha \in \gamma^*\gamma_1$$

□

Lemma 8. *If α and β are non-empty curves satisfying the equation*
$$\alpha^p = \beta^q \; p, q > 0$$
then there exists a curve γ such that $\alpha, \beta \in \gamma^+$.

Proof. If $|\alpha| = |\beta|$, then $p = q$ and $\alpha = \beta = \gamma$. Assume that $|\alpha| > |\beta|$. Then $\alpha = \beta^k \beta_1$ with $|\beta_1| < |\beta|$. If $\beta_1 = 1$, the proof is achieved. Otherwise, β_1 is a left factor and a right factor of α. In the same way, we can write $\alpha = \beta_2 \beta^k$ with $|\beta_2| < |\beta|$. And β_2 is also a left and a right factor of α. But $|\beta_2| = |\beta_1|$. So $\beta_2 = \beta_1$, and the following equality holds:
$$\beta_1 \beta^k = \beta^k \beta_1$$
By Lemma 7, β_1 and β^k are powers of the same curve γ, and $\alpha = \beta_1 \beta^k \in \gamma^+$. Now we have an equation $\gamma^{p_1} = \beta^q$ and $|\gamma| < |\beta|$ where $p_1 > p$. We iterate the process—if it does not stop, this means that α^p is a power of infinite number of curves, so by Lemma 6, α^p is a line segment and the result is proved in all the cases. □

Lemma 9. *If $\alpha\beta = \beta\alpha$ then α and β are both powers of some curve γ, or $\alpha\beta$ is a line segment.*

Proof. If $\alpha = 1$ or $\beta = 1$ it is clear (if we put $\alpha^0 = 1$). Let us suppose α and $\beta \neq 1$. If $|\alpha| = |\beta|$ then $\alpha = \beta = \gamma$. Let us suppose $|\alpha| > |\beta|$. Then $\alpha = \beta^k \beta_1 = \beta_1 \beta^k$ and $0 \leq |\beta_2| < |\beta_1|$. If $|\beta_1| = 0$, the result is proved. In the other case, we iterate the process for β_1 and β. We have $\beta_1 \beta = \beta \beta_1$. So, $\beta = \beta_1^{k'} \beta_2$ with $0 \leq |\beta_2| < |\beta_1|$. We can observe that $|\beta_2| \leq |\beta|/2$. If the process does not stop, the sequence $|\beta_1|$ is converging to zero, and so $\alpha\beta$ satisfies the hypothesis of Lemma 6. □

Lemma 10. *If α and β are primitive curves such that α^+ and β^+ have a common element γ of length greater than or equal to $|\alpha| + |\beta|$, then $\alpha = \beta$.*

Proof. Let γ be a curve such that $\gamma = \alpha^p = \beta^q$, with $|\gamma| \geq |\alpha| + |\beta|$. Suppose that $|\alpha| > |\beta|$; then $1 < p < q$. So we have $\alpha = \beta^k \beta_1$ with $0 < |\beta_1| < |\beta|$ because α is a primitive curve. Let us write $\beta = \beta_1 \beta_2$. But β_1 is a left and right factor of β and also β_2 because α^2 is a left factor of γ. So, $\beta_1 \beta_2 = \beta_2 \beta_1$ and there is a contradiction with Lemma 9. □

Lemma 11. *Let α be a primitive curve. If a conjugate β of α has two different writings $\beta = \beta_1 \beta_2 = \beta_1' \beta_2'$ such that $\alpha = \beta_2 \beta_1 = \beta_2' \beta_1'$, then $\beta_1 = 1$ and $\beta_1' = \beta$ or vice versa.*

Proof. The equality $\beta_2 \beta_1 = \beta_2' \beta_1'$ implies $\beta_1 \beta_2 \beta_1 \beta_2' = \beta_1 \beta_2' \beta_1' \beta_2'$. Whence $(\beta_1' \beta_2')(\beta_1 \beta_2') = (\beta_1 \beta_2')(\beta_1' \beta_2')$. By Lemma 9, there exists γ such that $\beta_1 \beta_2'$ and $\beta_1' \beta_2'$ belong to γ^+. In the same way, $\beta_1' \beta_2 \beta_1 \beta_2 = \beta_1' \beta_2' \beta_1' \beta_2 = \beta_1' \beta_2' \beta_1' \beta_2$ and so $(\beta_1' \beta_2)(\beta_1 \beta_2) = (\beta_1' \beta_2)(\beta_1' \beta_2)$. So we have $\beta_1 \beta_2 = \beta_1' \beta_2' = \gamma$. And $\beta_1' \beta_2, \beta_1 \beta_2' \in \gamma^*$. But one of the two curves $\beta_1' \beta_2$ or $\beta_1 \beta_2'$ has a length strictly less than γ because $\beta_1 \neq \beta_1'$ and so $\beta_1' \beta_2 = 1$ or $\beta_1 \beta_2' = 1$. □

12.4 Tilings

Our purpose is to characterize the exact tiles by a simple property, and to describe all the tilings we can obtain with an exact tile.

First of all we can observe that the following properties hold.

Lemma 12. *Let q be an exact tile and U be a tiling by q. Then two instances $q(u), q(v)$ $(u, v \in U)$ are disjoint or simply neighbouring.*

Proof. If two instances are neighbouring, but not simply neighbouring, their union forms a hole and by Lemmas 4 and 5 they cannot belong to a tiling of the plane. □

Lemma 13. *Let U be a tiling of the plane by an exact tile q. If $q(u)$ is an instance of q in the tiling, every surrounding of $q(u)$ with tiles of U has at least four tiles.*

Proof. It is an immediate consequence of Lemmas 4 and 5. □

Triads and contacts

A *triad* is a triple $(q(u), q(v), q(w))$ of tiles which are pairwise simply neighbouring and $[q(u), q(v)], [q(u), q(w)]$ are consecutive curves on the boundary of $q(u)$ in this order; moreover, the union of the three tiles has no hole. This implies the existence of a unique common point to the three tiles.

If we have

$$[q(u), q(v)] = [A_1 A]$$
$$[q(v), q(w)] = [B_1 B]$$
$$[q(w), q(u)] = [C_1 C]$$

then (A, B, C) will be called the *contact* of the triad $(q(u), q(v), q(w))$.

We can observe that if (p, q, r) is a triad with contact (A, B, C), then (q, r, p) and (r, p, q) are also triads with respective contacts (B, C, A) and (C, A, B).

A contact is said to be *exact* if there is at least one tiling of the plane by the corresponding tile in which this contact appears. In Figure 12.18, (a) and (c) but not (b) are exact contacts.

We shall give now a characterization of exact tiles.

Theorem 14. *A tile q is exact if and only if it admits a surrounding.*

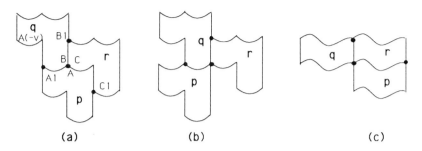

Fig. 12.18 $q(u) = p$, $q(v) = q$, and $q(w) = r$

Proof. Clearly, if U is a tiling of the plane by a tile q, then if $q(u)$ is a tile of U, the set of tiles of U which intersect $q(u)$ are simply neighbouring to $q(u)$ (Lemma 12) and form a surrounding of $q(u)$ when correctly ordered according to their common edge with $q(u)$.

Conversely, let us suppose that $(q(u_0),\ldots,q(u_{k-1}))$ is a surrounding of q. Because of Lemma 13, k is greater than or equal to 4. Let us write $e_i = [q, q(u_i)]$ and $e'_i = [q(u_i), q(u_{i+1})]$ (indices are defined modulo k).

Using translations we have

$$[q, q(u_i)] = [q(u_i), q(2u_i)] = [q(u_{i+1}, q(u_i + u_{i+1})] = e_i$$
$$[q, q(u_{i+1})] = [q(u_i), q(u_i + u_{i+1})] = [q(u_{i+1}), q(2u_{i+1})] = e_{i+1}$$
$$[q(u_i), q(u_{i+1})] = [q(2u_i), q(u_i + u_{i+1})] = [q(u_i + u_{i+1}), q(2u_{i+1})] = e'_i$$

In Figure 12.19 the tiles q, $q(u_i)$, $q(u_{i+1})$, $q(u_{i-1})$, $q(u_{i-1} + u_i)$, $q(2u_i)$, $q(u_i + u_{i+1})$, $q(2u_{i+1})$, and their common edges are represented.

The triads $(q, q(u_i), q(u_{i+1}))$, $(q(u_i), q(2u_i), q(u_i + u_{i+1}))$, and $(q(u_{i+1}), q(u_i + u_{i+1}), q(2u_{i+1}))$ are translated one of the other. So their contacts are the same.

But it is not obvious that $(q(u_{i+1}), q(u_i), q(u_i + u_{i+1}))$ is a triad. Actually the tiles are pairwise simply neighbouring, but perhaps the union has a hole h as in Figure 12.19.

Let us write r_i (or s_i, t_i) to denote the common edge of $q(u_i)$ (or $q(u_i+u_{i+1}), q(u_{i+1})$) with h. Let us observe that if $|r_i| = 0$ then $|s_i| = |t_i| = 0$; otherwise $q(u_i + u_{i+1})$ and $q(u_{i+1})$ would not be simply neighbouring. More generally, if one of these three curves is of length zero, so are the others. Now let us compute the length of the boundary of q, looking at the tile $q(u_i)$ (or $q(u_i + u_{i+1})$). We have

$$|b(q)| = |b(q(u_i))| = |b'_{i-1}| + |t_{i-1}| + |b_{i-1}| + |b_i| + |b_{i+1}| + |r_i| + |b'_i| + |b_i|$$

If we sum these equalities for $i = 0$ to $k-1$, we obtain

$$k|b(q)| = 2\sum |b'_i| + 4\sum |b_i| + \sum |t_i| + \sum |r_i| \qquad (1)$$

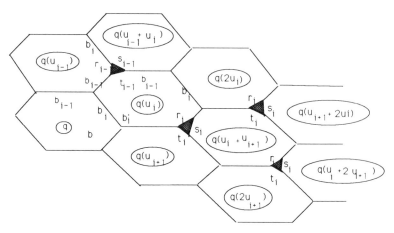

Fig. 12.19

On the other hand, computing the length of $b(q)$ looking at $q(u_i + u_{i+1})$ (Figure 12.19), we obtain

$$|b(q)| = |b_i| + |s_i| + |b_{i+1}| + |b'_i| + |t_i| + |b_i| + |b_{i+1}| + |r_i| + |b'_i|$$

Also

$$k|b(q)| = 4\sum |b_i| + 2\sum |b'_i| + \sum |s_i| + \sum |t_i| + \sum |r_i| \quad (2)$$

Comparing (1) and (2), we have

$$\sum |s_i| = 0$$

So for each i, $|s_i| = |t_i| = |r_i| = 0$, and the hole h is reduced to a point. So $(q(u_{i+1}), q(u_i), q(u_i + u_{i+1}))$ is a triad for each i. Moreover,

$$(q(u_i - u_{i+1}), q(u_i), q(u_{i+1}), q(u_{i+1} - u_i), q(-u_i), q(-u_{i+1})) \quad (3)$$

is a surrounding of q (because of Lemma 3).

This surrounding has the property that it can be translated, so the set $U = \{nu_i + n'u_{i+1}/n, n' \in \mathbb{Z}\}$ is a regular tiling of the plane. So the proof is achieved and it contains some more properties which we state below. □

Corollary 15. *If a tile q is exact, there exists a regular tiling of the plane by q.*

Corollary 16. *Every surrounding of a tile can be extended into a tiling of the plane, and every contact appearing in a surrounding is exact.*

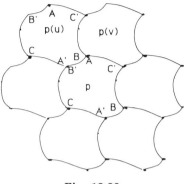

Fig. 12.20

Proof. Keeping the notations of the proof of Theorem 14, the set $U = \{ku_i + k'u_{i+1}/i = 0,\ldots, k-1,\ n, n' > 0\}$ is a tiling of the plane. □

We can now give another characterization of exact tiles which is a consequence of (3).

Lemma 17. *Let (p, q, r) be an exact triad (that is, a triad appearing in a tiling of the plane) with contact (A, B, C). Then we have the following.*

1. *Recalling that A' represents the symmetric point of A on the boundary of the tile,*

$$[p, q] = [B'A]$$
$$[q, r] = [C'B]$$
$$[r, p] = [AC']$$
$$[q, p] = [BA']$$
$$[r, q] = [CB']$$
$$[p, r] = [AC']$$

2. *Moreover, if we write $q = p(u)$, $r = p(v)$, then $(p(u), p(v), p(v-u), p(-u), p(-v), p(u-v))$ is a surrounding of p (Figure 12.20).*

Corollary 18. *If $[p, q] = [AB]$ is an edge in a tiling U, then $[q, p] = [A'B']$.*

Let q be a tile, and A, B two points of its boundary. From now on, the equivalence class in $\mathbf{C} = \mathbf{C}_0/\sim$ of the curve $[AB]$ will be denoted by $\langle [AB] \rangle$.

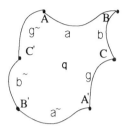

Fig. 12.21

Definition 19. A tile q is a pseudo-hexagon $ABAA'B'C'$ if there exist three points A, B, and C on the boundary of q such that $B \in [AC]$ and $\langle[A'B']\rangle = \langle[AB]\rangle$, $\langle[B'C']\rangle = \langle[BC]\rangle$, and $\langle[C'A]\rangle = \langle[CA']\rangle$.

This is equivalent to saying that there exists a point A of the boundary of q such that the equivalence class of the boundary of q considered as a Jordan curve starting and ending at A can be written as $\alpha\beta\gamma\tilde{\alpha}\tilde{\beta}\tilde{\gamma}$ (Figure 12.21).

The previous result can be stated in a new way.

Theorem 20. A tile q is exact if and only if it is a pseudo-hexagon.

Proof. The 'only if' part is proved by using Lemma 17. Conversely, suppose that q is a pseudo-hexagon $ABCA'B'C'$. Let $u = \vec{AC}, v = \vec{BA'}$. Then $(q(u), q(v), q(v - u), q(-u), q(-v), q(u - v))$ is a surrounding of q and so q is an exact tile. $\square$

As an immediate consequence of Lemma 17, we have the following proposition.

Proposition 21. If (A, B, C) is an exact contact for the tile q, then q is a pseudo-hexagon $(B'AC'BA'C)$ and (A', B', C') is also an exact contact.

In Theorem 20 we give an 'if and only if' condition for a tile to be exact. The question arising now is the effectiveness of this condition. Actually this condition is decidable for a large class of tiles. Let us consider a tile as defined by two real functions f and g such that $x = f(s)$ and $y = g(s)$ are the parametric equations of the boundary (where s is the curvilinear abscissa of the boundary); let us denote the tile by $q(f, g)$. Then we have the following theorem.

Theorem 22. Let P be the class of real functions of a real variable which are piecewise polynomial functions. Let $q(f, g)$ be a tile such that f and g belong to P. Then the exactness of q is decidable.

Proof. Let l be the length of q. Then f and g are defined on the segment $[0,l]$. There exists a sequence $(s_1 = 0, \ldots, s_n = l)$ such that f and g are polynomial functions in each segment $[s_i, s_{i+1}]$. So, there exists a finite number of choices for three numbers a, b, c in $[0, l/2]$ if we consider only the choice of the segments $[s_i, s_{i+1}]$ in which the numbers are located. Given this choice, we can decide whether $M(a)M(b)M(c)M(a+l/2)M(b+l/2)M(c+l/2)$ is a pseudo-hexagon for some value of a, b, c. □

We have proved that if a tile q is exact there exists a regular tiling of the plane by q. But a stronger property can be proved. Actually, *every* tiling by an exact tile is half-periodic. This is discussed in the next section.

12.5 Half-periodicity

In this section, q is assumed to be an exact tile. We first establish a main lemma which is quite important for the proof of half-periodicity.

Lemma 23. *If (A, C, B') and (A, D, B') are two different exact contacts $(C \neq D)$ such that $C \in [BD]$ and $(B \neq C$ or $A' \neq D)$ then one of the following properties is satisfied.*

1. *There exists a primitive curve $\alpha = \alpha_1 \alpha_2 \in C$ and a conjugate $\alpha_c = \alpha_2 \alpha_1$ of α such that (Figure 12.22)*

$$\langle [BA'] \rangle \in \alpha \alpha^+$$
$$\langle [B'A] \rangle \in \tilde{\alpha}^+$$
$$\langle [CD] \rangle \in \alpha_c^+$$
$$\langle [C'D'] \rangle \in \widetilde{\alpha^+}$$
$$\langle [BC] \rangle \in \alpha^* \alpha_1$$
$$\langle [B'C'] \rangle \in \widetilde{\alpha_1 \alpha^*} = \widetilde{\alpha_c^* \alpha_1}$$
$$\langle [DA'] \rangle \in \alpha_2 \alpha^*$$
$$\langle [D'A] \rangle \in \widetilde{\alpha^* \alpha_2} = \widetilde{\alpha_2 \alpha_c^*}$$

2. *$[BA']$ is a line segment (Figure 12.23).*

Proof. Let $\langle [BC] \rangle = a$, $\langle [CD] \rangle = b$ and $\langle [DA'] \rangle = c$. Then $\langle [CA'] \rangle = \langle [CD] \rangle \langle [DA'] \rangle = bc$. We have $\langle [C'A] \rangle = \langle [\widetilde{CA'}] \rangle = \widetilde{(bc)} = \tilde{c}\tilde{b}$. On the other hand,

$$\langle [C'A] \rangle = \langle [C'D'] \rangle [D'A] = \langle [C'D] \rangle \langle [\widetilde{DA'}] \rangle = \langle [C'D'] \rangle \tilde{c}$$

310   Tiling the plane with one tile

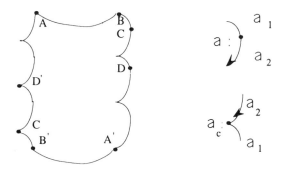

Fig. 12.22

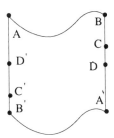

Fig. 12.23

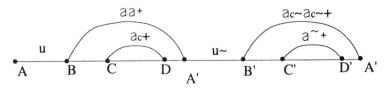

Fig. 12.24

Let $\langle [C'D'] \rangle = d$, when $d \neq 1$. We obtain the equation

$$\tilde{c}\tilde{b} = d\tilde{c}$$

By Lemma 7, there exists $\alpha = \alpha_1 \alpha_2 \in C$ and a conjugate $\alpha_c = \alpha_2 \alpha_1$ of α such that

$$\tilde{c} \in \alpha^* \alpha_1 \quad \tilde{b} \in \alpha_c^+ \quad d \in \alpha^+ \qquad (4)$$

On the other hand

$$\begin{aligned}\langle [BD] \rangle &= \langle [BC] \rangle \langle [CD] \rangle = ab \\ \langle [B'D'] \rangle &= \langle [\widetilde{BD}] \rangle = \tilde{b}\tilde{a}\end{aligned}$$

but also $\langle [B'D'] \rangle = \langle [B'C'] \rangle \langle [C'D'] \rangle = \tilde{a} \langle [C'D'] \rangle$. So $\tilde{b}\tilde{a} = \tilde{a}d$. Hence there exists a curve $\alpha' = \alpha'_1 \alpha'_2 \in C$ and a conjugate $\alpha'_c = \alpha'_1 \alpha'_2$ such that

$$\tilde{b} \in \alpha'^+_c \quad \tilde{a} \in \alpha_c^* \alpha'_2 = \alpha'_2 \alpha'^* \quad d \in \alpha'^+ \qquad (5)$$

If α is a line segment, so is α' and property 2 holds.

From (4) and (5) α and α' are primitive curves. Therefore $\alpha = \alpha'$ by Lemma 11, and consequently $\alpha_c = \alpha'_c$. by Lemma 11, this implies that $\alpha_1 = \alpha'_1$ and $\alpha_2 = \alpha'_2$, or $\alpha_1 = \alpha'_2 = 1$ or $\alpha_2 = \alpha'_1 = 1$.

In the three cases, we have

$$\langle [BA'] \rangle = \langle [BC] \rangle \langle [CA'] \rangle \in \widetilde{\alpha'_2 \alpha_c^* \alpha_c^+ \alpha_1 \alpha^*} = \widetilde{\tilde{\alpha}\alpha^+}$$

In the same way we have $\langle [B'A] \rangle \in \alpha_c \alpha^+$.

Therefore property 1 is proved (changing α in α_c) and can be made explicit by the following scheme obtained by 'unfolding' the boundary of q (Figure 12.24). □

We complete this result with the following obvious remark.

Lemma 24. *If (A, B, B') is an exact contact, then we have $\langle [AB] \rangle = \langle [\widetilde{A'B'}] \rangle$ and $\langle [BA'] \rangle = \langle [\widetilde{B'A}] \rangle$ (Figure 12.25), and (A, A', B') is also an exact contact.*

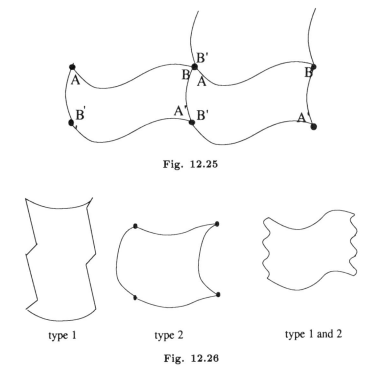

Fig. 12.25

type 1 type 2 type 1 and 2

Fig. 12.26

Definition 25. A tile q is a pseudo-parallelogram $ABA'B'$ if there are two points A, B on the boundary of q such that $B \in [AA']$ and either

1. $\langle[AB]\rangle = \langle[\widetilde{A'B'}]\rangle$ and there is a curve α and a conjugate α_c such that $\langle[BA']\rangle \in \alpha\alpha^+$ and $\langle[B'A]\rangle \in \widetilde{\alpha_c}\alpha_c^+$ (the tile will be called a pseudo-parallelogram of type 1, or

2. $\langle[AB]\rangle = \langle[\widetilde{A'B'}]\rangle$ and $\langle[BA']\rangle = \langle[\widetilde{B'A}]\rangle$ (the tile will be called a pseudo-parallelogram of type 2) (Figure 12.26).

We can see that a pseudo-parallelogram is a special kind of exact tile, and that a pseudo-parallelogram be of type 1 or type 2.

Lemma 26. If a tile q is a pseudo-parallelogram $ABA'B'$ and also a pseudo-parallelogram $ACA'C'$, then $B = C$.

Proof. Let us write $\langle[AB]\rangle = u$, $\langle[BC]\rangle = v$, and $\langle[CA']\rangle = w$. Suppose that $|[AB]| < |[AC]|$. There exist two curves w_1 and v_1 such that $vw = w_1v_1$ and $|v| = |v_1|$.

Since q is a pseudo-parallelogram $ABA'B'$ and also $ACA'C'$, the following equalities hold: $\langle\widetilde{[AC]}\rangle = \langle[A'C']\rangle$, so that $\tilde{v}\tilde{u} = \tilde{u}\tilde{v}_1$. Also, $\langle\widetilde{[BA']}\rangle = \langle[B'A]\rangle$, so that $\tilde{w}\tilde{v} = \tilde{v}_1\tilde{w}$.

Using Lemmas 7 and 11, we know there exists a curve $\gamma = \gamma_1\gamma_2$ such that $u \in \gamma^*\gamma_1$, $v \in (\gamma_2\gamma_1)^+$, $v_1 \in \gamma^+$, and $w \in (\gamma_2\gamma_1)^*\gamma_2$. Now we have $uvw \in \gamma\gamma^+$. So $uvw\tilde{u}\tilde{v}\tilde{w}$ admits a strict factor which is a closed curve, and so there is a contradiction. □

Lemmas 23 and 24 have a corollary.

Corollary 27. *If (A, C, B') and (A, D, B') are two different contacts then q is a pseudo-parallelogram and conversely.*

We now give four technical lemmas about pseudo-parallelograms which will be needed later. The first three lemmas deal with pseudo-parallelograms of type 1.

Lemma 28. *Let $ABA'B'$ be a pseudo-parallelogram p of type 1, and let U be a tiling of the plane by p. Let r, q, s be three instances of p in the tiling such that*

$$[r,q] = [CA'] \quad C \in [BA']$$
$$[q,s] = [DA'] \quad D \in [BA'] \quad D \neq B$$
$$C \neq D \quad ||[B'C']|| \geq |\alpha|$$

where α is a curve such that $\langle[BA']\rangle \in \alpha\alpha^+$ (Figure 12.27).

Under these hypotheses, the tile $r(A\vec{B'})$ belongs to the tiling of Figure 12.27.

Proof. Let t be the tile adjacent to r in the surrounding of q (before r) in the tiling U. If $t \neq r(A\vec{B'})$, then the contact of the triad (t, r, q) is (X, A', C') with $X \neq B$. There are two cases to examine according to $X \in [AB]$ or $X \in [BC]$.

1. $X \in [BC]$ (Figure 12.28)

    The tile p admits two different contacts (C', X, A') and (C', B, A'), and $B \in [AX]$. So, we apply Lemma 23 ($B \neq A$). If 2 holds then $\langle[AC]\rangle$ is a line segment with factor α, so $\langle[AA']\rangle$ is a line segment, which is impossible. So, 1 holds. There exists a primitive curve β and a conjugate β_c such that $\langle[AC]\rangle \in \beta\beta^+$ and $\langle[BX]\rangle \in \beta_c^+$. So we obtain the scheme in Figure 12.29.

314   Tiling the plane with one tile

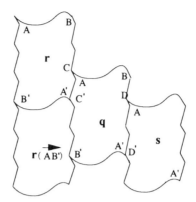

Fig. 12.27

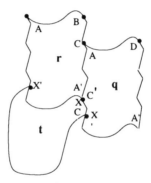

Fig. 12.28

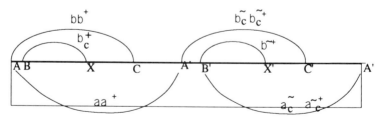

Fig. 12.29

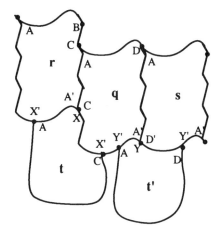

Fig. 12.30

But let us recall that $|[BC]| \geq |\alpha|$. If $|\beta| \geq |\alpha|$, then β_c, which is a left factor of $\langle[BX]\rangle$, has all its factors of length $|\alpha|$ which are conjugate curves of α. So the curve of length $2|\alpha|$ centred in A' is a closed curve because its equivalence class is α, and it cannot be a factor of the boundary of p.

If $|\alpha| \geq |\beta|$ the argument is symmetric.

2. $X \in [AB]$ (Figure 12.30)

   This case is less easy than the previous one. Let us observe that $X \neq A$, otherwise r and t would no longer be adjacent. So there exists an instance t' adjacent to s in the surrounding of q (after s). Let (Y, A', D') be the contact of the triad (t', q, s). But Y is distinct from B, and so there are two distinct contacts (D', Y, A') and (D', B, A'), with $D \neq B$. If 1 of Lemma 23 holds $\langle[AD]\rangle$ is a line segment, and so is α because $|[BC]| \geq |\alpha|$, which is impossible because $[AA']$ is not a segment line. This implies that $\langle[AD]\rangle \in \gamma\gamma^+$ (γ is primitive). So we have (assuming $D \in [BC]$) the situation in Figure 12.31.

   Let $\gamma = \gamma_1\gamma_2$ and $\gamma_c = \gamma_2\gamma_1$. Then $|\langle[AY]\rangle| \geq |\gamma_1|$ and $|\langle[BD]\rangle| \geq |\gamma_2|$. So $|\langle[AD]\rangle| \geq |\gamma_1|+|\beta|+|\gamma_2| = |\gamma|+|\beta|$. But $\langle[AD]\rangle \in F(\gamma\gamma^+) \cap F(\beta\beta^+)$, and so (Lemma 10) $\beta = \gamma$. This implies $\langle[DC]\rangle \in \beta^+$. So $|\langle[BC]\rangle| \geq |\beta|$ and $|\langle[BC]\rangle| \geq |\alpha|$.

   With the same kind of argument as the first case, we conclude that there exists a strict factor of the boundary of p which is a closed curve: the curve of length $2\inf(|\beta|,|\alpha|)$ centred in A'. If we suppose $C \in [BD]$, permuting C and D gives the result.

   $\square$

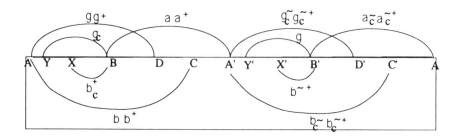

Fig. 12.31

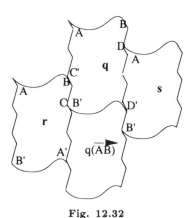

Fig. 12.32

Lemma 29. *Let $ABA'B'$ be a pseudo-parallelogram of type 1 and U a tiling of the plane by p. Let r, q, and s be three instances of p in the tiling such that*

$$[r, q] = [BC'] \quad C \neq A' \quad C \neq B$$
$$[q, s] = [DA'] \quad D \in [BA'] \quad C \neq D$$

Then the tile $q(\vec{AB'})$ belongs to the tiling U (Figure 12.32).

Proof. Let us assume that the tile adjacent to r in the surrounding of q (before r) is not $q(\vec{AB'})$, but a tile t such that the contact (t, r, q) is (X, C, B') with $X \neq A$ (Figure 12.33). Necessarily $X' \in [A'B']$, and so $X \in [AB]$.

So, there exists a tile t' adjacent to t (before t) in the surrounding of q with a contact (Y, B, X') for the triad (t', t, q). We cannot have $Y = C'$ because $Y' \in [A'X']$. So we have the following inequalities about exact

Half-periodicity 317

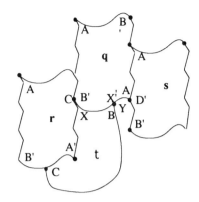

Fig. 12.33

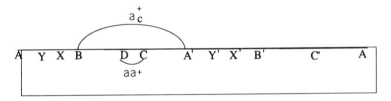

Fig. 12.34

contacts:

$$(B', A, C) \neq (B', X, C)$$
$$C' \neq A$$
$$(X', Y, B) \neq (X', C', B)$$
$$C \neq B$$

We apply Lemma 12.22 at these two inequalities. In both cases, 2 cannot hold because it implies $[BB']$ is a line segment (Figure 12.34).

In the same way, we can prove that we cannot have property 1 for one inequality and property 2 for the other. So, property 1 holds for the two inequalities (Figure 12.35).

Then by an argument we have already used, we observe that the curve centred in B with length $2\inf(|\beta|, |\gamma|)$ is closed and it is impossible. □

Lemma 30. *Let $ABA'B'$ be a pseudo-parallelogram of type 1 and U a tiling of the plane by p. Let $r, q,$ and s be three instances of p in the tiling such that*

$$[r, q] = [CA'] \quad C \in [BA']$$

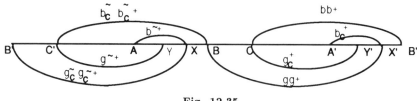

Fig. 12.35

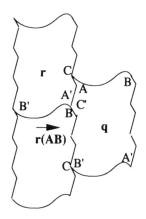

Fig. 12.36

and $|[BC]|$ is greater than or equal to the upper bound of the lengths of the edges of the tiling. Then $r(\vec{AB'})$ belongs to the tiling U (Figure 12.36).

Proof. The proof is similar to that of Lemma 28. But the hypothesis about $|[BC]|$ implies that tile t cannot satisfy the hypothesis of the second case ($X \in [AB]$), because in that case the edge $[XC] = [t, q]$ would have a length greater than the upper bound, and so t satisfies the hypothesis of the first case ($X \in [BC]$) and provides a contradiction (without using the tile s). □

Lemma 31. *Let $ABA'B'$ be a pseudo-parallelogram of type 1, and U a tiling such that U contains the bi-infinite band $B = \{q(k\vec{AB'})/k \in \mathbb{Z}\}$. Then U is invariant in the translation of vector $\vec{AB'}$.*

Proof. Let r be a tile adjacent to q in the surrounding of $q(-\vec{AB'})$ (before q) (Figure 12.37) such that the contact of the triad $(r, q, q(-\vec{AB'}))$ is (X, B, A'). Then, by Lemma 17, $[r, q] = [B'X]$ and $[r, q(-\vec{AB'})] = [XA]$. So, every tile of U which is adjacent to B has an edge with B of type $[B'A]$

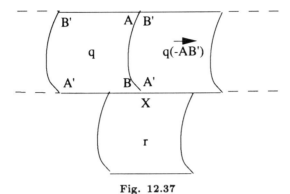

Fig. 12.37

on one side and $[AB']$ on the other side (symmetrically). This proves that all the tiles $r(k\vec{AB'})$ belong to U. So U is a union of bands which are a translation of B and hence U is invariant by the translation of vector $\vec{AB'}$. □

Lemma 32. *Let $ABA'B'$ be a pseudo-parallelogram p where $[BA']$ is a line segment. If a tiling U contains two adjacent instances of p, q, and r such that (Figure 12.38a)*

$$[q,r] = [CA'] \quad C \in [BA'] \quad C \neq B$$

then $q(\vec{AB'})$ belongs to the tiling or there exists a primitive curve $\gamma = \gamma_1\gamma_2$ and a conjugate $\gamma_c = \gamma_2\gamma_1$ ($\gamma_1, \gamma_2 \neq 1$) such that $\langle[AC]\rangle \in \gamma\gamma^+$ and $\langle[BC]\rangle = \gamma_2$.

Proof. Let s be the tile adjacent to r and neighbouring q (before q) in U. Let (C', X, A') be the contact of the triad (r, s, q). If $X \neq B$, Lemma 23 can be applied, so if 2 holds then $[AC]$ is a line segment, and this is a contradiction. Hence 1 holds: there is a primitive curve $\gamma = \gamma_1\gamma_2$ and a conjugate $\gamma_c = \gamma_2\gamma_1$ of γ such that $\langle[AC]\rangle$ belongs to $\gamma\gamma^+$ and $\langle[XB]\rangle$ (or $\langle[BX]\rangle$ depending on the location of X) belongs to γ_c^+.

If $X \in [BC]$ then $[AC]$ is a line segment, and so we have a contradiction.

If $X \in [AB]$ then we have

$$\langle[AX]\rangle \in \gamma^*\gamma_1$$
$$\langle[XB]\rangle \in (\gamma_2\gamma_1)^+$$
$$\langle[BC]\rangle = \gamma_2$$

So the result is proved. □

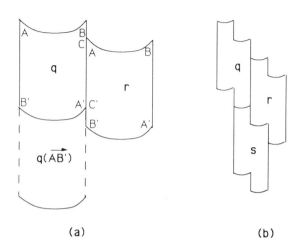

Fig. 12.38

Theorem 33. *If p is an exact tile, every tiling of the plane by p is half-periodic.*

Proof. Let U be a tiling of the plane by p. Let l be the upper bound of the lengths of the edges in U. There are two possibilities: there exists an edge of length l, or all the edges have a length strictly less than l.

We first treat the former case.

1. There exists in the tiling an edge of length l.

    The first step is to prove that the maximum edge is 'propagated' in the tiling, and that this occurs in two possible ways.

    Lemma 34. *Let q be a tile in U and $q(u_1), q(u_2), q(u_3), q(u_4)$ for consecutive tiles in a surrounding of q in U such that $[q(u_2), q(u_3)]$ has length l (the maximum length), and $q(u_1), q(u_4)$ are adjacent to q (Figure 12.39). Then U satisfies the following properties:*

    (a) $u_1 = u_2 - u_3$ or $u_4 = u_3 - u_2$;

    (b) if one of both equalities does not hold, then q is a pseudo-parallelogram.

    Part 1a proves that the maximum edge between $q(u_2)$ and $q(u_3)$ is propagated at least on one side between q and $q(u_1)$ or $q(u_4)$ and perhaps on the two sides; moreover, if it is propagated on only one side, then q is a pseudo-parallelogram.

Half-periodicity

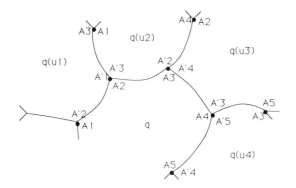

Fig. 12.39

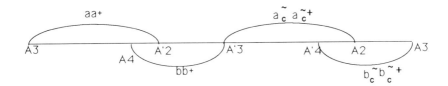

Fig. 12.40

Proof. First let us observe that $q(u_1), \ldots, q(u_4)$ exist by Lemma 13. Let $[q, q(u_i)] = [A_i A_{i+1}]$. Then by Lemma 17, we have

$$[q(u_i), q(u_{i+1})] = [A_i A_{i+1}]$$
$$[q(u_{i+1}), q(u_i)] = [A'_{i+2} A_i]$$

Assume that $u_1 \neq u_2 - u_3$ and $u_4 \neq u_3 - u_2$. Since $[q, q(u_1)] = [A_1 A_2]$ and $[q(u_3), q(u_2)] = [A'_4 A_2]$, this implies $A_1 \neq A'_4$.

In the same way $u_4 \neq u_3 - u_2$ implies $A_2 \neq A'_5$. Since (A'_2, A'_4, A_3) is a contact, so is (A_2, A_4, A'_3). And we have two distinct contacts (A_2, A'_1, A'_3) and (A_2, A_4, A'_3). We have also two distinct contacts (A'_3, A'_5, A_4) and (A'_3, A_2, A_4). We have $A_1 \neq A_2$, and so Lemma 23 can be applied to the first distinct contacts. Also, $A_4 \neq A_5$ implies that Lemma 23 can be applied to the last distinct contacts. So we have the scheme of Figure 12.40.

But $[q(u_2), q(u_3)]$ is of length l. So $|[A_3 A_4]| \leq |[A_4 A'_2]|$, whence $|\alpha| \leq |[A_4 A'_2]|$.

Also $|[A'_2 A'_3]| \leq |[A_4 A'_2]|$, whence $|\beta| \leq |[A_4 A'_2]|$. This implies that the curve centred in A'_3 with length $2 \inf(|\alpha|, |\beta|)$ is a closed curve,

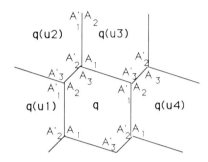

Fig. 12.41

and this is a contradiction. So 1a is proved. Now, 1b is an immediate consequence of Corollary 27. □

Definition 35. Let e be the edge $[q(u_2), q(u_3)]$.

If $u_1 = u_2 - u_3$ and $u_4 = u_3 - u_2$, we shall say that the maximum edge is propagated *two-sidedly* over q (Figure 12.41).

If $u_1 = u_2 - u_3$ and $u_4 \neq u_3 - u_2$, the maximum edge e is propagated *on the left* over q (Figure 12.42).

If $u_1 \neq u_2 - u_3$ and $u_4 = u_2 - u_3$ the edge e is propagated *on the right* over q (Figure 12.43).

At this step, two cases have to be considered.

(a) Every time the edge e appears in the tiling, it is two-sidedly propagated *at the two extremities of* e (and so four new occurrences of e appear).

(b) There exists an occurrence of e in the tiling, and an extremity of e such that e is propagated only on the left, or on the right.

(a) We can use the scheme in Figure 12.41. The curve $[A'_2 A_1]$ has a strictly positive length. So there exists in U, in a surrounding of q, a tile $q(u_0)$ neighbouring $q(u_1)$ and adjacent to q (before $q(u_1)$). Let (A'_2, A_1, X) be the contact of the triad $(q(u_1), q, q(u_0))$. Let us observe that

$$[q(u_0), q(u_1)] = [A_2 X]$$
$$[q, q(u_2)] = [A_2 A_3]$$

Half-periodicity 323

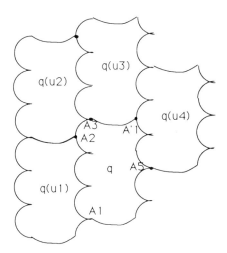

Fig. 12.42

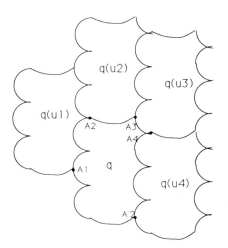

Fig. 12.43

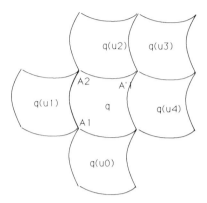

Fig. 12.44

- If $X \neq A_3$, then (A'_1, A_3, A'_2) and (A_1, X, A'_2) are two distinct contacts. So, using Lemmas 23 and 24, two situations are available.
    * If $X = A_2$, then q is a pseudo-parallelogram $A_1 A_2 A'_1 A'_2$ of type 2 ($A_3 = A'_1$), and we have the scheme of Figure 12.44.
    Now considering the edge $e = [q, q(u_4)]$ and the tile $q(u_0)$, this edge is propagated two-sidedly over $q(u_0)$ so that $q(u_4 + u_0)$ belongs to U. In the same way, $q(u_1 - u_0)$ and $q(u_1 + u_0)$ belong to U. Iterating this argument, we prove that all the tiles $q(ku_0 + k'u_4)$, $k, k' \in \mathbb{Z}$ belong to U, but these tiles realize a regular tiling. So, U is regular.
    * If $X \neq A_2$, then q is a pseudo-parallelogram $A_1 A_2 A'_1 A'_2$ of type 1 or of type 2 with a line segment $[A_2 A'_1]$ and two possible schemes corresponding to Figures 12.45 and 12.46.
    In the two cases, we iterate the process of propagation of the edge $[q(u_2), q(u_3)]$ and we obtain that the tiling is invariant in the translation of vector $u_3 - u_2 = u_4$.
- If $X = A_3$ we obtain Figure 12.47.
By the hypothesis of propagation of the edge $[q(u_1), q]$, $q(u_4 + u_0)$ belongs to the tiling, and iterating the process, we obtain that U is half-periodic for vector $u_4 = u_3 - u_2$ (U is not necessarily regular because $[A_2 A'_1]$ can be a non-primitive curve).

(b) We can use the scheme of Figure 12.43. Somewhere in the tiling this configuration appears where $[q(u_2), q(u_3)]$ has length l, $u_1 \neq$

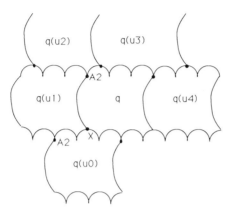

Fig. 12.45

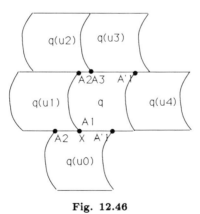

Fig. 12.46

Tiling the plane with one tile

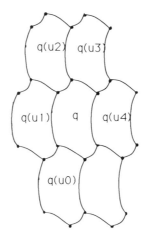

Fig. 12.47

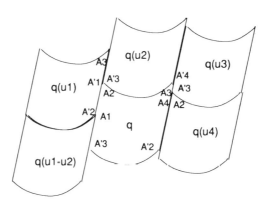

Fig. 12.48

$u_2 - u_3$ and $u_4 = u_2 - u_3$.

The triad $(q(u_1), q(u_2), q)$ has contact (A'_1, A'_3, A_2). So, since $A'_1 \neq A_4$, there are two different contacts (A_2, A'_1, A'_3) and (A_2, A_4, A'_3). And, looking at the proof of Lemma 34, we can conclude that p is a pseudo-parallelogram $A_2 A_3 A'_2 A'_3$ of type 1. Because of the maximality of $[q, q(u_4)]$, we can apply Lemma 28, and so $q(u_1 - u_2)$ belongs to the tiling (Figure 12.48). (We have represented $[A_3 A'_2]$ as a line segment to simplify the picture, but it is only assumed to be a non-primitive curve and $\langle [A'_3 A_2] \rangle$ is assumed to be the mirror image of a conjugate of $\langle [A_3 A'_2] \rangle$).

We have to discuss the location of point A'_1.

Half-periodicity

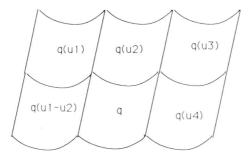

Fig. 12.49

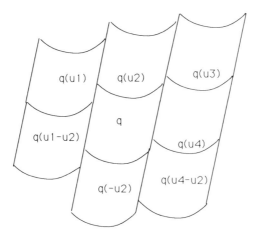

Fig. 12.50

- If $A'_1 = A'_2$, then by maximality of l, we have $A_3 = A_4$. And $u_2 - u_1 = u_3 - u_2 = u_4$ (Figure 12.49). This is a contradiction because in that case e is propagated two-sidedly. Therefore, we have the following.
- $A'_1 \neq A'_2$, and so we can apply Lemma 29 to Figure 12.48. Therefore $q(-u_2)$ belongs to U. By Lemma 30, $q(u_4 - u_2)$ belongs to U (Figure 12.50).

Iterating the process, we obtain that $q(u_1 - ku_2), q(-ku_2), q(u_4 - ku_2)$ belong to U for every $k \geq 0$.

In the same way that $q(u_1 + u_2)$ belongs to U by Lemma 28, then $q(u_2)$ belongs to U by Lemma 29. Finally, by Lemma 30 $q(u_3 + u_2)$ belongs to U, and we can iterate the process. Thus the bi-infinite band $B = \{q(ku_2)/k \in \mathbb{Z}\}$ belongs to U.

Lemma 31 provides the result in case 1.

2. Every edge in the tiling has a length strictly less than l.

It is possible that in the tiling the upper bound of the lengths of the edges is not reached. For example, with a rectangle, we can realize a tiling where the upper bound of lengths of edges is equal to the larger side of the rectangle, but where no edge has this size. So there exists a point A of the boundary of p which is an accumulation point for the left extremity of the edges of the tiling (because l is not attempted, there is an infinite number of edges). Let E be the set of edges of U. From E we can extract an infinite sequence S_1 of edges $S_1 = ([A_n B_n])$ such that their left extremity A_n converges monotonously to A (by the left or by the right) and such that the upper bound of lengths of edges in S_1 is equal to l. Now, looking at the right extremity B_n of edges of S_1 we can extract a subsequence S_2 such that this right extremity converges to a point B monotonously (by the left or by the right) and again keeps an upper bound of lengths equal to l. So it implies that $|\langle [AB] \rangle| = l$.

Finally, the contacts in points A_n in tiling U can be written (A_n, C_n, B'_n), and once more we extract a subsequence S_3 of S_2 so that the sequence C_n converges monotonously to a point C.

The sequence (A_n) and (B_n) cannot be ultimately constant because there is no edge of length l.

But we shall prove now the following statement.

Lemma 36. *Among the three sequences* $(A_n), (B_n), (C_n)$, *we have that* (A_n) *and* (C_n) *are trivial and* (B_n) *is not trivial or* (B_n) *and* (C_n) *are trivial and* (A_n) *is not trivial.*

Proof. The two sequences (A_n) and (B_n) cannot both be trivial because of the hypothesis about l. Let us suppose, for example, that (B_n) is trivial but (A_n) is not. Then A_n converges to A monotonously by the right side. We have to prove that (C_n) is trivial. If not, (C_n) converges to C by the right side (Figure 12.51); otherwise, since A'_n converges to A on the right we would have a sequence of edges $[C_n A'_n]$ which are nested and this is impossible because of Corollary 18.

So if (C_n) is not trivial, (C_n) converges to C by the right side and the edges $[C_n A'_n]$ are overlapping.

Let $n > m$. Corollary 18 provides the following equation. Let $\langle [C_n A'_m] \rangle = v$. Since $\langle [\widetilde{C_m A'_m}] \rangle = \langle [C'_m A_m] \rangle$ and $\langle [\widetilde{C_n A'_n}] \rangle = \langle [C'_n A_n] \rangle$, the curve $< C'_n A'_m]>$ admits $\tilde{v}$ as left and right factors. So we have an equation

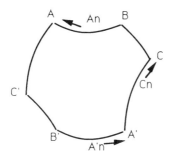

Fig. 12.51

$$\alpha \tilde{v} = \tilde{v} \beta$$

But n and m cannot be chosen large enough to have $|\alpha| < |v|$. So $\tilde{v}$ belongs to $\alpha^+ \alpha'$ where α' is a left factor of α. When n and m grow up, the curve $[C_n A'_m]$ converges to $[CA']$ and $|\alpha|$ converges to zero. This implies that $[CA']$ is a line segment (Lemma 6).

But if $[CA']$ is a line segment, $[C_n A'_n]$ is also a line segment, for n large enough. Considering three edges $[C_{n_1} A'_{n_1}], [C_{n_2} A'_{n_2}], [C_{n_3} A'_{n_3}]$ with $n_1 < n_2 < n_3$, we obtain a contradiction. $[C_{n_3} A'_{n_1}]$ is a line segment containing strictly $[C_{n_2} A'_{n_2}]$, and $[C'_{n_3} A A_{n_1}]$ is a line segment containing strictly $[C'_{n_2} A_{n_2}]$, and so $[C_{n_2} A'_{n_2}]$ cannot be an edge between two instances of p. The edge is necessarily larger.

Finally, we have proved that (C_n) is trivial. □

Let us assume that (A_n) and (C_n) are trivial and (B_n) is not. Then we have the following theorem.

Lemma 37. *$[AC]$ is a line segment.*

Proof. By the same argument as Lemma 36, we obtain that $[AB]$ and $[BC]$ are line segments, and also $[AB_n]$ and $[B_nC]$ so that $[AC]$ is a segment line. So, somewhere in the tiling we have the following situation: a contact (A, C, B'_n) in a triad (q, r, s) (Figure 12.52) and B_n is as near as we want to B; $[AC]$ is a segment line. □

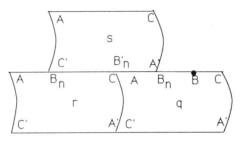

Fig. 12.52

If $s(\vec{CA})$ does not belong to the tiling, by Lemma 32 there exists a primitive curve $\gamma_n = \gamma_{1,n}\gamma_{2,n}$ such that

$$\langle [CA']\rangle \in \gamma_n^+ \gamma_{1,n}$$
$$\langle [A'B'_n]\rangle = \gamma_{2,n}$$

But this cannot happen for another value $m > n$. Indeed, we would have $\langle [A'B'_m]\rangle = \gamma_{2,m}$. And $|\gamma_{2,m}| > |\gamma_{2,n}|$. So $\gamma_{1,n}$ would end with a line segment oriented as $[A'B'_n]$. Then there is an impossibility for p in point C': the boundary of p would admit a factor centred in C' which is a closed curve. So, changing the value of n, $s(\vec{CA})$ belongs to the tiling. For the same reason, $r(\vec{CA})$ belongs to U. And now, iterating the process, $s(k\vec{CA})$ and $r(k\vec{CA})$ belong to U for $k \in \mathbb{Z}$.

We now just have to apply Lemma 31.

Hence Theorem 33 is proved. □

We can now give some complements about exact tiles and their surroundings which are deduced from the proof of the main theorem.

12.6 Surroundings of exact tiles

The different lemmas established in the previous section are now very useful for describing all the complete surroundings of an exact tile.

Theorem 38. *Every complete surrounding of an exact tile contains six, seven, or eight tiles, and the minimal surrounding extracted from the complete surrounding contains respectively six, five, or four tiles.*

Proof. First, by Corollary 15 every surrounding can be extended in a tiling of the whole plane, and so we have just to look at the surroundings appearing in the tilings of the plane, and observe what kinds of surrounding appear in the proof of Theorem 33. □

Surroundings of exact tiles   331

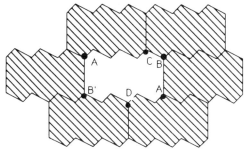

Fig. 12.53

We give below the different complete and associated minimal surroundings of an exact tile.

Complete 6-surroundings

These surroundings are also minimal.

- $\langle [BA'] \rangle = \langle [B'A] \rangle$
- There exists a curve $\alpha = \alpha_1 \alpha_2$ such that (Figure 12.53)

$$\langle [AC] \rangle \in \alpha^* \alpha_1 \qquad \langle [CB] \rangle \in \alpha_2^*$$
$$\langle [A'D] \rangle \in \tilde{\alpha}_1 \alpha^* \qquad \langle [DB'] \rangle \in \alpha^* \tilde{\alpha}_2$$

Complete 7-surroundings

The associated minimal surroundings contain five tiles.

- $\langle [BA'] \rangle = \langle \widetilde{[B'A]} \rangle$
- There exists a curve a such that (Figure 12.54)

$$\langle [AB] \rangle \in \alpha \alpha^+ \qquad \langle [A'C] \rangle \in \widetilde{\alpha^+} \qquad \langle [CB'] \rangle \in \widetilde{\alpha^+}$$

Complete 8-surroundings

The associated minimal surroundings contain four tiles (Figure 12.55).

- $\langle [AB] \rangle = \langle \widetilde{[A'B']} \rangle$, $\langle [BA'] \rangle = \langle \widetilde{[B'A]} \rangle$

From this description we immediately deduce the following property.

Proposition 39. *If a tile has a complete surrounding with seven tiles, then it also has a complete surrounding with six or eight tiles.*

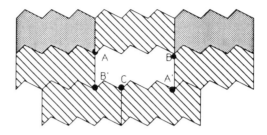

Fig. 12.54

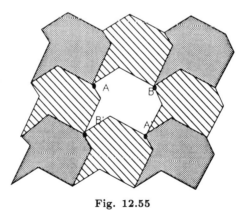

Fig. 12.55

References

Berger, M., and Gostiaux, B. (1972). *Géométrie différentielle.* Armand Colin.

Berger, M. (1966). The undecidability of the domino problem. *Mem. Amer. Math. Soc.* **66**.

Grunbaum, B., and Shephard, G. C. (1986). *Tilings and Patterns.* W. H. Freeman, New York.

Golomb, S. W. (1966). *Polyominoes.* George Allen and Unwin, London.

Harel, D. (1986). Effective transformations on infinite trees, with applications to high undecidability, dominoes and fairness. *J. Assoc. Comput. Mach.* **33** pp. 224–248.

Penrose, R. (1978). Pentaplexity. *Eureka* **39**.

Robinson, R. M. (1971). Undecidability and nonperiodicity of tilings of the plane. *Inventione Math.* **12** pp. 177–909.

Shapiro, H. D. (1978). Theoretical limitations on the efficient use of parallel memories. *I.E.E.E. Trans. Computing.*

Wang, H. (1965). Games, logic and computers. *Scientific American.*

Wijshoff, H. A. J., and van Leeuwen, J. (1984). Arbitrary versus periodic storage schemes and tessellations of the plane using one type of polymino. *Information and Control* **62** pp. 1–25.

13
An algebraic axiomatization of linear logic models

NARCISO MARTÍ-OLIET AND JOSÉ MESEGUER

Abstract

A new algebraic axiomatization of linear logic models is presented. The axioms directly reflect at the model-theoretic level the de Morgan duality exhibited by linear logic, and are considerably simpler than previous axioms. Several equationally defined classes of models are studied. One such class suggests a new variant of linear logic, called *cancellative linear logic*, in which it is always possible to cancel a proposition p (viewed as a resource) and its negation $p^\perp$ (viewed as a debt). This provides a semantics for a generalization of the usual token game on Petri nets, called the *financial game*. Poset models, called *Girard algebras*, are also defined equationally; they generalize for linear logic the Boolean algebras of classical logic, and contain the quantale models as a special case. The proposed axiomatization also provides a simple set of *categorical combinators* for linear logic, extending those previously proposed by Lafont.

13.1 Introduction

Girard's (1987b) linear logic has from its inception been recognized as a logic of action which seems particularly well suited to computer science applications. Girard and his coworkers have initiated some of these applications in the areas of parallelism (Girard 1987a), operational semantics (Lafont 1988a), and logic programming (Girard 1989), and there are at present interesting new developments including, among others, a logical treatment of polynomial time computability by means of bounded linear logic (Girard et al. 1989), new proposals for functional and logic programming languages (Lafont 1988a, 1990), and applications to concurrency (Asperti 1987, Asperti et al. 1990, Brown 1989, Engberg and Winskel 1990, Gunter and Gehlot 1989, Martí-Oliet and Meseguer 1989).

For many of these applications, as well as for the development of the subject itself, models are of great importance. Although several very useful models such as coherent spaces and phase semantics have been known from the beginning (Girard 1987b), the development of a general model theory for linear logic is more recent. In this context, the algebraic methods of category theory have been recognized as very useful by several authors, including Lafont (1988a), de Paiva (1988), and Seely (1989). Seely proposed using Barr's (1979) notion of the ∗-autonomous category to give a general categorical semantics for classical linear logic. Our own work (Martí-Oliet and Meseguer 1989) has refined and further developed Seely's proposal, and has established a systematic correspondence between this categorical semantics and the one for Petri nets developed by Meseguer and Montanari (1988a, 1988b). However, the technical details of ∗-autonomous categories are quite involved, and this presents an obstacle when explaining the basic ideas. This chapter presents a new algebraic axiomatization of linear logic models that is much simpler than, yet equivalent (Martí-Oliet and Meseguer 1990) to, our previous axiomatization in terms of ∗-autonomous categories.

An important aspect exploited in this work is the algebraic nature of the axioms, which define models in terms of their algebraic structure. This provides the usual benefits of equational logic[1] such as free models and equationally definable classes of models. In fact, one of the proposals made in this chapter is the introduction of a new variant of linear logic, called *cancellative linear logic*, which corresponds to imposing the equations $X \otimes Y = X \wp Y$ and $I = \bot$.

Linear logic is a resource-conscious logic in which a proposition p can be understood as a resource, and its negation $p^\perp$ as a *debt* of such a resource. However, owing to the distinction between $\otimes$, which accumulates resources, and $\wp$, which accumulates debts, in standard linear logic debts cannot, in general, be cancelled; cancellative linear logic makes this possible. This logic and its associated model theory suggest a generalization of the usual token game in a Petri net by what we call *financial games*, where a player can make progress by going into debt, but is kept honest about his finances.

Another important equationally definable class of models is that where the categories used as models are forced to be posets. We call a model of this kind a *Girard algebra*. Such algebras are for linear logic what Boolean algebras are for classical logic. The work on quantale models (Abramsky and Vickers 1988, Yetter 1990), which generalize Girard's (1987b) phase semantics, is contained within this framework.

Another advantage of our approach is that it naturally provides a set of *categorical combinators* for linear logic which extend those of Lafont

---

[1] Specifically, the underlying equational logic is that of essentially algebraic theories (Freyd 1972) or sketches (Barr and Wells 1985). Familiarity with this version of equational logic is not necessary for reading this chapter.

(1988a), and which are particularly simple because of our choice of axioms.

For expository reasons, we do not discuss exponentials in the body of this chapter. However, there is no problem in adding them to our treatment. This is explained in Section 13.7, where we also give a brief discussion of how our ideas could be extended to the case of *non-commutative* linear logic.

13.2 Dualizing objects and linear categories

We briefly recall the notion of a *symmetric monoidal category* $\mathcal{C}$ without giving all the details than can be found in MacLane (1971). The basic idea is that we have a 'binary product' $\_ \otimes \_$ defined as a functor $\_ \otimes \_ : \mathcal{C}^2 \longrightarrow \mathcal{C}$ and a 'unit object' I in $\mathcal{C}$ making $\mathcal{C}$ into a commutative monoid 'up to coherent isomorphism.' A good example is cartesian product in the category <u>Set</u> of sets and functions. Indeed, cartesian product is associative and commutative up to isomorphism in the sense that we have natural bijections

$$a_{A,B,C} : A \times (B \times C) \xrightarrow{\simeq} (A \times B) \times C$$
$$c_{A,B} : A \times B \xrightarrow{\simeq} B \times A$$

The one-point set 1 acts as a neutral element up to isomorphism by means of a natural bijection

$$e_A : 1 \times A \xrightarrow{\simeq} A$$

All these isomorphisms satisfy simple 'coherence conditions' as explained in (MacLane 1971).

More generally, any category $\mathcal{C}$ with finite products is an example of a symmetric monoidal category, but in other examples $A \otimes B$ is not a categorical product; the tensor product of vector spaces is one such example. As we shall see later, the linear logic connective $\otimes$ is interpreted in the models as a monoidal product of this kind, whereas the connective & is interpreted as a categorical product.

The well-known concept of a cartesian closed category can be recovered from the more general concept of a closed symmetric monoidal category defined below by just adding the additional requirement that $\_ \otimes \_$ is a categorical product and I is a final object.

In the following we use diagrammatic notation $f; g$ for the composition of a morphism f followed by a morphism g.

Definition 1. (MacLane 1971) *A closed symmetric monoidal category is a symmetric monoidal category* $(\mathcal{C}, \otimes, I, a, c, e)$ *such that, in addition, for each object A of $\mathcal{C}$, the functor* $\_ \otimes A : \mathcal{C} \longrightarrow \mathcal{C}$ *has a (chosen) right adjoint*

$A \multimap \_ : \mathcal{C} \longrightarrow \mathcal{C}$, that is, for all objects A, B, C in $\mathcal{C}$, we have a natural (in B and C) isomorphism

$$\varphi_{B,C}^A : Hom_\mathcal{C}(B \otimes A, C) \longrightarrow Hom_\mathcal{C}(B, A \multimap C)$$

The intuitive interpretation of $A \multimap B$ is of course the internalization of the morphisms from A to B as an object of $\mathcal{C}$. We have chosen the notation $A \multimap B$ to suggest that linear implication will be interpreted in the models by the functor $\_ \multimap \_$.

If $f : B \otimes A \to C$ is a morphism in $\mathcal{C}$ we write $f^\dagger$ for $\varphi_{B,C}^A(f)$, which is called the *currying* of f. The counit of this adjunction is a morphism

$$\varepsilon_{A,C} : (A \multimap C) \otimes A \longrightarrow C$$

called *evaluation*. The classical example of a closed symmetric monoidal category is the category of vector spaces over a field K with linear maps as morphisms, with $A \otimes B$ the usual tensor product, $I = K$, and $A \multimap B$ the space of linear maps from A to B.

By currying we can also view the evaluation map as a morphism

$$d_{A,B} : A \longrightarrow (A \multimap B) \multimap B$$

where $d_{A,B} = (c_{A, A \multimap B}; \varepsilon_{A,B})^\dagger$. Intuitively, $d_{A,B}$ corresponds to the map that is expressed in lambda notation as $\lambda x. \lambda f. f(x)$.

In the case of finite-dimensional vector spaces over a field K, the particular instance of the above morphism for $B = K$ is a natural isomorphism, usually written

$$A \xrightarrow{\simeq} A^{**}$$

This isomorphism expresses the well-known *duality* in finite-dimensional vector spaces. Here A^* denotes the vector space $A \multimap K$ dual of A. It is important to note that, although A and A^* have the same dimension, they are *not* naturally isomorphic. This can be generalized as follows.

Definition 2. *Given a closed symmetric monoidal category $(\mathcal{C}, \otimes, I, a, c, e, \multimap)$, an object $\bot$ in $\mathcal{C}$ is a dualizing object if, for every object A in $\mathcal{C}$, the natural morphism*

$$d_{A, \bot} : A \longrightarrow (A \multimap \bot) \multimap \bot$$

is an isomorphism.

Then, we just say that $(\mathcal{C}, \otimes, I, a, c, e, \multimap, \bot)$ is a category with a dualizing object.

For an object A in a category with a dualizing object we define $A^\perp = A \multimap \bot$. As this notation suggests, linear logic negation will be interpreted in the models by the functor $\_ \multimap \bot$.

Fact 3. *In a category with a dualizing object, the involution functor $(\_)^\perp : \mathcal{C}^{op} \longrightarrow \mathcal{C}$ is an equivalence of categories. Therefore it preserves limits and colimits. In particular, if $A \mathbin{\&} B$ is a cartesian product in $\mathcal{C}$, and $\top$ is a final object in $\mathcal{C}$, and if we define $A \oplus B = (A^\perp \mathbin{\&} B^\perp)^\perp$ on objects A, B, and $0 = \top^\perp$, then $A \oplus B$ is a coproduct in $\mathcal{C}$ and 0 is an initial object in $\mathcal{C}$. This is a categorical version of de Morgan's laws.*

The linear logic connectives $\&$ and $\oplus$ will be interpreted by products and coproducts respectively.

Fact 4. (Martí-Oliet and Meseguer 1989, 1990) *In a category with a dualizing object, we can define a new functor $\_ \wp \_ : \mathcal{C}^2 \longrightarrow \mathcal{C}$ by means of the formula $A \wp B = (A^\perp \otimes B^\perp)^\perp$. Then $\mathcal{C}$, together with $\wp, \bot$, and suitable coherence isomorphisms, is also a symmetric monoidal category.*

Again, our notation is chosen to suggest that the 'par' connective $\wp$ will be interpreted in the models by the functor $\_ \wp \_$.

Definition 5. *A* linear category *is a category with a dualizing object $(\mathcal{C}, \otimes, I, a, c, e, \multimap, \bot)$ and chosen finite products, that is, a final object $\top$ and for any objects A, B, a binary product denoted $A \mathbin{\&} B$ (therefore, by Fact 3, it has also finite coproducts).*

A linear functor *between two linear categories $\mathcal{C}$ and $\mathcal{C}'$ is a functor $F : \mathcal{C} \longrightarrow \mathcal{C}'$ that preserves all the additional structure in the category.*

The category LinCat *has linear categories as objects and linear functors as morphisms.*

We should like to emphasize that, since linear categories are just categories with additional algebraic structure, they can be fully axiomatized in a finitary first-order equational way. This is well known in categorical logic and we refer to (Freyd 1972, Lambek and Scott 1986, Barr and Wells 1985) for details about this style of axiomatization. This point is important because, since we shall use linear categories as models for linear logic, we can better understand and classify such models (and even find new ones) by considering additional equational axioms that can be imposed on them.

Example 6. (Martí-Oliet and Meseguer 1989) *The following list provides examples of linear categories. Additional examples will be discussed in Sections 13.4 and 13.5.*

- The category <u>Cohl</u> of coherent spaces and linear maps (Girard 1987b), that provided the first semantics for linear logic.

- For K a field, the category <u>FdVect$_K$</u> of finite-dimensional vector spaces over K and linear maps.

- Generalizing the previous example, the category <u>FSmod$_R$</u>, whose objects are free finitely generated R-semimodules over a commutative semiring R, and whose morphisms are linear maps.

- The category <u>Games$_K$</u> defined by Lafont (1988c).

13.3 Linear logic models

In a linear category we have the precise categorical structure needed to interpret linear logic. A *linear S-formula* is generated by the binary connectives $\otimes, \wp, \&, \oplus,$ and $\multimap$ and by the unary operation $(\_)^\perp$ from a collection S of propositional constants and the logical constants $I, \perp, \top, 0$. A *linear S-sequent*, written $\Gamma \vdash \Delta$, is an ordered pair of finite sequences of linear S-formulae, although the order of the formulae on both sides will be immaterial. A *linear theory* T is given by a collection of propositional constants S and a collection Ax of S-sequents, called *axioms*.

Given a linear theory $T = (S, Ax)$, an S-sequent $\Gamma \vdash \Delta$ belongs to the *closure of* T, denoted T^*, if it can be derived from the axioms Ax by using the axiom schemes and inference rules of linear logic[2].

Definition 7. *An interpretation $\mathcal{I}$ of a linear theory T in a linear category $\mathcal{C}$ is given by:*

1. *the assignment of an object $\mathcal{I}(s) \in Ob(\mathcal{C})$ to each constant $s \in S$, extended freely to S-formulas by interpreting the logical constants $I, \perp, \top, 0$ by the corresponding objects in $\mathcal{C}$, and the connectives $\otimes, \multimap, \wp, \&, \oplus, (\_)^\perp$ by their corresponding functor in $\mathcal{C}$;*

2. *the assignment of a morphism*

$$\mathcal{I}(A_1) \otimes \cdots \otimes \mathcal{I}(A_n) \longrightarrow \mathcal{I}(B_1) \wp \cdots \wp \mathcal{I}(B_m)$$

   in $\mathcal{C}$ to each S-sequent $A_1, \ldots, A_n \vdash B_1, \ldots, B_m$ in Ax.

A model of T is a linear category $\mathcal{C}$ together with an interpretation $\mathcal{I}$ of T in $\mathcal{C}$.

---

[2] We prefer a style of rules with standard sequents, as in (Martí-Oliet and Meseguer 1989) and (Seely 1989), over the more economic system of Girard (1987b) because it is more suitable for a categorical semantics.

Cancellative linear logic

We can define a category $\underline{LinTh}$ whose objects are linear theories, and also assign a linear theory $\mathcal{C}^\circ$ to a linear category $\mathcal{C}$ in such a way that an interpretation of T in $\mathcal{C}$ can be alternatively regarded as a morphism $T \to \mathcal{C}^\circ$ in $\underline{LinTh}$; see (Martí-Oliet and Meseguer 1989) for details.

Definition 8. *Given a linear theory $T = (S, Ax)$, a model $(\mathcal{C}, \mathcal{I})$ of T and an S-sequent $A_1, \ldots, A_n \vdash B_1, \ldots, B_m$, we say that $(\mathcal{C}, \mathcal{I})$ satisfies this sequent and write*

$$(\mathcal{C}, \mathcal{I}) \models A_1, \ldots, A_n \vdash B_1, \ldots, B_m$$

if there is in $\mathcal{C}$ a morphism $\mathcal{I}(A_1) \otimes \cdots \otimes \mathcal{I}(A_n) \longrightarrow \mathcal{I}(B_1) \wp \cdots \wp \mathcal{I}(B_m)$.

With these notions of model and satisfaction for linear logic sequents we can prove soundness and completeness theorems. We refer to (Martí-Oliet and Meseguer 1989) for details. The completeness theorem is a consequence of the following definition of a linear category generated by a linear theory.

Definition 9. *Given a linear theory $T = (S, Ax)$, there is a linear category $\mathcal{L}[T]$ whose objects are S-formulae, and whose morphisms are equivalence classes of derivations of S-sequents $\Gamma \vdash \Delta \in T^*$ with respect to the congruence generated by the equations that a category needs to satisfy in order to be a linear category.*

This construction provides a functor $\mathcal{L}[\_] : \underline{LinTh} \longrightarrow \underline{LinCat}$, left adjoint to the functor $(\_)^\circ : \underline{LinCat} \longrightarrow \underline{LinTh}$; for details see (Martí-Oliet and Meseguer 1989, Seely 1989). It is important to note that the equations that define a linear category are greatly simplified by using dualizing objects, and in particular this gives a much simpler construction of $\mathcal{L}[N]$.

13.4 Cancellative linear logic

Martí-Oliet and Meseguer (1989) established a triangular correspondence between computations in a Petri net N (generated from the basic transitions and idle transitions by sequential and parallel composition,) proofs in the tensor ($\otimes$) fragment of linear logic from a theory associated to N, and morphisms in a symmetric monoidal category $T[N]$ generated from N. In this way, a systematic connection was developed between linear logic models and categorical models of Petri nets in the sense of Meseguer and Montanari (1988a). The proof-theoretic correspondence between computations and proofs was first pointed out by Asperti (1987), and has been further developed by Gunter and Gehlot (1989). This correspondence has also been studied for the implicative ($\otimes, \multimap$) fragment in (Asperti et al. 1990).

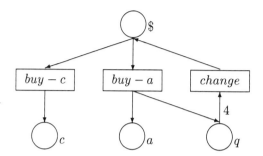

Fig. 13.1 A Petri net for buying cakes and apples

Since a Petri net N can be regarded as a linear theory, it has an associated linear category $\mathcal{L}[N]$ that can be viewed as a 'category of computations' containing all the computations in N (that is, $\mathcal{T}[N]$) and additional 'idealized' computations corresponding to possibilities of external choice (&), internal choice ($\oplus$), etc. (Martí-Oliet and Meseguer 1989). In particular, we proposed interpreting linear negation $(\_)^\perp$ as a *debt* of tokens or resources and $\wp$ as accumulation of debts, in the same way that $\otimes$ is interpreted as resource accumulation. However, the interaction between resources and debts, as well as the meaning of $\otimes$ or $\wp$, becomes complicated when both are mixed.

Let us consider an example. Figure 13.1 shows a Petri net N that represents a machine to buy cakes and apples; a cake costs a dollar and an apple costs three quarters. Owing to an inadequate design, the machine only accepts dollars, and it returns a quarter when the user buys an apple; to alleviate this problem in part, the machine can change four quarters into a dollar.

The $\otimes$-theory associated with this net is given by the following axioms:

$$buy - c : \$ \vdash c \qquad buy - a : \$ \vdash a \otimes q \qquad change : q^4 \vdash \$$$

Consider a user who wants to buy an apple but has only three quarters. Since the category $\mathcal{L}[N]$ represents all the computations of the machine as well as more general 'idealized' computations, the user could perform a *Gedankenexperiment* in $\mathcal{L}[N]$ to see how he could possibly get an apple with his three quarters. Using the idea of debts, he could borrow a quarter, creating at the same time the corresponding debt; this is reflected in $\mathcal{L}[N]$ by the existence of the morphism $I \longrightarrow q \wp q^\perp$. Then he could change the four quarters into a dollar, buy an apple, and cancel the debt with the quarter he gets back; this last step would be reflected in the morphism $q \otimes q^\perp \longrightarrow \perp$, dual to the previous one. If we try to do this inside $\mathcal{L}[N]$, we can attempt the following composition:

$$q^3 \longrightarrow q^3 \otimes I \longrightarrow q^3 \otimes (q \wp q^\perp) \longrightarrow (q^3 \otimes q) \wp q^\perp \longrightarrow \$ \wp q^\perp \longrightarrow$$
$$(a \otimes q) \wp q^\perp \stackrel{?}{\longrightarrow} a \wp (q \otimes q^\perp) \longrightarrow a \wp \perp \longrightarrow a$$

Everything seems right, except for the morphism marked with ?, because it simply *does not exist* in $\mathcal{L}[N]$. Notice that this problem has to do with the reorganization of resources and debts; otherwise, the debt is maintained until one acquires the necessary resource to cancel it. Another related problem is the asymmetry between $I \longrightarrow q \wp q^\perp$ and $q \otimes q^\perp \longrightarrow \perp$; in the former the absence of resources and debts is indicated by I, and the composition of a resource with the corresponding debt is represented by $\wp$, while in the latter these are represented by $\perp$ and $\otimes$ respectively.

The problem is that in standard linear logic, since $\otimes$ and $\wp$ are different connectives, the idea of *cancelling a debt* in general cannot be realized. We could have foreseen this from the beginning, since the user cannot get an apple in the original machine, whose computations correspond to deductions in the $\otimes$ fragment, and the extension to the whole of linear logic is conservative over the $\otimes$ fragment (Asperti 1989, Asperti et al. 1990).

The solution to this problem is very simple. It is enough to identify the connectives $\otimes$ and $\wp$ and their respective neutral elements I and $\perp$. Then the asymmetry noted above disappears, and the reorganization of resources is just associativity of $\otimes$. The computation in which we are interested becomes (forgetting associativity)

$$q^3 \longrightarrow q^3 \otimes I \longrightarrow q^3 \otimes q \otimes q^\perp \longrightarrow \$ \otimes q^\perp \longrightarrow$$
$$a \otimes q \otimes q^\perp \longrightarrow a \otimes I \longrightarrow a \quad (1)$$

In linear logic, the axioms and rules for the constants $I, \perp$ and the connectives $\otimes, \wp$ are the following:

(IR)	$\vdash I$	$(\perp L)$	$\perp \vdash$
$(neg\perp)$	$I \vdash \perp^\perp$	$(negI)$	$I^\perp \vdash \perp$
(IL)	$\dfrac{\Gamma \vdash \Delta}{\Gamma, I \vdash \Delta}$	$(\perp R)$	$\dfrac{\Gamma \vdash \Delta}{\Gamma \vdash \perp, \Delta}$
$(\otimes L)$	$\dfrac{\Gamma, A, B \vdash \Delta}{\Gamma, A \otimes B \vdash \Delta}$	$(\otimes R)$	$\dfrac{\Gamma \vdash A, \Delta \quad \Gamma' \vdash B, \Delta'}{\Gamma, \Gamma' \vdash A \otimes B, \Delta, \Delta'}$
$(\wp L)$	$\dfrac{\Gamma, A \vdash \Delta \quad \Gamma', B \vdash \Delta'}{\Gamma, \Gamma', A \wp B \vdash \Delta, \Delta'}$	$(\wp R)$	$\dfrac{\Gamma \vdash A, B, \Delta}{\Gamma \vdash A \wp B, \Delta}$

We shall call linear logic with the identifications $I = \perp$, $\otimes = \wp$ *cancellative linear logic*. The rules for this logic can be obtained from the rules for linear logic by substituting I for $\perp$ and $\otimes$ for $\wp$ in the rules above. In particular, the two rules resulting from this substitution on the rules $(\otimes R)$ and $(\wp L)$ become equivalent to the single rule

$$\frac{\Gamma \vdash \Delta \qquad \Gamma' \vdash \Delta'}{\Gamma, \Gamma' \vdash \Delta, \Delta'}$$

that is called (mix) by Girard (1987b, 1989). This rule says that $\otimes$ is stronger than $\wp$, which is true in *Cohl*; however, we show below that *Cohl* does *not* provide a model of cancellative linear logic.

The categorical models of cancellative linear logic are linear categories with natural isomorphisms $(A \otimes B)^\perp \cong A^\perp \otimes B^\perp$, $I \cong \perp$, making the symmetric monoidal structures of $\otimes$ and $\wp$ isomorphic. Therefore they constitute an equationally axiomatizable class of models called *cancellative linear categories*.

Fact 10. *The linear category $FdVect_K$ of finite-dimensional vector spaces over a field K and linear maps is cancellative. More generally, the same holds true for the category $FSmod_R$, whose objects are free finitely generated R-semimodules over a commutative semiring R, and whose morphisms are linear maps.*

Fact 11. *As already mentioned, the category Cohl of coherent spaces and linear maps is an example of a non-cancellative linear category, although in this case $I = \perp$.* We recall that a coherent space X is characterized by a reflexive unoriented graph, called the web of X, which represents the coherence relation on the underlying set $|X|$; moreover, if two coherent spaces are isomorphic in *Cohl*, their corresponding webs are isomorphic as graphs. Now, if X is represented by

$$\begin{array}{c} \bullet 2 \\ 1\bullet\!\!-\!\!\!-\!\!\bullet 3 \end{array}$$

and Y is the discrete space

$$1\bullet \quad \bullet 2$$

we have

$$X \otimes Y = \begin{array}{cc} (1,1) \bullet & \bullet(1,1) \\ (2,1) \bullet & \bullet(2,2) \\ (3,1) \bullet & \bullet(3,2) \end{array}$$

with an edge between $(1,1)$ and $(3,1)$ on the left side.

while

$X \wp Y = (2,1)$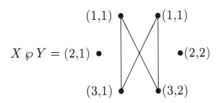

Therefore, in general, $X \otimes Y \not\cong X \wp Y$ in *Cohl*.

Proposition 12. *Let CLinCat denote the category of cancellative linear categories and structure-preserving functors. Then the forgetful functor CLinCat $\longrightarrow$ LinCat has a left adjoint.*

We shall denote by $\mathcal{C}[T]$ the image under this left adjoint of the linear category $\mathcal{L}[T]$ associated with a linear theory T.

A *cancellative model* of a theory T is a model $(\mathcal{C}, \mathcal{I})$ of T, as in Definition 7, such that $\mathcal{C}$ is a cancellative linear category. With a notion of satisfaction analogous to the one in Definition 8, we have a soundness and completeness theorem.

Theorem 13. *A sequent $\Gamma \vdash \Delta$ is satisfied in all cancellative models of a theory T if and only if it is provable from T by using the axioms and rules of cancellative linear logic.*

As before, the completeness part of this theorem is proved by using the category $\mathcal{C}[T]$.

Consider the cancellative linear category $\mathcal{C}[N]$ associated with a Petri net N. In this category we have the desired morphisms $I \longrightarrow a \otimes a^\perp$ and $a \otimes a^\perp \longrightarrow I$ for any place a in the net N. They correspond respectively to computations that 'borrow' a token, creating the corresponding debt at the same time, and 'cancel a debt' with a corresponding token. Of course, we also have the basic computations $t : {}^\bullet t \longrightarrow t^\bullet$ for each transition t. In addition, we have computations that 'transfer a debt' $t^\perp : (t^\bullet)^\perp \longrightarrow ({}^\bullet t)^\perp$; for instance, in the example above the transition *change* : $q^4 \longrightarrow \$$ also provides a computation *change*$^\perp$: $\$^\perp \longrightarrow (q^4)^\perp$ that transfers a debt of one dollar to a debt of four quarters (or equivalently four debts of one quarter each, because of the condition $(A \otimes B)^\perp \cong A^\perp \otimes B^\perp$).

This suggests a generalized notion of the token game that we shall call a **financial game**. Besides the standard moves in the usual game, one can also carry out the following.

1. *Create a debt*, thus obtaining simultaneously a positive token (•) and a negative token (∘) in the desired place. For example, if we begin with three quarters, we can borrow another by going into debt:

346   An algebraic axiomatization of linear logic models

2. *Transfer a debt* by playing the token game in reverse for negative tokens. For example, we can transform a debt of a dollar into a debt of four quarters by applying the *change* transition in reverse:

3. *Cancel a debt* by annihilating positive and negative tokens in the same place; thus

Therefore, although our user would never be able to buy an apple in the original machine, he could nevertheless obtain one in a more sophisticated machine allowing financial games. Figure 13.2 shows some snapshots of the financial game where the apple is now bought by borrowing a dollar, which corresponds to the following morphism[3] in $\mathcal{C}[N]$ (we again forget associativity and commutativity):

$$q^3 \longrightarrow q^3 \otimes \$ \otimes \$^\perp \xrightarrow{id \otimes buy - a \otimes change^\perp} q^4 \otimes a \otimes (q^4)^\perp \longrightarrow a$$

A categorical model for all the financial games of a Petri net N is provided by the subcategory of $\mathcal{C}[N]$ consisting of objects and morphisms generated only by $\otimes$ and $(\_)^\perp$ from the original set of places and the basic transitions in N[4].

13.5 Girard algebras and quantale models

Recently, a new kind of model for linear logic has been proposed in the form of quantales (Abramsky and Vickers 1988, Yetter 1990) (the references in these papers provide more sources on the subject of quantales) that are also suitable for the non-commutative case (Yetter 1990). Indeed, the original

---

[3]The reader can play the game for the morphism (1) presented before.

[4]In order to avoid the extra structure created by the coherence isomorphisms, it could be convenient to consider a *strict* category where they are just identities, as is done for the category $T[N]$ in (Meseguer and Montanari 1988a).

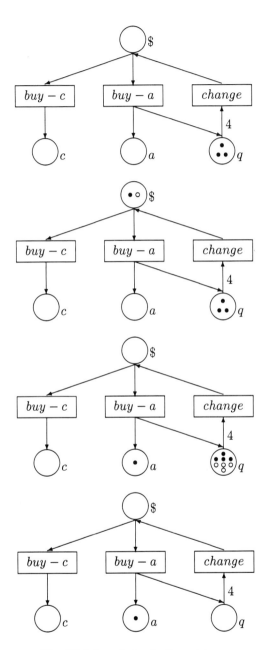

Fig. 13.2 Snapshots of a financial game

phase semantics for linear logic given by Girard (1987b) is based on the free quantale over a monoid. In this section, we relate this kind of model to the categorical models presented in Sections 13.2 and 13.3. We restrict ourselves to the commutative case. The relationship is very simple. It is just an inclusion, in the sense that Girard quantales are a special case of linear categories whose category structure[5] is a poset, which we call Girard algebras.

Definition 14. *A Girard algebra is a linear category that is a poset.*

This notion corresponds to restricting ourselves to an equationally defined class of models satisfying the two additional equations

$$\forall f, g \in Mor, \ \text{dom}\, f = \text{dom}\, g \wedge \text{cod}\, f = \text{cod}\, g \implies f = g$$
$$\forall f, g \in Mor, \ \text{dom}\, f = \text{cod}\, g \wedge \text{cod}\, f = \text{dom}\, g \implies f = g$$

The first equation forces a category to be a preorder, and the addition of the second requires this preorder to be a poset.

Proposition 15. *Let $\underline{GirAlg}$ denote the full subcategory of $\underline{LinCat}$ whose objects are Girard algebras. Then the inclusion functor $\underline{GirAlg} \hookrightarrow \underline{LinCat}$ has a left adjoint.*

In a poset category, products and coproducts correspond to meets and joins respectively; in particular, final and initial objects correspond to top and bottom elements. A functor $\_ \otimes \_ : P^2 \to P$ is exactly a monotone function, and it endows P with a symmetric monoidal structure if we have an element $I \in P$ such that $(P, \otimes, I)$ is a commutative monoid. Given an element $a \in P$, a right adjoint $a \multimap \_$ to $\_ \otimes a$ is given by a monotone function $a \multimap \_ : P \to P$ such that

$$c \otimes a \leqslant b \iff c \leqslant a \multimap b$$

Finally, an element $\bot \in P$ is dualizing if and only if, for all $a \in P$,

$$(a \multimap \bot) \multimap \bot = a$$

Therefore a Girard algebra can be equivalently defined as a lattice $(P, \oplus, \&)$ with top $\top$ and bottom 0, together with a commutative monoid structure $(\otimes, I)$, a 'relative pseudo-complement' $\multimap$, and a dualizing object $\bot$. This is a generalization of the well-known fact that a Heyting algebra is a cartesian closed category that is a poset and has finite coproducts. In fact, since linear negation is classical, Girard algebras generalize Boolean algebras, which can be equationally characterized as follows.

---

[5] Any poset $(P, \leqslant)$ can be seen as a category whose objects are the elements of P and which has a morphism $a \to b$ if and only if $a \leqslant b$ in $(P, \leqslant)$.

Proposition 16. *The category* BoolAlg *of Boolean algebras is the full subcategory of* GirAlg *defined by the equations*

$$X \otimes Y = X \,\&\, Y \tag{2}$$
$$I = \top \tag{3}$$

Moreover, any linear category $\mathcal{C}$ *satisfying equations (2) and (3) is a preorder, equivalent as a category to a Boolean algebra.*

Proof. First, notice that imposing equations (2) and (3) means that the linear category is a cartesian closed category with finite coproducts in which the dualizing object is initial. Then we have

$$Hom(A, B) \cong Hom(A, B^{\perp\perp}) \cong Hom(A \otimes B^{\perp}, \perp)$$

and, since $\perp$ is initial, the last hom-set has at most one morphism (Lambek and Scott 1986, Proposition I.8.3); hence, the category is a preorder and any preorder is equivalent, as a category, to a poset. Of course, a poset category that is cartesian closed, has finite coproducts, and has a dualizing initial object is exactly a Boolean algebra. □

In the same way that a complete Heyting algebra is a cartesian closed category that is a poset and has all (not only finite) products and coproducts, we can make the following definition.

Definition 17. *A Girard algebra is complete if it has all products (coproducts exist automatically by Fact 3, as well as by general arguments in lattice theory; we denote the coproduct of* $\{a_i\}$ *by* $\bigoplus\{a_i\}$.)

Analogously to the way in which implication is defined in a complete Heyting algebra, in a complete Girard algebra linear implication $\multimap$ can be defined by $a \multimap b = \bigoplus\{c \mid c \otimes a \leqslant b\}$. Therefore we have the following proposition.

Proposition 18. *A complete Girard algebra is a complete join semilattice* $(Q, \bigoplus)$ *equipped with a commutative monoid structure* $(\otimes, I)$ *that distributes over arbitrary joins*

$$\forall a \in Q \,\forall S \subseteq Q \quad a \otimes (\bigoplus S) = \bigoplus\{a \otimes b \mid b \in S\}$$

and with a dualizing object $\perp$. *Therefore a complete Girard algebra is exactly a Girard quantale as defined in (Yetter 1990).*

Of course, as a Girard algebra is a particular case of a linear category, we can give semantics to linear logic in this setting. Yetter (1990) shows that the quantale semantics is sound and complete for the linear sequent calculus. In this case, satisfaction of a sequent $A_1, \ldots, A_n \vdash B_1, \ldots, B_m$ means that

$$\mathcal{I}(A_1) \otimes \cdots \otimes \mathcal{I}(A_n) \leqslant \mathcal{I}(B_1) \wp \cdots \wp \mathcal{I}(B_m)$$

If we drop the requirement of a dualizing object in the definition of a complete Girard algebra we obtain a quantale as defined in (Abramsky and Vickers 1988, Yetter 1990).

Fact 19. *We can obtain a Girard quantale from a quantale $(Q, \bigoplus, \otimes, I)$ and an arbitrary element $d \in Q$ by restricting to the set of elements $a \in Q$ that satisfy $(a \multimap d) \multimap d = a$.*

It should be remarked that in an arbitrary quantale we can give semantics to *intuitionistic* linear logic, that is, the fragment of linear logic that does not include either negation $(\_)^\perp$ or $\wp$. It is the dualizing element that allows the classical interpretation of negation. Of course, this is also true for the categorical semantics.

Given a commutative monoid $(M, \cdot, i)$, the *free quantale* over M is given by the lattice $\mathcal{P}(M)$ together with the operation $\otimes$ on sets $A, B \subseteq M$ defined by $A \otimes B = \{a \cdot b \mid a \in A, b \in B\}$. If we apply the general method indicated in Fact 19 to obtain a Girard quantale, we obtain the phase semantics in terms of *facts* as defined by Girard (1987b).

Engberg and Winskel (1990) associate with a given Petri net N a quantale $Q[N]$ whose elements are the sets of markings downwards closed with respect to the reachability relation. This allows them to define an interpretation of intuitionistic linear logic, and consequently a relation of satisfaction of an intuitionistic linear logic sequent by a net. However, this relation of satisfaction is not complete with respect to the inference rules for the linear sequent calculus: there are sequents that are satisfied by the net and are not derivable in the calculus from some axioms associated to the net. This approach can be extended to treat negation intuitionistically by fixing an element of the quantale as the interpretation of $\perp$; they choose as $\perp$ the set of unreachable markings from an initial state. This permits them to state some negative information about the net, for example, the satisfaction of a mutual exclusion property. Nevertheless, the element chosen as $\perp$ is not dualizing and there does not seem to be an easy way to regain duality in this set-up; for example, the restriction to a Girard quantale in Fact 19 does not seem to provide a meaningful interpretation.

13.6 Categorical combinators

Lafont (1988a) gave a categorical semantics of intuitionistic linear logic in terms of closed symmetric monoidal categories, and used it to develop a *Linear Abstract Machine*.

The general idea of this passage from logic to categories and from categories to machine code is as follows. First, a translation is defined from a logical calculus to a collection of special morphisms, called *categorical combinators*, subject to a set of equations, that define a class of categories; then, an abstract machine is defined whose programs are expressions denoting morphisms in a free category, and whose computation is reduction of such expressions to canonical form using the equations. This generalizes to closed symmetric monoidal categories the translation between natural deduction proofs, written as typed λ-calculus terms, and categorical combinators for cartesian closed categories, studied by Curien (1986) and used in the *Categorical Abstract Machine* (Cousineau et al. 1985). The parallelism between both ideas is made more explicit in Lafont's (1988b) thesis.

Our construction of a free linear category, presented in detail in (Martí-Oliet and Meseguer 1989, Appendix), can be seen as an extension of Lafont's work that provides categorical combinators for linear categories, that is, for linear logic including negation[6]. However, the first presentation in terms of $*$-autonomous categories was too complicated; in addition to the categorical combinators that define a closed symmetric monoidal category, we needed an explicit negation $(\_)^\perp$ on objects, and natural isomorphisms $s_{A,B} : (A \multimap B) \longrightarrow (B^\perp \multimap A^\perp)$ and $d_A : A \longrightarrow A^{\perp\perp}$ subject to four somewhat unintuitive equations. The axiomatization in terms of dualizing objects presented in this chapter provides a much simpler, yet equivalent, set of combinators. We *only* need to add a special object $\perp$ (the dualizing one) and an inverse to the already existing morphism $(c; \varepsilon)^\dagger : A \longrightarrow A^{\perp\perp}$ where $A^\perp = A \multimap \perp$. It would be interesting to investigate appropriate notions of reduction for these combinators as well as the possibility of extending the Linear Abstract Machine to this more general context.

13.7 Concluding remarks

We have not yet discussed the exponentials *of course* (!) and its dual *why not* (?). Our reasons for postponing their discussion have been both simplicity of the exposition and the lack at present of good applications

---

[6] Although (Martí-Oliet and Meseguer 1989) does not treat the exponential !, which is treated in (Lafont 1988a), this can also be included within our framework by adding the additional categorical structure discussed in Section 13.7.

for these connectives in terms of Petri nets[7]. Nevertheless, the categorical interpretation of linear logic can be extended to include the exponentials also, both in the intuitionistic and the classical cases, along the lines of Lafont (1988a), de Paiva (1988), and Seely (1989).

Without going into details, the interpretation of the connective ! is given by a *comonad* $! : \mathcal{C} \longrightarrow \mathcal{C}$ that maps the comonoid structure

$$\top \longleftarrow A \longrightarrow A \& A$$

(given by the 'diagonal' map $\Delta : A \to A \& A$) into a *comonoid* structure

$$I \longleftarrow !A \longrightarrow !A \otimes !A$$

via isomorphisms $!\top \cong I$ and $!(A \& A) \cong !A \otimes !A$. Lafont (1988a) also asks this comonoid structure to be *cofree*. (We refer the reader to MacLane (1971) for the definitions of comonad, comonoid, cofree, etc.) To obtain the interpretation of the dual ? it is enough to apply the duality $(\_)^\perp$.

What is important to realize is that this amounts simply to adding more structure to a linear category, and that this can be done equationally. Therefore our framework can be extended to include the exponentials without any change in the algebraic point of view.

A generalization of linear logic that has been mentioned (Girard 1989, Yetter 1990), but has not yet been well studied, is *non-commutative* linear logic, where the connective $\otimes$ is no longer commutative. Yetter (1990) considers a variant where some permutations are valid and other conditions are added; moreover, he does not include the connective $\multimap$ in his system. To study the categorical semantics of non-commutative linear logic we need a monoidal category $(\mathcal{C}, \otimes, I)$ (note that the symmetry condition has been dropped) together with *two* right adjoints $\_ \multimap A$ to $\_ \otimes A$ and $A \circ\!\!\!-\, \_$ to $A \otimes \_$, since we no longer have an isomorphism $\_ \otimes A \cong A \otimes \_$. Such categories have already been studied by Lambek (1989). An interesting problem is finding the right notion of dualizing object in this setting. Another issue worth exploring is the relationship to the notion of *braided monoidal category* studied by Joyal and Street (1988, 1989).

Acknowledgements

We wish to thank Ross Street for stimulating us to seek a formulation of linear categories in terms of dualizing objects. We also thank Andrea Asperti, Gianluigi Bellin, David de Frutos, Jean-Yves Girard, Carl Gunter, and Robert Seely for helpful discussions.

---

[7]Note that, intuitively, !a could be interpreted as arbitrary generation of more tokens in place a.

This work has been supported by Office of Naval Research Contract N00014-88-C-0618, NSF Grant CCR-8707155, and by a grant from the System Development Foundation. The first author is supported by a Research Fellowship of the Spanish Ministry for Education and Science.

References

Abramsky, S., and Vickers, S. (1988). *Linear Process Logic*, Notes by Steve Vickers.

Asperti, A. (1987). *A Logic for Concurrency*, unpublished manuscript.

Asperti, A. (1989). Personal communication.

Asperti, A., Ferrari, G. L., and Gorrieri, R. (1990). Implicative formulae in the 'proofs as computations' analogy, in: *Proc. 17th. Annual ACM Symposium on Principles of Programming Languages*, San Francisco, California, pp. 59–71.

Barr, M. (1979). *∗-Autonomous Categories*, Volume 752 of Lecture Notes in Mathematics, Springer-Verlag, Berlin.

Barr, M., and Wells, C. (1985). *Toposes, Triples and Theories*, Volume 278 of Grundlehren der mathematischen Wissenschaften, Springer-Verlag, New York.

Brown, C. (1989). *Relating Petri Nets to Formulae of Linear Logic*, Technical Report ECS-LFCS-89-87, Laboratory for Foundations of Computer Science, University of Edinburgh.

Cousineau, G., Curien, P.-L., and Mauny, M. (1985). The categorical abstract machine, in: J.-P. Jouannaud (Ed.), *Functional Programming Languages and Computer Architecture, Nancy, France, September 1985*, Volume 201 of Lecture Notes in Computer Science, Springer-Verlag, Berlin, pp. 50–64.

Curien, P.-L. (1986). *Categorical Combinators, Sequential Algorithms and Functional Programming*, Research Notes in Theoretical Computer Science, Pitman, London.

Engberg, U., and Winskel, G. (1990). On Linear Logic and Petri Nets, to appear in *Proc. CAAP'90*, Copenhagen.

Freyd, P. (1972). Aspects of topoi, *Bulletin Australian Mathematical Society*, **7**, pp. 1–76 and 467–480.

Girard, J.-Y. (1987a). Linear logic and parallelism, in: Marisa Venturini Zilli (Ed.), *Mathematical Models for the Semantics of Parallelism, Advanced School, Rome, Italy, September 1986*, Volume 280 of Lecture Notes in Computer Science, Springer-Verlag, Berlin, pp. 166–182.

Girard, J.-Y. (1987b). Linear logic, *Theoretical Computer Science*, **50**, pp. 1–102.

Girard, J.-Y. (1989). Towards a geometry of interaction, in: J. W. Gray and

A. Scedrov (Eds.), *Categories in Computer Science and Logic, Boulder, June 1987*, Volume 92 of Contemporary Mathematics, American Mathematical Society, Providence, pp. 69–108.

Girard, J.-Y., Scedrov, A., and Scott, P. J. (1989). *Bounded Linear Logic*, invited talk by Andre Scedrov at the *Conference on Category Theory and Computer Science*, Manchester, UK.

Gunter, C., and Gehlot, V. (1989). *Nets as Tensor Theories*, Technical Report MS-CIS-89-68 Logic & Computation 17, Department of Computer and Information Science, University of Pennsylvania.

Joyal, A., and Street, R. (1988). *Braided Tensor Categories*, Macquarie Mathematics Reports.

Joyal, A., and Street, R. (1989). *The Geometry of Tensor Calculus I*, Macquarie Mathematics Reports.

Lafont, Y. (1988a). The linear abstract machine, *Theoretical Computer Science*, **59**, pp. 157–180.

Lafont, Y. (1988b). *Logiques, Catégories & Machines*, Thèse de Doctorat, Université de Paris 7.

Lafont, Y. (1988c). *From Linear Algebra to Linear Logic*, Preliminary Draft.

Lafont, Y. (1990). Interaction nets, in: *Proc. 17th. Annual ACM Symposium on Principles of Programming Languages*, San Francisco, California, pp. 95–108.

Lambek, J. (1989). Multicategories revisited, in: J. W. Gray and A. Scedrov (Eds.), *Categories in Computer Science and Logic, Boulder, June 1987*, Volume 92 of Contemporary Mathematics, American Mathematical Society, Providence, pp. 217–239.

Lambek, J., and Scott, P. J. (1986). *Introduction to Higher Order Categorical Logic*, Volume 7 of Cambridge Studies in Advanced Mathematics, Cambridge University Press, Cambridge.

MacLane, S. (1971). *Categories for the Working Mathematician*, Volume 5 of Graduate Texts in Mathematics, Springer-Verlag, Berlin.

Martí-Oliet, N., and Meseguer, J. (1989). From petri nets to linear logic, in: D. H. Pitt et al. (Eds.), *Category Theory and Computer Science, Manchester, UK, September 1989*, Volume 389 of Lecture Notes in Computer Science, Springer-Verlag, Berlin, pp. 313–340. Full version submitted for publication.

Martí-Oliet, N., and Meseguer, J. (1990). *Duality in Closed and Linear Categories*, Technical Report SRI-CSL-90-01, Computer Science Laboratory, SRI International; submitted for publication.

Meseguer, J., and Montanari, U. (1988a). *Petri Nets Are Monoids*, Technical Report SRI-CSL-88-3, Computer Science Laboratory, SRI International; to appear in *Information and Computation*.

Meseguer, J., and Montanari, U. (1988b). Petri nets are monoids: a new algebraic foundation for net theory, in: *Proc. Logic in Computer*

Science, Edinburgh, pp. 155–164.

de Paiva, V. C. V. (1988). *The Dialectica Categories*, Ph. D. thesis, University of Cambridge.

Seely, R. A. G. (1989). Linear logic, *-autonomous categories and cofree coalgebras, in: J. W. Gray and A. Scedrov (Eds.), *Categories in Computer Science and Logic, Boulder, June 1987*, Volume 92 of Contemporary Mathematics, American Mathematical Society, Providence, pp. 371–382.

Yetter, D. N. (1990). Quantales and (non-commutative) linear logic, *Journal of Symbolic Logic*, **55**, pp. 41–64.

14
Types as theories

JOSEPH A. GOGUEN*

Abstract

There are many notions of type in computing. The most classical notion is 'types as sets', which has been extended to cover many features of modern programming languages. This chapter shows that such features are handled perhaps even more naturally by an extension of the 'types as algebras' notion to a 'types as theories' notion. This notion naturally supports object-oriented concepts, including inheritance and local state, as well as generic modules and dependent types. Moreover, it explains why polymorphic operations are natural transformations and provides a systematic foundation for a powerful form of programming in the large. This chapter also suggests a new semantics for dependent types based on strict fibrations.

14.1 Introduction

'Types' are used to classify programming entities. Because the entities involved in modern programming have many different kinds of structure and may be seen at many different levels of abstraction, it is not surprising that there are many different notions of type. For each notion of type, we should ask what it serves to classify. The two major established traditions are the 'types as sets' tradition, and the 'types as algebras' tradition; both these notions of type classify values. The first leads from classical programming languages to the lambda-calculus and so-called 'type theory', while the second leads from abstract data types into object-oriented concepts and programming in the large. Brief introductions to these two traditions are given in the following two subsections, with some indication of how the second extends to the 'types as theories' approach in which types classify models rather than values, and in particular classify software modules and objects.

---

*The research in this paper was supported in part by grants from the Science and Engineering Research Council, the National Science Foundation, and the System Development Foundation, plus contracts with the Office of Naval Research and the Fujitsu Corporation.

A basic distinction concerns the time at which types are checked against what they classify:

- Types that are checked at *compile time* are called 'static types'. One problem with static typing is that it can be too strong; for example, the expression $(-24/-8)!$ would fail a strong static type check under the assumption that factorial (that is, !) is only defined on natural numbers, because computation is needed to determine whether the division yields a natural number. The polymorphic types of ML (Milner 1978) can be checked at compile time.

- Types checked at *run time* are called 'dynamic types'. Phenomena in this area include coercions (as in many conventional programming languages), retracts (as in OBJ (Futatsugi et al. 1985, Goguen and Winkler 1988)), and some forms of polymorphism.

- Types checked at *design time* do not seem to have an established popular name: in this chapter we call them 'theories'. We believe that this kind of type is the most relevant to formal methods for program development, and to programming in the large.

This chapter assumes familiarity with category, functor, natural transformation, and colimit, for which see texts and papers such as (Goldblatt 1979, Mac Lane 1971, Burstall and Goguen 1982); Lawvere theories and indexed categories are also mentioned but not assumed. Section 14.2 assumes adjoints, and some other advanced concepts are defined as they arise. The identity morphism at A is denoted 1_A, and composition is indicated by semicolon, so that, for example, $(f;g)(x) = g(f(x))$.

This chapter uses OBJ (Futatsugi et al. 1985, Goguen and Winkler 1988) as a notation for examples. Hopefully, much of this notation is self-evident because it is based directly on algebra, and the rest is explained where it arises. Indeed, this chapter provides semantics for much of OBJ.

Types as sets

The oldest notion of type may be summarized by the slogan 'types as sets'. Types in this sense classify values. Cardelli and Wegner (1985) survey this tradition (but, despite the title of their paper, do not survey other approaches to types).

This notion is embodied in the most traditional programming languages, such as FORTRAN and BASIC, which have few types and no operations for combining types. More advanced languages may enrich this approach with some simple operations on sets, such as disjoint union and cartesian product ($\times$ and $\uplus$, for records and variant records), and possibly inclusion ($\subseteq$, for a somewhat restricted notion of subtype). Many more or less modern programming languages embody this approach, including Algol68, Pascal and Modula-2. More recent work tends to take a higher-order viewpoint, treating functions as values to be classified; ML and functional languages like Orwell (Bird and Wadler 1988) embody this approach.

Theoretical work in this tradition, such as domain theory and Scott's 'data types as lattices' (Scott 1972, 1976), tends to impose a partial ordering $\sqsubseteq$ on the entities in a type. Higher-order functions arise through an exponentiation operation on types. A variant models types by 'ideals' in domains (MacQueen et al. 1984). A major benefit of this work is its support for the lambda-calculus in programming, as anticipated by Landin (1964, 1965) and McCarthy et al. (1966). An important recent advance is the second-order polymorphic calculus, independently discovered by Reynolds (1974) and Girard (1972).

In the 'types as predicates' variant of the 'types as sets' approach, types are taken to be predicates, which therefore denote sets (or some variant thereof, such as domains). However, many advocates of this view are more proof-theoretically inclined, and hence might resist such denotations. Among the best known work along this line is Martin-Löf's (1982) 'type theory', which also provides dependent types, as implemented in Pebble (Burstall and Lampson 1984) and other languages. (Note that 'type theory' is not a general theory of types, but rather a specific intuitionistic logic which provides one specific notion of type.)

Although much effort has been put into the 'types as sets' tradition, perhaps because it seems easier to understand, it nonetheless has some defficiencies: it does not take sufficient account of the operations associated with types, and in particular, selectors (for constructors) are not treated in a natural way (see (Smyth and Plotkin 1982) for a treatment of some special cases, and (Goguen and Meseguer 1987b) for a general discussion of the problem); it does not deal with the correctness of implementations; it has a difficult logic, so that it is hard to prove properties of types and operations; it is difficult to apply to modules, exceptions, states, object-oriented concepts, and so on, often requiring complex *ad hoc* encodings to achieve such ends.

Types as algebras

The essential insight of the 'types as algebras' notion is that the *operations* associated with data are at least as important as the values. Thus, this approach generalizes from sets to *algebras*, which are just sets with some given operations. Early work anticipating or advocating this approach includes lectures by Goguen at ETH Zürich (Goguen 1973), Hoare's work on data representation (Hoare 1972), and work by Zilles (1974), Liskov and Zilles (1975), ADJ (Goguen et al. 1975a, 1976), and Guttag (1975) on data abstraction.

The 'initial algebra semantics' of ADJ says that a concrete data representation is a concrete (many-sorted) algebra, and an abstract data type is an abstract algebra, where, as usual in algebra, 'abstract' means 'unique up-to-isomorphism.' A convenient way to describe (or 'present') an abstract algebra is as 'the' initial algebra in the variety of all algebras that satisfy some equations; that is, one gives a signature Σ (of sorts and operation symbols) and a set E of (possibly conditional) Σ-equations, and then defines the class of the initial models of $\langle \Sigma, E \rangle$ to be the abstract data type. This works because any two such initial algebras are necessarily isomorphic. Such a pair $\langle \Sigma, E \rangle$ is called a *theory*, and serves to classify algebras by whether or not they satisfy it; thus theories are types.

For example, consider the natural numbers, where Σ has the one sort Nat, and the following two operations:

```
0 : -> Nat
succ : Nat -> Nat
```

There are no equations, and so $E = \emptyset$. Now let B be the concrete initial Σ-algebra of natural numbers represented in binary positional notation. Finally, let $\mathcal{B}$ be the class of all initial Σ-algebras, that is, of all Σ-algebras isomorphic to B. Thus the intended denotation of the theory NAT = $\langle \Sigma, \emptyset \rangle$ is $\mathcal{B}$. This way of characterizing the natural numbers is due to Lawvere (1964), and is shown to be equivalent to the Peano axioms in (Mac Lane and Birkhoff 1967).

We can extend this approach in many different ways. Section 14.2 develops it into the 'types as theories' approach, in which the theories themselves, and various operations upon them, come to play a dominant role. Figure 14.1 gives an overview of the various levels of abstraction that are involved. From this, we can see that the three approaches to type have a natural hierarchical order: the 'types as sets' approach is the most concrete, the 'types as algebras' approach is a modest abstraction of it, and the 'types as theories' approach supports genuine independence of representation, while still encompassing the other approaches.

Another natural step extends from many-sorted algebra to order-sorted algebra (Goguen and Meseguer 1989), in which a partial order is given on the set of sorts, interpreted semantically as subset inclusion. This permits a nice treatment of exception handling, and of partial and overloaded operations, as illustrated later in this chapter; see also the 'unified algebra' of Mosses (1989). Another extension admits relations as well as operations, leading to various forms of first-order logic. Horn clause logic with equality is especially convenient, because it still has initial models for all theories, as shown in work on the semantics of the language Eqlog, which combines functional and logic programming (Goguen and Meseguer 1986, 1987a).

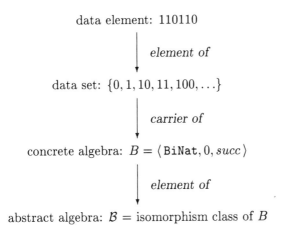

Fig. 14.1 Levels of abstraction of data

Yet another extension considers 'machines' that may have internal states, identifying two such machines iff their behaviour is equivalent (Goguen and Meseguer 1982, Meseguer and Goguen 1985). This gives a semantics for abstract objects which generalizes that of abstract data types, as in the semantics of FOOPS (Goguen and Meseguer 1987c). An early attempt to handle state appears in Guttag's (1975) thesis, later formalized by Wand (1979) using final algebras. However, approaches which admit a class of models, for example, those models that are behaviourally equivalent, seem more satisfactory. Section 14.3 gives another such approach based on the satisfaction of equations up to observability.

The existence of applications using many different logical systems suggests generalizing to an arbitrary logical system. This requires formalizing the notion of logical system. Fortunately, such a formalization is available in the form of *institutions* (Goguen and Burstall 1985), which arose in the semantics of the specification language Clear (Burstall and Goguen 1977, 1980). The framework of institutions is reviewed in Section 14.2, and then systematically used thereafter.

14.2 Types as theories

This section first gives a brief introduction to institutions, and then uses them to explicate theories, parameterized (also called 'dependent') theories, and parameterized theory application. Although these concepts are syntactic in nature, they have formal denotations using constructions on classes of models. Some novel ideas about fibrations and indexed categories arise in this connection, and many examples are given.

Institutions

The formal development in this section assumes a fixed but arbitrary logical system, having *signatures* denoted Σ, Σ', etc., having Σ-*models* denoted M, M', etc., having Σ-*sentences* denoted P, Q, etc., and having Σ-*satisfaction* relations denoted

$$M \vDash_\Sigma P$$

and indicating that the Σ-sentence P is satisfied by the Σ-model M.

For many purposes, we also need signature morphisms, denoted $\phi \colon \Sigma \to \Sigma'$, and we need to know how these morphisms act on sentences and on models. Thus, if P is a Σ-sentence, we let $\phi(P)$ denote the *translation* of P by ϕ, a Σ'-sentence (generally obtained by substituting $\phi(\sigma)$ into P for each $\sigma \in \Sigma$), and if M' is a Σ'-model, we let $\phi(M')$ denote the *translation* (or *reduct*) of M' by ϕ, a Σ-model (generally obtained by letting $\phi(\sigma)$ be the interpretation in M' for each $\sigma \in \Sigma$).

Imposing some simple conditions on this data leads to the notion of an *institution* (Goguen and Burstall 1985), details of which are omitted here. The main technical condition, called the *Satisfaction Condition*, expresses the invariance of satisfaction under change of notation:

$$\phi(M') \vDash_\Sigma P \iff M' \vDash_{\Sigma'} \phi(P)$$

where $\phi \colon \Sigma \to \Sigma'$ is a signature morphism, P is a Σ-sentence and M' is a Σ'-model. In addition, signatures and their morphisms form a category, denoted **Sign**: if $Sen(\Sigma)$ denotes the class of all Σ-sentences, then $Sen \colon$ **Sign** $\to$ **SET** is a functor; if $Mod(\Sigma)$ denotes the category of all Σ-models, then $Mod \colon$ **Sign** $\to$ **CAT**^{op} is a functor. Example institutions include many-sorted equational logic, Horn clause logic, order-sorted equational logic, and first-order logic, as shown in (Goguen and Burstall 1985).

The illustrations in this chapter mainly use many-sorted equational logic. In this institution, a signature is a pair $\langle S, \Sigma \rangle$ where S is a *sort set* and Σ is an $S^* \times S$-indexed family of sets. A signature morphism $\phi \colon \Sigma \to \Sigma'$ consists of a function $f \colon S \to S'$ and an $S^* \times S$-indexed family of functions $g_{w,s} \colon \Sigma_{w,s} \to \Sigma'_{f(w), f(s)}$. Σ-models are Σ-algebras, Σ-sentences are (conditional) Σ-equations, and Σ-satisfaction is as usual. (Actually, the main examples will use order-sorted equational logic (Goguen and Meseguer 1989).)

Theories

This subsection develops theories over an arbitrary institution, following the semantics of Clear (Burstall and Goguen 1980, Goguen and Burstall 1985).

Definition 1. *A theory consists of a signature Σ and a set E of Σ-sentences, that is, it is a pair $\langle \Sigma, E \rangle$. We may also call $\langle \Sigma, E \rangle$ a Σ-theory.*

Given a theory $T = \langle \Sigma, E \rangle$ and a Σ-model M, we say that M is a T-model, or that M satisfies T iff $M \vDash_\Sigma P$ for each $P \in E$, written $M \vDash_\Sigma E$.

A theory $\langle \Sigma, E \rangle$ is closed iff

$$M \vDash_\Sigma E \;\Rightarrow\; M \vDash_\Sigma P \;\Rightarrow\; P \in E$$

(Sometimes theories are called 'presentations' to emphasize that closure is not required.)

Thus a Σ-theory classifies Σ-models by whether or not they satisfy it.

Definition 2. *Given a set T of Σ-sentences and a class $\mathcal{M}$ of Σ-models, let*

$$T^* = \{M \mid M \vDash_\Sigma T\}$$

be the variety *or* denotation *of T, and let*

$$\mathcal{M}^* = \{P \mid M \vDash_\Sigma P, \text{ for all } M \in \mathcal{M}\}$$

be the theory *of $\mathcal{M}$, also denoted $\mathrm{Th}(\mathcal{M})$.*

Because an institution provides the category $Mod(\Sigma)$ of all Σ-models, we can let T^* be a full subcategory of $Mod(\Sigma)$.

Notice that satisfaction induces a duality (or more precisely, a Galois connection) between

- sets of sentences (that is, theories), and
- sets (or classes, to be more precise) of models

by the operations denoted * in the above definition. We might think that because of this duality we could just as well take model classes as basic, instead of theories. However, theories have the advantage over model classes that they are 'more presentable': we can often find a *finite* set of (finite) sentences that describes what we are interested in, whereas the model classes (as well as their elements) are much less likely to be finite.

Fact 3. Given a theory $T = \langle \Sigma, E \rangle$, there is a least closed set $\overline{E}$ of Σ-sentences containing E, defined by

$$P \in \overline{E} \iff M \vDash_\Sigma P, \text{ for all } M \text{ such that } M \vDash_\Sigma E$$

or equivalently, by

$$\overline{E} = E^{**}$$

We shall call $\overline{E}$ the closure of E, and $\langle \Sigma, \overline{E} \rangle$ the closure of T, written $\overline{T}$.

Let us now consider some examples, using the syntax of OBJ. The simplest non-void theory is

```
th TRIV is
  sort Elt .
endth
```

Notice that TRIV* = **SET**, the category of all sets. Next, we consider monoids and groups, two simple theories from abstract algebra:

```
th MONOID is
  sort Elt .
  op e : -> Elt .
  op _·_ : Elt Elt -> Elt .
  vars A B C : Elt .
  eq A · e = A .
  eq e · A = A .
  eq A · (B · C) = (A · B) · C .
endth

th GROUP is
  sort Elt .
  op e : -> Elt .
  op _·_ : Elt Elt -> Elt .
  op _-1 : Elt -> Elt .
  vars A B C : Elt .
  eq A · e = A .
  eq e · A = A .
  eq A · (B · C) = (A · B) · C .
  eq A · (A -1) = e .
  eq (A -1) · A = e .
endth
```

These are theories over the institution $\mathcal{EQ}$ of (conditional many-sorted) equational logic. The denotation MONOID* of the theory MONOID of monoids is the variety (or category) **MONOID** of all monoids; similarly, GROUP* = **GROUP**.

Dependent theories

In order to talk about relationships between theories — for example, inheritance, or that a certain theory serves as interface for a certain generic module — we need the following definition.

Definition 4. *A* view *or* theory morphism $\Phi\colon \langle \Sigma, E\rangle \to \langle \Sigma', E'\rangle$ *is a signature morphism* $\Phi\colon \Sigma \to \Sigma'$ *such that*

$$P \in E \;\Rightarrow\; \Phi(P) \in \overline{E'}$$

This gives a category **Theo**$(\mathcal{I})$ *of theories over an institution* $\mathcal{I}$.

Fact 5. $\Phi\colon \Sigma \to \Sigma'$ *is a theory morphism* $\Phi\colon \langle \Sigma, E\rangle \to \langle \Sigma', E'\rangle$ *iff*

$$M' \vDash_{\Sigma'} E' \;\Rightarrow\; M' \vDash_{\Sigma'} \Phi(E)$$

for all Σ'-*models* M', *where* $\Phi(E) = \{\Phi(P) \mid P \in E\}$.

A subtheory is a view $\langle \Sigma, E\rangle \to \langle \Sigma', E'\rangle$, where $\Sigma \subseteq \Sigma'$ and $E \subseteq E'$; although this only makes sense for institutions where the notion of subsignature is well defined, most cases of practical interest are included. For example, MONOID $\subseteq$ GROUP.

Definition 6. *The* denotation *of a theory morphism* $\Phi\colon T \to T'$ *is its* reduct *or* forgetful functor, $\Phi^*\colon T'^* \to T^*$, *defined to send* M' *to* $\Phi(M')$, *which is a* T-*model by Fact 5 plus the Satisfaction Condition, and to send* $f\colon M_0' \to M_1'$ *to* $\Phi(f)$.

Because TRIV is a subtheory of every theory (that has at least one sort), we have

$$\text{TRIV} \subseteq \text{GROUP}$$

and

$$\text{TRIV} \subseteq \text{MONOID}$$

which induce the familiar forgetful functors

$$\text{GROUP}^* \longrightarrow \textbf{SET}$$

and

$$\text{MONOID}^* \longrightarrow \textbf{SET}$$

For another example, the inclusion MONOID ⊆ GROUP induces the forgetful functor

$$\text{GROUP}^* \longrightarrow \text{MONOID}^*$$

Many institutions are 'liberal' in the sense that all their forgetful functors have 'left adjoints' which give the 'free things generated by'. It is possible to define this without a lot of category theory.

Definition 7. *Given a theory morphism* $\Phi: T \to T'$, *let*

$$U = \Phi^*: T'^* \to T^*$$

be its forgetful functor, and let M be a model in T^*. Then a model M' in T'^* is free (with respect to U) over M iff there is some morphism $i: M \to U(M')$ such that, given any $j: M \to U(N')$, there is a unique morphism $h: M' \to N'$ such that $i; U(h) = j$, that is, such that the diagram

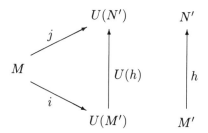

commutes, where i, j are Σ-morphisms and h is a Σ'-morphism. We may write $j^\#$ for h, because it is uniquely determined by j.

An insitution is liberal iff for every such theory morphism Φ and model M, there is some model M' that is free over M with respect to Φ^*.

Intuitively, M' is the best T'^*-model that 'extends' M. Examples include the following:

- the free group on a set of X of generators;
- the free monoid on a set of X of generators;
- the free group over a monoid M.

Liberal institutions include equational logic (in all the variations that we have mentioned), Horn clause logic, Horn clause logic with equality, and many others — but not first-order logic.

Proposition 8. *Given a theory morphism* $\Phi \colon T \to T'$ *in a liberal institution, the function* $M \mapsto M'$ *for* $M \in T^*$, *where* M' *is free (with respect to* Φ^**) over* M, *extends uniquely to a functor denoted*

$$\Phi^\$ \colon T^* \to T'^*$$

and called a free functor. *It is left adjoint to the forgetful functor* Φ^*.

For the 'types as theories' point of view to include the 'types as models' point of view, it is convenient to introduce a new kind of sentence which says that an *initial* model is intended for a theory (or, more generally, for part of a theory). The theory of institutions provides such sentences, called 'constraints' (Goguen and Burstall 1985); see also (Reichel 1980). In order to define these, we first formalize the idea that a model is free over part of a theory. Instead of just considering a theory inclusion, we generalize to an arbitrary theory morphism.

Definition 9. *Given a theory morphism* $\Phi \colon T \to T'$, *we say that a* T'-*model* M' *is* Φ-free *iff* M' *is free over* $\Phi^*(M')$ *and the morphism*

$$(1_{\Phi^*(M')})^\# \colon \Phi^\$(\Phi^*(M')) \longrightarrow M'$$

called the co-unit morphism *at* M' *and denoted* $\epsilon_{M'}$, *is an isomorphism.*

We now define the new kind of sentence.

Definition 10. *Given a signature* Σ, *then a* Σ-constraint *is a pair*

$$\langle \Phi \colon T'' \to T', \theta \colon Sign(T') \to \Sigma \rangle$$

where T', T'' *are theories and* $Sign(T')$ *is the signature of* T'. *Given a* Σ-*constraint* $c = \langle \Phi \colon T'' \to T', \theta \colon Sign(T') \to \Sigma \rangle$ *and a* Σ-*model* M, *we say that* M *satisfies* c *iff* $\theta(M)$ *satisfies* T' *and is* Φ-*free. Given signature morphism* $\phi \colon \Sigma \to \Sigma'$, *then the* ϕ-*translation of* $c = \langle \Phi, \theta \rangle$ *is the constraint* $\langle \Phi, \theta; \phi \rangle$, *denoted* $\phi(c)$.

Proposition 11. *Given an institution* $\mathcal{I}$, *a new institution denoted* $\mathcal{C}(\mathcal{I})$ *is obtained from* $\mathcal{I}$ *as follows: its signatures are the same as those of* $\mathcal{I}$; *its sentences include those of* $\mathcal{I}$, *plus constraints as new sentences, with translation and satisfaction as given in Definition 10.*

See (Goguen and Burstall 1985) for more detail, including a proof. In $\mathcal{C}(\mathcal{I})$, a Σ-theory contains both Σ-sentences from $\mathcal{I}$ and Σ-constraints, and is satisfied by a model M iff M satisfies both all the sentences and all the constraints. It is worth noting that some $\mathcal{C}(\mathcal{I})$ theories may be satisfied by no models, even when $\mathcal{I}$ is liberal, because of the possibility of inconsistent constraints.

Let us now consider some examples to show how these ideas are used in practice. Again we use OBJ as a notation. For the unparameterized case, the keyword th indicates that any model is acceptable, whereas the keyword obj indicates that the only acceptable models are Φ-free with respect to the theory inclusion $\Phi \colon \emptyset \to T$, where $\emptyset$ denotes the empty theory (with no sorts) and T is the body of the theory in question. In this way, unparameterized theories are a special case of parameterized theories. (Note that $\emptyset$ has just one model, having no carriers and no operations.) For example, the natural numbers may be specified as follows:[1]

```
obj NAT is
  sort Nat .
  op 0 : -> Nat .
  op s_ : Nat -> Nat [prec 5].
  op _+_ : Nat Nat -> Nat .
  vars M N : Nat .
  eq 0 + N = N .
  eq s M + N = s(M + N).
endo
```

This theory denotes the class of all initial algebras over the theory $T = \langle \Sigma, E \rangle$, where Σ has $S = \{\texttt{Nat}\}$ and the three operation symbols 0, s_, _+_, and where E contains the two equations given above, according to the following result.

Proposition 12. *Given an $\mathcal{I}$-theory $T = \langle \Sigma, E \rangle$, a model M is an initial model of T iff it satisfies the $\mathcal{C}(\mathcal{I})$-theory $T' = \langle \Sigma, E' \rangle$, where E' is E plus the constraint $\langle \emptyset \to T, 1_\Sigma \rangle$. We may call constraints of this form* initiality constraints.

Our machinery also provides a very natural semantics for parameterized theories, such as the following theory of actions of monoids on sets of states; intuitively, $A \cdot S$ is the new state that results from taking action A when in state S.

```
th ACT[M :: MONOID] is
  sort State .
  op _·_ : Elt State -> State .
```

---

[1] The attribute [prec 5] of the prefix operation s indicates that it is more tightly binding than +, which OBJ3 gives the default precedence for binary infix operations of 41; see (Goguen and Winkler 1988).

```
    var A A' : Elt .
    var S : State .
    eq A · (A' · S) = (A · A') · S .
    eq e · S = S .
endth
```

We take the forgetful functor $\Phi^*\colon \text{ACT}^* \longrightarrow \text{MONOID}^*$ to be the denotation of this parameterized theory, where ACT includes not only the operations and equations shown above, but also those in MONOID.

In much the same way, we can give a denotation for a constrained parameterized theory. It is traditional to illustrate this with the following example, which also makes good use of order-sorted algebra (for more information on order-sorted algebra, see (Goguen and Meseguer 1987b, 1989)):

```
obj STACK[X :: TRIV] is
    sorts Stack NeStack .
    subsort NeStack < Stack .
    op push : Elt Stack -> NeStack .
    op empty : -> Stack .
    op pop_ : NeStack -> Stack .
    op top_ : NeStack -> Elt .
    var E : Elt .
    var S : Stack .
    eq pop push(E,S) = S .
    eq top push(E,S) = E .
endo
```

This denotes the $\mathcal{C}(\mathcal{E}\mathcal{Q})$-theory with all the sorts, operations, and equations stated above, plus Elt from TRIV and the constraint

$$\langle \text{TRIV} \hookrightarrow \text{STACK}, 1_{Sign(\text{STACK})} \rangle$$

Letting $\Phi\colon Sign(\text{TRIV}) \to Sign(\text{STACK})$ be the signature inclusion and letting $\text{STACK}^* = \mathbf{STACK}$, then $\Phi^*\colon \mathbf{STACK} \longrightarrow \mathbf{SET}$ is the denotation for the parameterized theory STACK, whereas $\Phi^\$\colon \mathbf{SET} \longrightarrow \text{STACKEQ}^*$ would be its denotation in the style of Clear (Burstall and Goguen 1980, Goguen and Burstall 1985), where STACKEQ is the purely equational part of the theory STACK.

The following example, which given integers a, b, c, returns the 'type' of functions $f\colon \text{Int} \to \text{Int}$ which always lie within c of $ax + b$, is more similar to examples of dependent types in the usual literature on that subject:

```
th 3INT is
    pr INT .
    ops a b c : -> Int .
endth
```

Fig. 14.2 Application of a parameterized theory

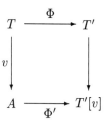

(The line pr INT . indicates that the theory INT of integers is being imported, that is, 'inherited', in the sense explained in Section 14.3 below.)

```
th WLIN[P :: 3INT] is
  op f : Int -> Int .
  var X : Int .
  eq f(X) - c < (a * X) + b = true .
  eq (a * X) + b < f(X) + c = true .
endth
```

These examples motivate the following definition.

Definition 13. *A* parameterized *(or* dependent*) theory is a theory morphism*
$$\Phi: T \to T'$$
The denotation of $\Phi: T \to T'$ *is its forgetful functor* $\Phi^*: T'^* \to T^*$, *as in Definition 6.*

Such a parameterized theory corresponds to a 'universal' or 'product' dependent type, as illustrated by the above examples, provided that the institution involved has the form $\mathcal{C}(\mathcal{I})$. (Note that either, or both, T and T' may include constraints.)

Theory instantiation

The specification language Clear provides a semantics for applying (also called 'instantiating') parameterized theories. This semantics uses colimits, and was inspired by some earlier ideas from general systems theory (Goguen 1971):

> Given a parameterized theory $\Phi: T \to T'$ and a view $v: T \to A$ of an 'actual' theory A as an 'instance' of T, then the result of 'applying' Φ to v is the pushout shown in Figure 14.2, whose pushout object is denoted $T'[v]$ and whose morphism opposite to Φ may be denoted Φ'. The denotation of $T'[v]$ is $(T'[v])^*$.

To know that we can always perform this construction, we need the following results from (Goguen and Burstall 1985).

Theorem 14. *If the category* **Sign** *of signatures of an institution* $\mathcal{I}$ *has all (finite) colimits, then so does the category* **Theo**$(\mathcal{I})$ *of all theories over* $\mathcal{I}$.

Corollary 15. *If the category* **Sign** *of signatures of an institution* $\mathcal{I}$ *has all (finite) colimits, then so does the category* **Theo**$(\mathcal{C}(\mathcal{I}))$ *of all theories over* $\mathcal{C}(\mathcal{I})$.

All this generalizes to the case where there are 'shared' subtheories, but it is convenient to defer this topic to Section 14.3, where it is discussed under the heading of 'inheritance'.

Now let us perform some simple theory instantiations, beginning with a view of the Booleans as a monoid, again using syntax from OBJ:

```
view BOOMON from MONOID to BOOL is
  sort Elt to Bool .
  op (_._) to (_and_) .
  op e to true .
endv
```

(Actually, the sort line could be omitted, because of default the view conventions that are built into OBJ (Goguen and Winkler 1988).) Notice that there is no 'preferred' view of the Booleans as a monoid; we could just as well have mapped '·' to disjunction, exclusive-or, or iff. Now we can apply ACT to BOOL, using the following syntax:

```
make BOOACT is ACT[BOOMON] .
```

which gives the theory BOOACT of actions of BOOL on sets, with conjunction as composition.

For another example,

```
make NATSTACK is STACK[NAT] .
```

gives the theory of stacks of natural numbers, exploiting OBJ's ability to infer the following default view:

```
view NATT from TRIV to NAT is
  sort Elt to Nat .
endv
```

Finally, let us instantiate WLIN as follows:

```
view MY3INT from 3INT to INT is
  op a to 2 .
  op b to 3 .
  op c to 3 .
endv

make MYWLIN is WLIN[MY3LIN] .
```

This yields the following theory:

```
th MYWLIN is
  pr INT .
  op f : Int -> Int .
  var X : Int .
  eq f(X) - 3 < (2 * X) + 3 = true .
  eq (2 * X) + 3 < f(X) + 3 = true .
endth
```

Fibration and indexed category semantics

Given a parameterized theory $\Phi \colon T \to T'^*$, notice that for each T^*-model M there is a full subcategory of T'^* whose objects are the models M' such that $\Phi^*(M') = M$. Examining this situation more closely suggests some interesting structure, first noticed in a more general form by Grothendieck (1963) in the context of algebraic geometry.

Definition 16. *A (strict) fibration is a functor $P \colon \mathbf{A} \to \mathbf{B}$ together with, for each morphism $b \colon B \to B'$ in $\mathbf{B}$, a functor $b^{\spadesuit} \colon B'^{\spadesuit} \to B^{\spadesuit}$ such that $(b;b')^{\spadesuit} = b'^{\spadesuit}; b^{\spadesuit}$ and $1_B^{\spadesuit} = 1_{B^{\spadesuit}}$, where $B^{\spadesuit}$ denotes the full subcategory of $\mathbf{A}$ with objects A such that $P(A) = B$. We may also write $P^{\spadesuit}(B)$ for $B^{\spadesuit}$, the fibre over B, and $P^{\spadesuit}(b)$ for $b^{\spadesuit}$. We call $\mathbf{B}$ the base of the fibration.*

For example, if Φ is the theory inclusion MONOID $\longrightarrow$ ACT, then Φ^* is a strict fibration, where, given a monoid homomorphism $h \colon M \to M'$, then $\Phi^{\spadesuit}(h) \colon \Phi^{\spadesuit}(M') \to \Phi^{\spadesuit}(M)$ sends an action $a \colon M' \times S \to S$ to the action $(h \times 1_S); a \colon M \times S \to S$ which sends $\langle m, s \rangle$ to $a(h(m), s)$. Here $\Phi^{\spadesuit}(M)$ is the category of all actions of M.

If $P \colon \mathbf{A} \to \mathbf{B}$ is a strict fibration, then $P^{\spadesuit} \colon \mathbf{B} \to \mathbf{CAT}^{op}$ is a strict indexed category (in the sense of (Tarlecki et al. 1989)). Conversely, any strict indexed category gives rise to a strict fibration, by the strict Grothendieck construction described in (Tarlecki et al. 1989).

The following concept (from (Goguen and Burstall 1985)) seems useful in giving a construction for Φ^* to be a strict fibration.

Object-oriented concepts

Definition 17. *A theory morphism* $\Phi\colon T \to T'$ *is persistent iff* Φ^* *has a left adjoint* $\Phi^\$$ *such that its unit* η *is an isomorphism, and is strictly persistent iff* η *is an identity.*

Thus when Φ is strictly persistent, $\Phi^*(\Phi^\$(M)) = M$ for each T-model M, and general results imply that if Φ is persistent, then it is also strictly persistent, by choosing a slightly different functor $\Phi^\$$. In a more geometrical language, we might say that such a $\Phi^\$$ is a *section* of Φ^*, in the sense that $\Phi^\$; \Phi^* = 1_{T^*}$.

Conjecture 18. *If* $\Phi\colon T \to T'$ *is strictly persistent and if* T'^* *has pullbacks, then* Φ^* *can be given the structure of a strict fibration.*

In such cases, we may take the fibration Φ^*, or equivalently the corresponding indexed category $\Phi^\spadesuit\colon T^* \longrightarrow \mathbf{CAT}^{op}$, as the denotation of the dependent theory Φ.

For example, given the $\mathcal{C}(\mathcal{E}\mathcal{Q})$-theory morphism $\Phi\colon \mathtt{TRIV} \to \mathtt{STACK}$, the indexed category $\Phi^\spadesuit\colon \mathbf{SET} \longrightarrow \mathbf{CAT}^{op}$ sends each set X to the isomorphism class of initial $\mathtt{STACK\,[X]}$-algebras.

It is natural to ask how the pushout semantics of theory application relates to the indexed category semantics, which might suggest that $v; \Phi^\spadesuit\colon T^* \longrightarrow \mathbf{CAT}^{op}$ should be the denotation of the application $T'[v]$.

Conjecture 19. *Let* $\Phi\colon T \to T'$, *let* M *be a* T-*model, and let* $\underline{M}$ *be the theory inclusion* $T \to \{M\}^*$ *(note that* $\{M\}^*$ *is the theory of* M). *Then*

$$\Phi^\spadesuit(M) = (T'[\underline{M}])^*$$

and

$$v^*; \Phi^\spadesuit = \Phi'^\spadesuit$$

14.3 Object-oriented concepts

This section applies the 'types as theories' viewpoint to object-oriented programming. Although algebra has already been applied to object-oriented programming (Goguen and Meseguer 1982, Meseguer and Goguen 1985, Sannella and Tarlecki 1987, Goguen and Meseguer 1987c, Hayes and Coleman 1989), the present approach goes further and is perhaps more elegant.

Inheritance

The basic issues concerning inheritance are orthogonal to issues concerning object and state, because they have to do with importing one module into another, that is, with theory morphisms, and make sense in any institution.

The keyword **protecting** (abbreviated **pr**) indicates that a theory is imported in a way that preserves its denotation. For example,[2]

```
th POSET is
  pr BOOL .
  sort Elt .
  op _<_ : Elt Elt -> Bool .
  vars E1 E2 E3 : Elt .
  eq E1 < E1 = false .
  cq E1 < E3 = true if E1 < E2 and E2 < E3 .
endth
```

Let Σ denote the signature of POSET, which includes the signature of BOOL plus the operation _<_. Then POSET is the $\mathcal{C}(\mathcal{EQ})$-theory which includes the two equations shown above, plus all the equations in BOOL, plus the Σ-constraint $\langle \emptyset \to \text{BOOL}, Sign(\text{BOOL}) \to \Sigma \rangle$. Thus, a Σ-algebra M satisfies this theory iff it satisfies all the equations in POSET, including those in BOOL, plus the constraint $\langle \emptyset \to \text{BOOL}, Sign(\text{BOOL}) \to \Sigma \rangle$, which just says that it contains an uncorrupted copy of the Booleans.

Now let us consider an example with the keyword **obj** instead of **th**:

```
obj STACK is
  pr NAT .
  sorts Stack NeStack .
  subsort NeStack < Stack .
  op push : Nat Stack -> Nestack .
  op empty : -> Stack .
  op pop_ : NeStack -> Stack .
  op top_ : NeStack -> Nat .
  var S : Stack .
  var N : Nat .
  eq pop push(N,S) = S .
  eq top push(N,S) = N .
endo
```

---

[2] Because OBJ assumes that BOOL is inherited into every module, the line **pr BOOL .** could be omitted, and future examples will do so. Also, note that the condition in a conditional equation is required to have sort Bool in OBJ, so that it is unnecessary to write **= true** there.

This denotes the $\mathcal{C}(\mathcal{EQ})$-theory containing all the equations stated above, plus those in NAT, and the two constraints

$$\langle \text{NAT} \hookrightarrow \text{STACK}, 1_{Sign(\text{STACK})} \rangle$$

and

$$\langle \emptyset \to \text{NAT}, Sign(\text{NAT}) \to Sign(\text{STACK}) \rangle$$

In general, the import declarations of a theory provide a base over which it is extended, freely if its keyword is **obj** (by adding a constraint over its base); also, all the imported constraints have their signature extended to that of the importing theory.

OBJ has two other forms of import declaration, **extending** and **using** (plus **including** in a more recent release), whose semantics can be handled in a similar way, making use of some more general forms of constraint which are discussed in (Goguen and Burstall 1985); in particular, the semantics of **extending** uses the version of hierarchy constraints developed in (Goguen and Burstall 1985).

Often a theory is imported into two (or more) other theories which are later combined, for example, through instantiating a parameterized theory, and we expect to see only one copy of this shared subtheory in the result theory; in particular, BOOL is imported into every theory in OBJ. The desired effect is obtained in a natural way by letting each specification determine a *diagram* in the category of theories, called its *environment*, such that the desired result theories are obtained by taking *colimits* of appropriate subdiagrams. Such environment diagrams in general have just one copy of fundamental theories such as BOOL, NAT and INT, shared among many other theories, while there may be many instances of useful dependent theories, such as LIST and SET. Colimits indeed behave this way, and the extended example in Section 14.5 presents several environment diagrams.

Thus, sharing is one reason why pushouts alone are not quite enough. Colimits are also needed to evaluate general *module expressions*, which may call for more complex ways of combining modules that just the instantiation of parameterized modules. One important additional operation is the *sum* of theories, denoted +, which is computed by coproducts; see (Burstall and Goguen 1980, Goguen and Burstall 1985).

The above considerations extend to what de Bruijn (1989) has called 'telescopes', which are sequences

$$T_1 \xrightarrow{\Phi_1} T_2 \xrightarrow{\Phi_2} \cdots \xrightarrow{\Phi_n} T_{n+1}$$

where each Φ_i is a theory morphism. These represent types with multiple nested dependencies. There are many interesting relationships among the various denotations of the parts of a telescope, such as $(\Phi_1; \ldots; \Phi_n)^*$

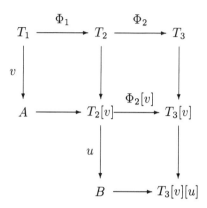

Fig. 14.3 A telescope diagram

and $(\Phi_1; \ldots; \Phi_n)^{\spadesuit}$. For example, Figure 14.3 shows how to instantiate a telescope of length 3: first instantiate T_2 with $v\colon T_1 \to A$, obtaining the object $T_2[v]$, and then instantiate T_3 with $u\colon T_2[v] \to B$. Alternatively, after instantiating with u, one could complete the top by taking a second pushout, yielding the object $T_3[v]$, and then do the bottom pushout. Well-known general results about pushouts guarantee that either route leads to the same result (up to isomorphism), denoted $T_3[v][u]$ in Figure 14.3.

A comparison

It is interesting to compare this notion of inheritance at the level of *theories* with the notion of inheritance at the level of *sorts* which is given by order-sorted algebra (Goguen and Meseguer 1989). Perhaps it is best to start by comparing the notions of type that each involves: the first is, of course, the 'types as theories' notion which is the focus of this chapter; the second lifts the inclusion notion of inheritance from the 'types as sets' to the 'types as algebras' level. Whereas inheritance for theories has to do with the re-use of code and the foundations of programming methodology, including programming in the large, inheritance by subsorts has to do with the hierarchical classification of values, whereby some value classes are contained in others. The level of theories has to do with the classification of implementations, in the sense of models defined by concrete code; here, inheritance has as denotation a map from some model class to another by a forgetful functor arising from a theory inclusion. (Note that many sorts of data may be involved simultaneously in a single relationship of theory inheritance.)

Objects and state

An appropriate institution for object-oriented programming is the *hidden-sorted* (conditional) *equational institution*, in which signatures are triples $\langle S, V, \Sigma \rangle$, where $\langle S, \Sigma \rangle$ is a many-sorted signature and $V \subseteq S$ is a subset of sorts called *visible*, such that

$$(s \leqslant v \lor v \leqslant s) \land v \in V \Rightarrow s \in V$$

that is, the visible and hidden sorts form disjoint hierarchies, where we call $s \in H = S - V$ a *hidden sort*.

A *hidden sort signature morphism* $\Phi \colon \langle S, V, \Sigma \rangle \to \langle S', V', \Sigma' \rangle$ is a many-sorted signature morphism $\Phi = \langle f, g \rangle \colon \langle S, \Sigma \rangle \to \langle S', \Sigma' \rangle$ such that $f(V) \subseteq V'$ and $f(H) \subseteq H'$ where $H' = S' - V'$, and such that if $\sigma' \in \Sigma'_{w',s'}$ and some sort in w' is hidden, then for each $\sigma \in \Sigma_{w,s}$ such that $f(w) = w'$, $f(s) = s'$, and $g_{w,s}(\sigma) = \sigma'$, there is a hidden sort in w. This last condition captures the intuition in object-oriented programming that objects are 'encapsulated' in the sense that a module cannot add new attributes or methods to classes that are defined in another module (here the morphism represents the relation between the two modules).

An $\langle S, V, \Sigma \rangle$-model is an $\langle S, \Sigma \rangle$-algebra, an $\langle S, V, \Sigma \rangle$-morphism is an $\langle S, \Sigma \rangle$-homomorphism, and an $\langle S, V, \Sigma \rangle$-sentence is an $\langle S, \Sigma \rangle$-equation. Similarly, the Φ-translations of $\langle S', V', \Sigma' \rangle$-models are their $\langle S, \Sigma \rangle$-translations, and the Φ-translations of $\langle S, V, \Sigma \rangle$-sentences are their $\langle S, \Sigma \rangle$-translations. However, the satisfaction relation is rather different from that of the ordinary equational institution. We define

$$M \vDash_{\langle S, V, \Sigma \rangle} (\forall X)\, t_0 = t_1$$

iff for all $v \in V$ and all $t \in T_{\Sigma \cup X \cup \{z\}, v}$ having just one occurrence of z, where z is a new variable having the same sort as t_0 and t_1, we have

$$M \vDash_{\langle S, \Sigma \rangle} (\forall X)\, t(z \leftarrow t_0) = t(z \leftarrow t_1)$$

where $t(z \leftarrow t_0)$ indicates the result of substituting t_0 for each occurrence of z in t. Let us call this *behavioural satisfaction*, and let us call equations that are intended to be satisfied in this way *behavioural equations*. Also, terms t of the form described above are called *contexts* for the equation.

Proposition 20. *Hidden-sorted equational signatures and morphisms, with ordinary (conditional) equations as sentences, with ordinary algebras as models, and with translation and satisfaction as defined above, form an institution.*

Proof. The Satisfaction Condition for this institution follows from that of the ordinary equational institution, by using the definition of hidden-sorted signature morphism to show that sufficiently many contexts can be translated. Given $\Phi\colon \langle S,V,\Sigma\rangle \to \langle S',V',\Sigma'\rangle$, given a Σ'-algebra A, and given a Σ-sentence $e = (\forall X)\ t_0 = t_1$, we must show that

$$\Phi(A) \vDash_\Sigma e \text{ iff } A \vDash \Phi(e)$$

that is, that
$$\Phi(A) \vDash_\Sigma (\forall X)\ t(z \leftarrow t_0) = t(z \leftarrow t_1)$$
iff
$$A \vDash_{\Sigma'} (\forall \Phi(X))\ t'(z' \leftarrow \Phi(t_0)) = t'(z' \leftarrow \Phi(t_1))$$

The key observation is that each context t' for $\Phi(e)$ is of the form $t''(\Phi(t))$, where t'' involves only operations with all arity and value sorts visible, and t is a context for e. □

It is sometimes convenient to extend the hidden-sorted equational institution to ordered sorts, Horn clauses, and/or continuity. This presents no difficulty, and in fact the following example uses the order-sorted version. Of course we can also add constraints.

```
th STACK is
  pr NAT .
  sorts Stack NeStack [hidden] .
  subsort NeStack < Stack .
  op push : Nat Stack -> NeStack .
  op empty : -> Stack .
  op pop_ : NeStack -> Stack .
  op top_ : NeStack -> Nat .
  var S : Stack .
  var N : Nat .
  eq pop push(N,S) = S .
  eq top push(N,S) = N .
endth
```

The attribute [hidden] after a sort declaration makes all the sorts in its list hidden; visible sorts are introduced by sort declarations that do not have the hidden attribute.

Notice that the keyword here is th rather than obj, so that there is no constraint for STACK over NAT; however, the constraint

$$\langle \emptyset \to \text{NAT}, Sign(\text{NAT}) \to Sign(\text{STACK}) \rangle$$

does appear, because of the line **pr NAT**. This constraint guarantees that the sort **Nat** really does represent the natural numbers.

Models of **STACK** need only 'appear' to satisfy its equations when observed through visible-valued terms, which necessarily have **top** as their head operation. Notice that the usual implementation of a stack by a pointer and an array does *not* actually satisfy the first equation above, although it does satisfy it behaviourally and hence satisfies **STACK**. Hence, the models of this theory are not all isomorphic to one another.

This theory defines the stack *class* (in the sense of object-oriented programming) by providing axioms that are the basis for reasoning about stack objects. If we let **STACKV** denote the above **STACK** theory without hidden sorts, then its initial algebra also satisfies **STACK**, and we can reason about it in all the usual ways, including induction, and be sure that any equation we prove is behaviourally satisfied by all models, because all models are behaviourally equivalent. However, this approach is not in general *complete*; that is, some equations may be behaviourally satisfied in all models of **STACK**, but cannot be proved in this way.

This specification uses a functional notation which may require us to write explicit calls for state, contrary to the usual notational conventions of imperative programming. However, imperative programs certainly do access state, and we can regard their notation as syntactic sugar for the more explicit notation used in our algebraic specifications. In this way, we can have our sweet syntactic cake and eat its healthy semantics too.

A theory like **STACK** above does not address the creation or deletion of objects, or concurrent interactions among objects; however, techniques such as process algebra (Milner 1980, Hoare 1978) and sheaf theory (Goguen 1991) may be useful. There is also relevant work by the ISCORE group, such as (Ehrich *et al.* 1989) and (Sernandas *et al.* 1989). Hidden sorts, but not behavioural satisfaction, were introduced in (Goguen and Meseguer 1982).

14.4 Polymorphism is natural

Following Milner (1978), polymorphism has become a hallmark of higher-order functional programming languages, such as ML (Harper *et al.* 1986), Miranda (Turner 1985) and Orwell (Bird and Wadler 1988). For this kind of polymorphism to work correctly, all the types involved must be 'trivial', so that any type constructor can be applied to any other. For example, *list list* α, *stack list* α, and *pair*$(\alpha, list\ \beta)$ are all polymorphic types, where α and β are type variables. In the language of this chapter, we would say that the parameterized theories which describe these sorts must all have **TRIV** as their interface theory. It is also necessary to have a designated

380   Types as theories

principal exported sort.[3] Thus, such a polymorphic type is a trivially parameterized theory with a designated sort, and it determines a functor **SET** ⟶ **SET**, because

$$\text{TRIV} \xrightarrow{\Phi} T \xleftarrow{\Psi} \text{TRIV}$$

gives rise to

$$\textbf{SET} \xrightarrow{\Phi^\$} T^* \xrightarrow{\Psi^*} \textbf{SET}$$

and these two functors can be composed to yield the desired denotation. More generally,

$$\text{TRIV}^n \xrightarrow{\Phi} T \xleftarrow{\Psi} \text{TRIV}$$

gives

$$\textbf{SET}^n \xrightarrow{\Phi^\$} T^* \longrightarrow \textbf{SET}$$

An observation which has aroused interest in the functional programming community, and has been proved there in some special cases, is that polymorphic operations are natural transformations. For example, the append function on lists is natural in the sense that, given any function $f: X \to Y$ (in **SET**), the following diagram commutes:

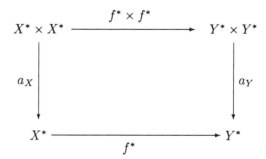

where X^*, Y^* are lists of X, Y, and where a_X, a_Y are their append operations.

---

[3] It is natural to take the principal sort to be the one that is first mentioned in a module, as in OBJ (Goguen and Winkler 1988).

It is easy to prove this fact in general, making use of Lawvere's (1963) view that an algebraic theory is a category with operations as its morphisms, an algebra is a product-preserving functor (to **SET**), and a homomorphism is a natural transformation. (Lawvere's insight that homomorphisms are natural transformations is at first surprising, but is actually fairly simple and rather useful; for example, see (Goguen et al. 1975b).)

Now given a trivially parameterized type

$$\text{TRIV} \xrightarrow{\Phi} T$$

we obtain a functor

$$\textbf{SET} \xrightarrow{\Phi^\$} [\mathsf{T} \xrightarrow{\pi} \textbf{SET}]$$

where T is T as a Lawvere theory and $[\_ \xrightarrow{\pi} \_]$ indicates the category of product-preserving functors with natural transformations as morphisms. Viewing this as a category with *graph* morphisms as objects and with natural transformations as their morphisms, let us pick out an edge, say $\sigma: s_1 \ldots s_n \longrightarrow s$ representing $\sigma \in \Sigma_{s_1 \ldots s_n, s}$. Then, by restriction, we obtain the functor

$$\textbf{SET} \xrightarrow{\Phi^\$} [(\cdot \xrightarrow{\sigma} \cdot) \longrightarrow \textbf{SET}]$$

where $\cdot \xrightarrow{\sigma} \cdot$ denotes the graph with two nodes and one edge, labelled σ. But a functor $\textbf{SET} \longrightarrow [(\cdot \rightarrow \cdot) \longrightarrow \textbf{SET}]$ is *exactly* the same thing as a natural transformation between two functors $\textbf{SET} \to \textbf{SET}$, which is what we set out to find.

In the usual functional programming context, the 'types' (that is, sets X, Y, etc.) are defined recursively by a given set of type constructors, including $\longrightarrow$ and *list* (that is, *); but unfortunately × is probably encoded into $\longrightarrow$ by currying.

Similarly, polymorphism can be provided for languages like OBJ by regarding it as a notation for functions from implicitly imported instantiations of parameterized theories. Under this view, polymorphism is not limited to theories whose parameter interface is trivial, and moreover it embodies the second-order capabilities (such as map functions) of parameterized modules in what would otherwise be a first-order language, using the notation of conventional functional programming but with an underlying logic that is entirely first order and hence is simpler to reason with. See (Goguen 1990a) for further discussion. The same conventions can be used for object-oriented programming, because of the developments in Section 14.3.

14.5 Programming in the large

The polymorphism for trivial types discussed in the previous section is rather limited, and is not useful for programming in the large (that is, for modules), where interface theories may involve many sorts and operations (as in Clear, Ada, OBJ, and Standard ML) and may also have laws (as in Clear and OBJ). We want a calculus for putting together theories (specifications), code (compiled or source), documentation (explanations), animations, management and accounting information, test cases, correctness proofs, and so on. Let us call such a capability *hyperprogramming* (Goguen 1986, 1990b). It provides an approach for integrating many diverse aspects of programming environments, and has been given a concrete syntax in the LIL system (Goguen 1986), originally intended for combining Ada modules.

Two key concepts in hyperprogramming are module expression and colimit. The first generalizes the UNIX **make** command, which actually constructs a (sub)system when evaluated, while the second provides its semantics, as in Section 14.2.

First a series of declarations build up an 'environment', which is a diagram in the category of theories of some institution; its denotation is its colimit. For example, the following code for SORTING[NAT] and SORTING[NATD] has the environment shown in Figure 14.4, in which the dotted lines indicate the relationship of 'instantiation' for parameterized theories:

```
obj LIST[X :: TRIV] is
  sorts List NeList .
  subsorts Elt < NeList < List .
  op nil : -> List .
  op __ : List List -> List [assoc id: nil prec 9]
  op head_ : NeList -> Elt .
  op tail_ : NeList -> List .
  var X : Elt .  var L : List .
  eq head(X L) = X .
  eq tail(X L) = L .
  op empty_ : List -> Bool .
  eq empty L = L =/= nil .
endo

obj SORTING[X :: POSET] is
  pr LIST[X] .
  op sorting_ : List -> List .
  op unsorted_ : List -> Bool .
  vars L L' L'' : List .  vars E E' : Elt .
```

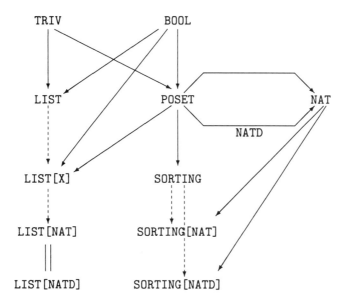

Fig. 14.4 SORTING[NATD]

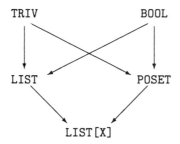

Fig. 14.5 LIST[X]

384 Types as theories

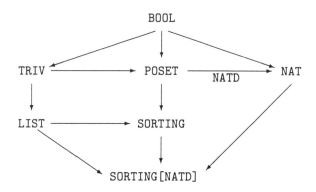

Fig. 14.6 SORTING[NATD] Again

```
  cq sorting L = L if unsorted L =/= true .
  cq sorting L E L' E' L'' = sorting L E' L' E L''
                             if E' < E .
  cq unsorted L E L' E' L'' = true if E' < E .
endo

make SORTING-NAT is SORTING[NAT].
reduce sorting 1 4 2 3 .

view NATD from POSET to NAT is
  vars L1 L2 : Elt .
  op L1 < L2 to L1 divides L2 and L1 =/= L2 .
endv

make SORTING-NATD is SORTING[NATD].
reduce sorting 1 4 2 3 .
```

The results of these two reductions are the lists 1 2 3 4 and 1 2 4 3 respectively.

Figure 14.5 shows the structure of LIST[X] inside SORTING[X :: POSET]. Figure 14.6 shows the diagram for the result of the second make, done all in one step.

14.6 Summary

We have explained a notion of 'types as theories' and shown that it:

- supports dependent types;

- supports a very general form of inheritance, based on theory morphisms;

- extends abstract data types to abstract objects, using an institution of hidden-sorted equational logic;

- explains the naturality of polymorphic types;

- provides a foundation for programming in the large;

- suggests a new way to unify programming environments.

In addition, we have suggested a new denotation for dependent theories based on strict Grothendieck fibrations and indexed categories.

It could be argued that the proper handling of abstraction is the most important problem in modern computer science, because it is the key to mastering the enormous complexity of modern software and software–hardware systems. It is hoped that the 'types as theories' viewpoint can contribute to this effort by clarifying the concepts involved, because it captures several different forms of abstraction. The ideas in this chapter direct applications to the design of programming languages and environments; in particular, they have been used in designing OBJ, FOOPS and Eqlog, and have influenced the design of other languages.

Acknowledgements

The approach in this paper evolved out of research on Clear (Burstall and Goguen 1977, 1980, Goguen and Burstall 1985) and its semantics; this joint work with Rod Burstall includes the concept of institution and the putting together of theories by colimits to form larger specifications. These ideas were further developed in joint work with José Meseguer on various topics, including OBJ (Futatsugi *et al.* 1985, Goguen and Winkler 1988), which can be seen as an implementation of Clear for the order-sorted (conditional) equational institution, as well as FOOPS (Goguen and Meseguer 1987c), which extends these ideas to object-oriented programming. Special thanks are due to Andrzej Tarlecki and Razvan Diaconescu for spotting some bugs in a draft of this chapter. My debt to the Venerable Chögyam Trungpa Rinpoche is not expressible in words, and his inspiration is pervasive.

References

Bird, R., and Wadler, P. (1988). *Introduction to Functional Programming.* Prentice-Hall.

Burstall, R., and Goguen, J. (1977). Putting theories together to make specifications. In Reddy, R., editor, *Proceedings, Fifth International Joint Conference on Artificial Intelligence,* pp. 1045–1058. Department of Computer Science, Carnegie-Mellon University.

Burstall, R., and Goguen, J. (1980). The semantics of Clear, a specification language. In Bjorner, D., editor, *Proceedings, 1979 Copenhagen Winter School on Abstract Software Specification,* pp. 292–332. Springer-Verlag. Lecture Notes in Computer Science, Volume 86; based on unpublished notes handed out at the Symposium on Algebra and Applications, Stefan Banach Center, Warszawa, Poland.

Burstall, R., and Goguen, J. (1982). Algebras, theories and freeness: An introduction for computer scientists. In Wirsing, M., and Schmidt, G., editors, *Theoretical Foundations of Programming Methodology,* pp. 329–350. Reidel. Proceedings, 1981 Marktoberdorf NATO Summer School, NATO Advanced Study Institute Series, Volume C91.

Burstall, R., and Lampson, B. (1984). A kernel language for abstract data types and modules. In *Proceedings, International Symposium on the Semantics of Data Types,* pp. 1–50. Springer-Verlag. Lecture Notes in Computer Science, Volume 173.

Cardelli, L., and Wegner, P. (1985). On understanding types, data abstraction, and polymorphism. *Computing Surveys,* 17(4):471–522.

de Bruijn, N. (1989). Telescope mapping in typed lambda calculus. Technical Report P161-M17, Department of Mathematics and Computing Science, Eindhoven University of Technology. Earlier versions from 1986 and 1987.

Ehrich, H.-D., Sernadas, A., and Sernadas, C. (1989). Objects, object types, and object identification. In Ehrig, H. et al., editors, *Categorical Methods in Computer Science with Aspects from Topology,* pp. 142–156. Springer-Verlag. Lecture Notes in Computer Science, Volume 393.

Futatsugi, K., Goguen, J., Jouannaud, J.-P., and Meseguer, J. (1985). Principles of OBJ2. In Reid, B., editor, *Proceedings, 12th ACM Symposium on Principles of Programming Languages,* pp. 52–66. Association for Computing Machinery.

Girard, J.-Y. (1972). Interprétation fonctionelle et élimination des coupures de l'arithmétique d'ordre supérieur. Thèse de Doctorat d'État, Université Paris VII.

Goguen, J. (1971). Mathematical representation of hierarchically organized systems. In Attinger, E., editor, *Global Systems Dynamics,* pp. 112–128. S. Karger.

Goguen, J. (1973). Some remarks on data structures. Abstract of 1973 Lectures at Eidgenoschiche Technische Hochschule, Zürich.

Goguen, J. (1986). Reusing and interconnecting software components. *Computer*, 19(2):16–28. Reprinted in *Tutorial: Software Reusability*, Peter Freeman, editor, IEEE Computer Society Press, 1987, pp. 251-263.

Goguen, J. (1990a). Higher-order functions considered unnecessary for higher-order programming. In Turner, D., editor, *Proceedings, University of Texas Year of Programming, Institute on Declarative Programming*. Addison-Wesley. Preliminary version in SRI Technical Report SRI-CSL-88-1, January 1988.

Goguen, J. (1990b). Hyperprogramming: A formal approach to software environments. In *Proceedings, Symposium on Formal Approaches to Software Environment Technology*. Joint System Development Corporation, Tokyo, Japan.

Goguen, J. (1991). Semantics of concurrent interacting objects using sheaf theory. Lecture given at U.K.-Japan Symposium on Computation, Oxford, September 1989; to appear in Springer-Verlag Lecture Notes in Computer Science.

Goguen, J., and Burstall, R. (1985). Institutions: Abstract model theory for specification and programming. To appear in *Journal of the Association for Computing Machinery*. Report ECS-LFCS-90-106, Computer Science Department, University of Edinburgh, January 1990; preliminary version, Report CSLI-85-30, Center for the Study of Language and Information, Stanford University, 1985, and remote ancestor in 'Introducing Institutions', in *Proceedings, Logics of Programming Workshop*, Edward Clarke and Dexter Kozen, editors, Springer-Verlag Lecture Notes in Computer Science, Volume 164, pp. 221-256, 1984.

Goguen, J., and Meseguer, J. (1982). Universal realization, persistent interconnection and implementation of abstract modules. In Nielsen, M., and Schmidt, E., editors, *Proceedings, 9th International Conference on Automata, Languages and Programming*, pages 265–281. Springer-Verlag. Lecture Notes in Computer Science, Volume 140.

Goguen, J., and Meseguer, J. (1986). Eqlog: Equality, types, and generic modules for logic programming. In DeGroot, D., and Lindstrom, G., editors, *Logic Programming: Functions, Relations and Equations*, pp. 295–363. Prentice-Hall. An earlier version appears in *Journal of Logic Programming*, Volume 1, Number 2, pp. 179-210, September 1984.

Goguen, J., and Meseguer, J. (1987a). Models and equality for logical programming. In Ehrig, H., Levi, G., Kowalski, R., and Montanari, U., editors, *Proceedings, 1987 TAPSOFT*, pp. 1–22. Springer-Verlag. Lecture Notes in Computer Science, Volume 250.

Goguen, J., and Meseguer, J. (1987b). Order-sorted algebra solves the constructor selector, multiple representation and coercion problems. In *Proceedings, Second Symposium on Logic in Computer Science*, pp. 18–29. IEEE Computer Society Press. Also Technical Report CSLI-87-92, Center for the Study of Language and Information, Stanford University, March 1987.

Goguen, J., and Meseguer, J. (1987c). Unifying functional, object-oriented and relational programming, with logical semantics. In Shriver, B., and Wegner, P., editors, *Research Directions in Object-Oriented Programming*, pp. 417–477. MIT Press. Preliminary version in *SIGPLAN Notices*, Volume 21, Number 10, pp. 153-162, October 1986.

Goguen, J., and Meseguer, J. (1989). Order-sorted algebra I: Equational deduction for multiple inheritance, overloading, exceptions and partial operations. Technical Report SRI-CSL-89-10, SRI International, Computer Science Lab. Given as lecture at Seminar on Types, Carnegie-Mellon University, June 1983; many draft versions exist.

Goguen, J., and Winkler, T. (1988). Introducing OBJ3. Technical Report SRI-CSL-88-9, SRI International, Computer Science Lab.

Goguen, J., Thatcher, J., and Wagner, E. (1976). An initial algebra approach to the specification, correctness and implementation of abstract data types. Technical Report RC 6487, IBM T.J. Watson Research Center. In *Current Trends in Programming Methodology, IV*, Raymond Yeh, editor, Prentice-Hall, 1978, pp. 80-149.

Goguen, J., Thatcher, J., Wagner, E., and Wright, J. (1975a). Abstract data types as initial algebras and the correctness of data representations. In Klinger, A., editor, *Computer Graphics, Pattern Recognition and Data Structure*, pp. 89–93. IEEE Press.

Goguen, J., Thatcher, J., Wagner, E., and Wright, J. (1975b). An introduction to categories, algebraic theories and algebras. Technical report, IBM Watson Research Center, Yorktown Heights NY. Research Report RC 5369.

Goldblatt, R. (1979). *Topoi, the Categorial Analysis of Logic*. North-Holland.

Grothendieck, A. (1963). Catégories fibrées et descente. In *Revêtements étales et groupe fondamental, Séminaire de Géométrie Algébraique du Bois-Marie 1960/61, Exposé VI*. Institut des Hautes Études Scientifiques. Reprinted in Lecture Notes in Mathematics, Volume 224, Springer-Verlag, 1971, pp. 145-194.

Guttag, J. (1975). *The Specification and Application to Programming of Abstract Data Types*. PhD thesis, University of Toronto. Computer Science Department, Report CSRG-59.

Harper, R., MacQueen, D., and Milner, R. (1986). Standard ML. Technical Report ECS-LFCS-86-2, Department of Computer Science, University of Edinburgh.

Hayes, F., and Coleman, D. (1989). Objects and inheritance: An algebraic view. Technical report, Hewlett-Packard Labs, Bristol.

Hoare, C. (1972). Proof of correctness of data representation. *Acta Informatica*, 1:271–281.

Hoare, C. (1978). Communicating sequential processes. *Communications of the ACM*, 21:666–677.

Landin, P. (1964). The mechanical evaluation of expressions. *Computer Journal*, 6:308–320.

Landin, P. (1965). A correspondence between ALGOL60 and Church's lambda notation. *Communications of the Association for Computing Machinery*, 8(2):89–101.

Lawvere, F. W. (1963). Functorial semantics of algebraic theories. *Proceedings, National Academy of Sciences, U.S.A.*, 50:869–872. Summary of Ph.D. Thesis, Columbia University.

Lawvere, F. W. (1964). An elementary theory of the category of sets. *Proceedings, National Academy of Sciences, U.S.A.*, 52:1506–1511.

Liskov, B., and Zilles, S. (1975). Specification techniques for data abstraction. *IEEE Transactions on Software Engineering*, SE-1(1):7–19.

Mac Lane, S. (1971). *Categories for the Working Mathematician*. Springer-Verlag.

Mac Lane, S., and Birkhoff, G. (1967). *Algebra*. Macmillan.

MacQueen, D., Sethi, R., and Plotkin, G. (1984). An ideal model for recursive polymorphic types. In *Proceedings, Symposium on Principles of Programming Languages*, pp. 165–174. Association for Computing Machinery.

Martin-Löf, P. (1982). Constructive mathematics and computer programming. In *Logic, Methodology and Philosophy of Science VI*, pages 153–175. North-Holland.

McCarthy, J., Levin, M., et al.(1966). *LISP 1.5 Programmer's Manual*. MIT Press.

Meseguer, J., and Goguen, J. (1985). Initiality, induction and computability. In Nivat, M., and Reynolds, J., editors, *Algebraic Methods in Semantics*, pp. 459–541. Cambridge University Press.

Milner, R. (1978). A theory of type polymorphism in programming. *Journal of Computer and System Sciences*, 17(3):348–375.

Milner, R. (1980). *A Calculus of Communicating Systems*. Springer-Verlag. Lecture Notes in Computer Science, Volume 92.

Mosses, P. (1989). Unified algebras and institutions. Technical Report DAIMI PB-274, Computer Science Department, Aarhus University.

Reichel, H. (1980). Initially restricting algebraic theories. In Dembinski, P., editor, *Mathematical Foundations of Computer Science*, pp. 504–514. Springer-Verlag. Lecture Notes in Computer Science, Volume 88.

Reynolds, J. (1974). Towards a theory of type structure. In *Colloquium sur la Programmation*, pp. 408–423. Springer-Verlag. Lecture Notes in Computer Science, Volume 19.

Sannella, D., and Tarlecki, A. (1987). On observational equivalence and algebraic specification. *Journal of Computer and System Science*, 34:150–178. Earlier version in *Proceedings, Colloquium on Trees in Algebra and Programming*, Lecture Notes in Computer Science, Volume 185, Springer-Verlag, 1985.

Scott, D. (1972). Lattice theory, data types and semantics. In Rustin, R., editor, *Formal Semantics of Algorithmic Languages*, pp. 65–106. Prentice Hall.

Scott, D. (1976). Data types as lattices. *SIAM Journal of Computing*, 5(3):522–586.

Sernadas, A., Fiadeiro, J., Sernandas, C., and Ehrich, H.-D. (1989). The basic building block of information systems. In Falkenberg and P. Lindgreen, editors, *Proceedings IFIP 8.1 Working Conference on Information Systems Concepts: an In-Depth Analysis*, pp. 225–246. North Holland.

Smyth, M., and Plotkin, G. (1982). The category-theoretic solution of recursive domain equations. *SIAM Journal of Computation*, 11:761–783. Also Technical Report D.A.I. 60, University of Edinburgh, Department of Artificial Intelligence, December 1978.

Tarlecki, A., Burstall, R., and Goguen, J. (1989). Some fundamental algebraic tools for the semantics of computation, part 3: Indexed categories. Technical Report PRG-77, Programming Research Group, University of Oxford. To appear in *Theoretical Computer Science*.

Turner, D. (1985). Miranda: A non-strict functional language with polymorphic types. In Jouannaud, J.-P., editor, *Functional Programming Languages and Computer Architectures*, pp. 1–16. Springer-Verlag. Lecture Notes in Computer Science, Volume 201.

Wand, M. (1979). Final algebra semantics and data type extension. *Journal of Computer and System Sciences*, 19:27–44.

Zilles, S. (1974). Abstract specification of data types. Technical Report 119, Computation Structures Group, Massachusetts Institute of Technology.